T0239457

# Bioinformatics

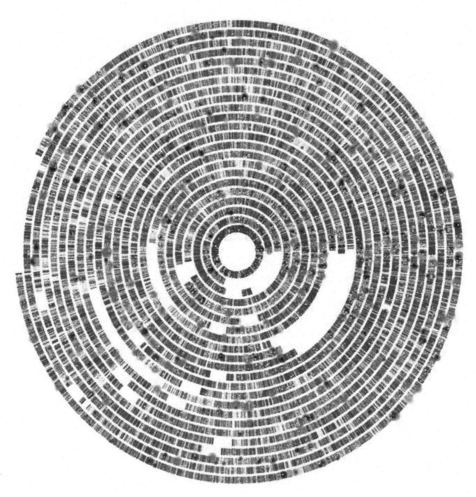

No black and white: Shown are the fascinating shades of individuality. In this artistic representation, all the variants of a healthy human being (NIH assembly identifier: NA12878) are displayed. They are organized on several circles, representing the different chromosomes, according to their position on the chromosome. The size and color of the variants were chosen according to the severity of the impact on the function of the genome. For example, you can see the many gray variants that do not fall on any gene and are therefore difficult to classify. This contrasts with the black and dark variants, which cause a severe defect in the affected genes. This shows how a considerable number of gene defects can be found even in healthy people as they are compensated by the healthy gene copy from the other parent.

Thomas Dandekar  •  Meik Kunz

# Bioinformatics

## An Introductory Textbook

Springer

Thomas Dandekar
Department of Bioinformatics
University of Würzburg
Würzburg, Germany

Meik Kunz
Chair of Medical Informatics
Friedrich-Alexander University
Erlangen-Nürnberg, Germany

ISBN 978-3-662-65035-6      ISBN 978-3-662-65036-3   (eBook)
https://doi.org/10.1007/978-3-662-65036-3

# Contents

# Part I

# How Does Bioinformatics Work?

## Access

We are searching the key to understand life – this is how bioinformatics is oriented nowadays! It has evolved from data processing, just the assistant and auxiliary science for large amounts of data, to now establish *quantitative* theoretical biology. For the first time, theories about something as complex as living beings no longer remain pure theory, but are directly verifiable and measurable, and have already led to remarkable results and progress – from drugs against cancer and HIV to new insights, for example into the exciting question of why our cells and we age.

Nevertheless, my main motivation for studying medicine and later becoming a bioinformatician was not so much the prospect of ploughing through large amounts of data, but the fascination that biology has always had for people, the eternal questions about the key to the language of life, about the "water of life" that heals everything. I wanted to recognize and understand what holds us together in our innermost self, that is, how our consciousness and our brain function. Tracing these great questions is precisely the purpose of this book. Because today's bioinformatics is doing this to an increasing extent, and because one can also start from very small, simple examples, we will begin with these. We provide case-based examples for each chapter and a tutorial in the appendix for you to play with and discover for yourself. The new English edition 2021 brings everything up to date and adds further important aspects.

The unbelievable has happened silently: Whereas before the computer was just a stupid data storage device, new insights into life and the world and ourselves are now emerging in simulations. This is only possible because life itself is not dead and is permeated by numerous recognition processes. These are, for example, key-lock relationships between molecules, but also memory and molecular languages at all levels of life. We want to explore this in more detail here, first looking at the "how" of bioinformatics, in order to then better understand in Part II why bioinformatics is so successful right now – similar to theoretical physics in the first half of the last century. This will also prepare us to explore the fascination of information processing in living beings and its reflection in the computer model (Part III), whether we want to better fight infections, understand cancer, or even understand ourselves.

## Short Instructions for Usage of the Book

A classical textbook should (i) teach you the practice of bioinformatics and (ii) provide accurate definitions. For these two points, we have (i) prepared not only exercises in each chapter, but also tutorials for the most important software examples along with tips for use, and (ii) included a number of definitions in the glossary so that important terms are defined and explained.

Nevertheless, the book here is deliberately not a classical textbook. We want to convey joy and interest in bioinformatics. You can and are welcome to read the examples and chapters at your leisure and then, if you are interested in certain analyses in more detail, to practice them, work through the questions, look at the tutorials and do everything in more detail. Systematically, all current areas of bioinformatics are presented in a broad over-view, and each end of chapter briefly summarizes the presented area again in a conclusion. We can only provide a stimulating introduction here. Without practicing and working through several examples for each of the software, it is not possible to gain sufficient expe-rience for your own analyses. A sound knowledge of biology is also important, since you should be able to critically examine the program outputs with your knowledge. A number of suggested books on molecular biology but also on the national research data and medi-cal informatics initiative are listed in the chapters. For students who enjoy programming, appropriate references for further reading are given in the introduction to the tutorials. Since bioinformatics lives on databases and software, we have summarized databases and programs and their basic use in the chapters and in the appendix.

# Sequence Analysis: Deciphering the Language of Life

**Abstract**

Sequence analysis is a central tool of bioinformatics with relevant databases (NCBI, GenBank, Swiss-Prot) and software to detect sequence similarity (BLAST) and domain databases (Pfam, SMART). Crucial is the ability to know and use such software on the web, the tutorials and exercises encourage this. Programming sequence comparison software and databases only makes sense if it enables a better analysis of the biological question, in particular for large-scale analysis – in all other cases, it is better to use the numerous software that already exist, the internet is only a mouse click away.

Bioinformatics requires data on living organisms, processes them and then designs a corresponding model of the living process that is thereby mapped. A good simple example is when a polymerase chain reaction (*PCR*) is used to detect a virus in the blood. Polymerases copy DNA (*deoxyribonucleic* acid) and were originally derived from bacteria. Hereby they also duplicate their genetic information. PCR is a modern method of molecular biology. Using such a chain reaction, so much of a molecule (if, for example, there is only one virus molecule in the blood) is produced by constant doubling of the molecules with the help of polymerase that it can be easily detected in the laboratory and, above all, the sequence can be read.

Nowadays, this can be deciphered quite easily by a sequencing machine. However, this initially leaves us with a salad of letters that lists the nucleotides, i.e. the genetic material, of the virus in sequence, such as tgtcaacata ... (Fig. 1.1).

© Springer-Verlag GmbH Germany, part of Springer Nature 2023
T. Dandekar, M. Kunz, *Bioinformatics*,
https://doi.org/10.1007/978-3-662-65036-3_1

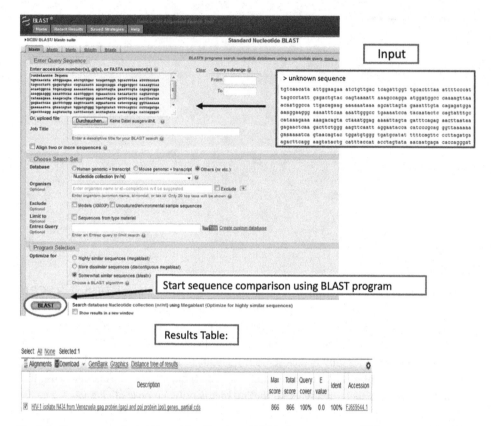

**Fig. 1.1** Sequence analysis allows identification of HIV virus sequences. HIV sequence identification using BLAST (https://blast.ncbi.nlm.nih.gov/Blast.cgi). Shown is the sequence comparison of an initially unknown sequence against a database using the program BLAST. The result line indicates that the unknown sequence is an HIV-1 N434 retrovirus strain from Venezuela (result line: Venezuela gag coat protein and pol polymerase protein; the result link then leads to the detailed sequence comparison)

**Collect, Compare and Understand Data**   In order to now know which virus we have in front of us (in practice, usually even much more precisely, namely which virus strain), we have to let the computer identify this sequence.

**Collect Data**   This is particularly easy if you have created a database of virus sequences. You already know their sequence because you have sequenced them before. As an example, let us consider HIV, the human *immunodeficiency* virus. With the help of the database, it is easy to find out whether the sequence found by PCR for a virus in the blood matches one of the entries in the database. Databases are fundamental in bioinformatics. They store all the information and can then be used for further investigations.

**Analyze and Compare Data**

So this is how you do a sequence comparison (also called sequence analysis). You look to see which sequence in the **database** is most similar to the new sequence. This can be done over the entire length of the sequence, i.e. globally. However, because a virus can be relatively strange and one would then usually like to know whether it is not at least similar in sections, one typically performs a section-by-section local comparison, which thereby yields the most similar sequence section (Fig. 1.1). But in order for the computer to do anything at all, you have to tell it what to do down to the last detail, until it finally presents a result of the computation. All the instructions for this, e.g. to perform such a comparison up to the final result, are together a program. In the past, **programs** were written using instructions that the machine understood particularly well. But these could only be very short, because they were written in machine language, which essentially contained simple register instructions (clear 1 bit, write, move or check). Today, however, a richer language is used that contains far more complicated instructions, which is therefore called a higher programming language (e.g. Perl, Java, Python, C++ or R, currently the most popular programming languages in bioinformatics).

Let us return to our sequence example: What do we see as a result in Fig. 1.2? This is a so-called *Basic Local Alignment*, the corresponding tool in bioinformatics is called BLAST, for *Basic Local Alignment Search Tool* (Altschul et al. 1990), where the result indicates a veritable diagnosis for the patient.

The sequence comparison shows that it is an HIV strain from Venezuela. It becomes clear that one can actually make a diagnosis (HIV infection, probably acquired in South

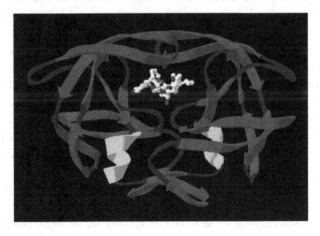

**Fig. 1.2** *Drug design*, example of HIV infection. The HI virus is blocked in its activities (dark molecule around the drug) by a drug (centre, white). Computer representation of the three-dimensional structure of the HIV-1 protease (molecular structure consisting of leaflets [red], loop regions [blue] and helices [yellow]) and its inhibitor ritonavir (shown as a sphere and edge model). The aim of such bioinformatic *drug designs* is to design a suitable therapy on the computer, in this case, for example, the inhibition of the protease for the treatment of an HIV-1 infection, so that the virus can no longer produce new viral envelopes - its protease no longer functions

America) with this computer program, which only writes letters as optimally as possible among each other (hence sequence comparison or alignment). The decisive prerequisite for this is that one knows and understands the results correctly in their biological meaning - and this is precisely the work of the bioinformatician.

**Understanding Data**

Finally, there is a third area of work in bioinformatics: "understanding data". In addition to collecting data (databases) and comparing data (e.g. using BLAST), one ultimately wants to understand the data and use it appropriately, for example to develop new therapeutic approaches. This can happen, among other things, by integrating the data in a suitable **bioinformatics model** and then modelling it. This modelling can be a simulation, for example if I am looking for new drugs against HIV and want to destroy the sequence of the virus. Since the virus consists of nucleic acids, as we have already seen above, I can, for example, insert the wrong nucleotides into the virus and thus also destroy its polymerase (the copying enzyme with which the virus reproduces). A complex but highly successful modelling technique consists of reproducing the three-dimensional structure of this polymerase in the computer and then selecting from a database of molecules which best fits into the polymerase in such a way that it is blocked, i.e. the virus can no longer reproduce (Fig. 1.2 shows an example of this *drug design*). Such methods have been very successful with HIV in particular. There are now more than 20 drugs that target the virus with the wrong nucleotides, by inhibiting its nucleic acid or its enzymes. The result is remarkable, the combination therapy (*highly active* antiretroviral therapy; HAART, Antiretroviral Therapy Cohort Collaboration 2008) works so well that one has an almost normal life expectancy, while one can only withstand the viral infection for a few years without therapy (Hoog et al. 2008). This illustrates that bioinformatics can strongly support medicine for instance regarding therapy.

What would you actually have to pay special attention to if, for example, you now perform such sequence comparisons yourself? It is important to know that the BLAST search is not completely accurate (heuristic), but it delivers faster results than a 1:1 comparison over the entire sequence length against the database. Therefore, such hits are only credible if the probability of getting such a hit by chance is low enough. As a first rule of thumb you can remember: The *E-Value* (i.e. the expected value of a random hit) should be less than 1 in one million. This is then already a very convincing value. In borderline cases (random expectation value at 1 in 1000), you can also take the hit sequence and see if you can find the initial sequence again (called "reverse search" in technical jargon). If we keep in mind that this is a local search, then we also understand why we should search the whole hit length (given in the example, sequence similarity over the whole sequence length). But there are also BLAST results where only one subsequence in the protein has high similarity and the rest instead shows no similarity. In this case, the BLAST search turned up only one protein domain, the one with the highest similarity in the whole database. To determine the remaining parts of the sequence in terms of function as well, you then need to use only those domains that do not yet have database hits again, without the first sequence part for the search. In this way, you can match domain by domain in the protein with a new BLAST search each time for the sequence portion that has not yet been matched by the search. Finally, in difficult cases, the BLAST search may only reveal a similarity to a database entry that has no clear function. In this case (protein sequence), you can use the "position-specific iterative BLAST", or Psi-BLAST for short, which then searches with all the still unrecognized sequences at the same time (a so-called "profile") until it lands a hit to which a sequence can be assigned. This almost always works, but may take several repetitions. You should also only continue searching with Psi-BLAST if something changes in the repeat search, otherwise the search is "converged" in vain.

However, the drug search shown in Fig. 1.2 is a somewhat involved process, requiring many intermediate results to be obtained and calculations and comparisons to be made. What can be done, on the other hand, is to perform direct databases that provide additional secondary information besides the primary sequence information. These are also called secondary databases. An example would be to search for the HIV protease in the protein database PDB (https://www.rcsb.org/pdb/home/home.do). In addition to the protein sequence, this database also holds the coordinates of the protein's three-dimensional structure, as well as other details about its structure and function. There is a great deal of further information available on the HIV structure in particular, including information on the *drug design*.

## 1.1   How Do I Start My Bioinformatics Analysis? Useful Links and Tools

Generally speaking, we first look at the function of the molecule we want to bioinformatically determine by comparing it directly to a database. The best known example is the direct sequence comparison with BLAST, which we have already discussed in detail. The next step is to use other databases or programs for analyses and comparisons to obtain additional information. A simple example is to search for secondary data, and our first example of this was the protein database. As a primary database, it contains the three-dimensional coordinates of protein structures, but it also contains a lot of secondary data about these proteins where this structure determination was successful. As a third step, we can finally follow up with detailed analyses.

In the following, useful supporting sites for these steps are briefly presented. The BioNumbers database describes number relationships in biology (https://bionumbers.hms.harvard.edu). This was established at Harvard University by students who first calculated these biological problems and then made these numbers available to the interested reader.

Unfortunately, most bioinformatics websites are in English, including this book. This is due to the fact that the Anglo-Americans were simply faster with many initial developments than German bioinformatics. In addition, English is now the language of science, and the creator of a bioinformatics website would like everyone to be able to use this site.

**Already Prepared Results: "BioNumbers"**

▶       https://bionumbers.hms.harvard.edu/

So here you can find out how different sizes and numbers are related in biology. Just look it up and learn about the exciting world of sizes and numbers in different organisms and diseases, but also in humans.

For a better understanding, we would like to show a simple *screenshot* of a list of useful biological quantities and numbers from the BioNumbers database (Fig. 1.3). It is best to simply look at it yourself and be amazed at the interesting correlations and differences.

**MEDLINE as a Large Online Library**
One of the main problems in all bioinformatics work is to get a quick overview of the knowledge that exists about the object of study. This is the only way to assess the accuracy

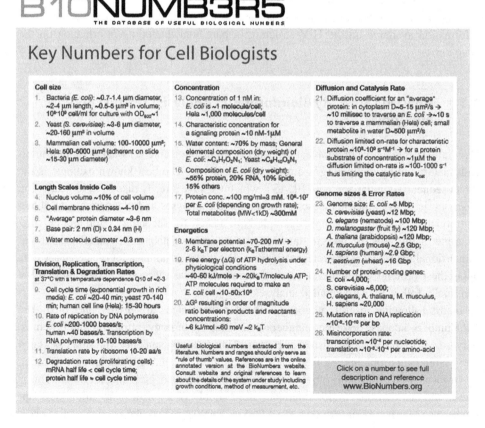

**Fig. 1.3** Listing of useful biological quantities and numbers from the literature in the BioNumbers database (for details see text)

and also the value of your results. For this purpose MEDLINE, the online version of the library at the National Institute of Health, is an indispensable tool. A large, worldwide open library about medicine and biology:

▶    MEDLINE (or also PubMed)

▶    https://www.ncbi.nlm.nih.gov/pubmed

It is the online version of the library. Only here, in Betheds (near Washington), the Health Research Center of the United States of America, has it been possible to keep a sufficiently large staff of service scientists permanently on hand to ensure easy use of the web pages and to keep the data constantly up to date. This is a truly extraordinary achievement, which is precisely why it looks and feels child's play to use.

Here you can search for keywords ("HIV", "*sequence analysis*", "*aging*"), for authors ("Dandekar-T", "Kunz-M"), journals ("Nature", "Science"). For each article found, a table of contents will then appear, but also links to related articles (including search options). A steadily increasing number of articles also offer a directly readable full-text link ("*Open Access* ", even for current articles already more than 30%, for articles one to 2 years old it is now even the majority). It is possible for the experienced to search for an article much more precisely and with many more criteria ("*advanced search*"). It is helpful to have a look at the PubMed tutorials or our tutorial in the appendix. In addition, PubMed also provides important textbooks online and a variety of other resources.

### How Do I Get the Sequence to My Molecule?

Many bioinformatics studies start with the sequence of a molecule and analyze it. Interestingly, this important starting information, i.e. what sequence the molecule I am interested in has, is already known for many millions of entries. This is especially true for important organisms such as humans, the bacterium *Escherichia coli (E. coli)*, plants such as *Arabidopsis*, mice, the worm *Caenorhabditis elegans (C. elegans)*, and the fruit fly *Drosophila melanogaster*. To check if my sequence for this protein or term is already known, look it up at NCBI in particular. If it is known, the sequence for DNA, RNA (option "*nucleotide*" or "*gene*") or proteins (option "*protein*") can easily be found here, e.g. for "HIV" there are hundreds of thousands of entries:

▶    https://www.ncbi.nlm.nih.gov/protein/?term=hiv

One of the first offers from the long list of hits is an artificial sequence for the "TAR protein":

▶    https://www.ncbi.nlm.nih.gov/protein/AAX29205.1

The now mostly quite long header entry explains already existing information about the respective protein:

```
LOCUS          AAX29205      367 aa linear      SYN 29-MAR-2005
DEFINITION     TAR, partial [synthetic construct].
ACCESSION      AAX29205
VERSION        AAX29205    .1 GI:60653021
DBSOURC        Eaccession      AY892288.1
KEYWORDS       Human      ORF project
SOURCE         synthetic      construct
ORGANISM       synthetic      construct
```

... and so on. In particular, you can find information about the authors of the sequence, journal articles about it and the exact properties of the sequence, that is, from where to where, for example, the protein, the region and specific binding sites go:

```
Protein      1..>367
             /product="TAR"
Region       30  ..95
             /region_name="DSRM"
             /note="Double-stranded RNA binding motif. Binding is not
             sequence specific but is highly specific for double stranded
             RNA. Found in a variety of proteins including dsRNA depen-
             dent protein kinase PKR, RNA helicases, Drosophila staufen
             protein, E. coli RNase III; cd00048"
             /db_xref="CDD:238007"
Site         order(30,36..37,78..81,84)
             /site_type="other"
             /note="dsRNA binding site [nucleotide binding]"
             /db_xref="CDD:238007"
Region       159      ..222
             /region_name="DSRM"
             /note="Double-stranded RNA binding motif. Binding is not
             sequence specific but is highly specific for double
             stranded RNA. Found in a variety of proteins including
             dsRNA dependent protein kinase PKR, RNA helicases,
             Drosophila staufen protein, E. coli RNase III; cd00048"
             /db_xref="CDD:238007"
             Siteorder(159,165..166,208..211,214)
             /site_type="other"
             /note="dsRNA binding site [nucleotide binding]"
```

Finally, this is followed by the original sequence as determined by the authors and used in their research. In the example:

```
ORIGIN
        1  mseeeqgsgt ttgcglpsie qmlaanpgkt pisllqeygt rigktpvydl
           lkaegqahqp
```

```
      6   lnftfrvtvgd ·tsctgqgpsk kaakhkaaev alkhlkggsm lepaledsss
          fspldsslpe
     12   ldipvftaaaa atpvpsvvlt rsppmelqpp vspqsecnp vgalqelvvq
          kgwrlpeytv
    181   tqesgpahrk eftmtcrver fieigsgtsk klakrnaaak mllrvhtvpl
          dardgnevep
    241   dddhfsigvg srldglrnrg pgctwdslrn svgekilslr scslgslgal
          gpaccrvlse
   3011   seeqafhvs yldieelsls glcqclvels tqpatvchgs attreaarge
          aarralqylk
    361   imagskl
```

The NCBI site brings a lot more information for bioinformatics:

https://www.ncbi.nlm.nih.gov/guide/ ...

| All resources | A detailed overview of molecular and literature analysis and data banks |
|---|---|
| Chemicals and bioassays | Bioinformatic analyses should eventually lead to new experiments to confirm the results; the necessary ingredients and measurement methods are collected here: Chemicals and biological measurement methods (*bioassays*) |
| Data and software | Here we find numerous databases and programs |
| DNA and RNA | Software and tools for the analysis of DNA and RNA |
| Domains and structures | Analysis of protein domains (small folding units) and large structures |
| Genes and expression | Analysis of the transcription of genes under different conditions |
| Genetics and medicine | Numerous genetic information |
| Genomes and maps | Useful *maps to* find your way around genomes |
| Homology | Similarity comparisons to proteins, but at the structural level. In particular, it is thus possible to calculate one's own protein structure by pointing out a similar three-dimensional structure |
| Literature | In addition to MEDLINE (see above), there are many articles that can be found on the site and read online, as well as important textbooks |
| Proteins | General analyses of protein sequence, structure and function. In particular, the protein domains, i.e. the functional building units in the protein, are also examined in more detail |
| Sequence analysis | Other programs besides BLAST that examine the sequence of a protein or a nucleic acid |
| Taxonomy | Classification of a sequence in a catalogue of all species. Many of the results are presented as phylogenetic trees |
| Training and tutorials | Highly recommended for a first start, see: https://www.ncbi.nlm.nih.gov/guide/training-tutorials/ Especially the BLAST search and the *taxonomy* are explained in a very nice beginner tutorial |
| Variation | How do I do justice to biodiversity and variety? |

In addition to the NCBI site, which is certainly the best known bioinformatics site, there are also good introductory sites at the European Bioinformatics Institute (EBI). These are especially helpful for those people who also like programming modules and are looking for information at an advanced level:

▶    https://www.ebi.ac.uk

For example:

▶    https://www.ebi.ac.uk/services

*"We maintain the world's most comprehensive range of freely available and up-to-date molecular databases."* This refers to the wealth of data that the EBI site offers. The difference to the NCBI website is that it is easier to download the entire data of the database and not only to perform individual queries via the web interface.

It is also important that the EMBL database is located here, which provides comparably detailed sequence information as GenBank at the NIH. However, there are small differences in the preferences and the offer, but also in the preparation of the entries. In addition, there is somewhat more and somewhat faster information on new sequences identified in Europe (NCBI is more detailed and faster for American sequences).

Other important sites can be found at the Swiss Bioinformatics Institute (see next chapter) and at the Japanese gene bank DDBJ (DNA Data Bank of Japan).

▶    https://www.ddbj.nig.ac.jp

Again, there is a daily comparison with the EMBL and NCBI databases in order to keep "all known" sequences available. This time, however, this is done from the Japanese point of view; it is precisely the sequences from Japan that are particularly complete and quickly recorded here.

Finally, reference should also be made to the new German National Research Data Infrastructure, in which targeted digitisation and infrastructure is being promoted in numerous areas.

▶    https://www.nfdi.de, https://www.nfdi.de/konsortien-2

For biology, for example, DataPlant (plant databases), the German Human Genome-Phenome Archive, NFDI4BioDiversity and NFDI4microbiota. This is also where very useful data for bioinformatics analysis is concentrated and made available as an infrastructure for all.

▶   https://nfdi4microbiota.de (Dandekar is an affiliate).

In addition, within the framework of the Medical Informatics Initiative of the Federal Ministry of Education and Research, there are several Germany-wide consortia to which university hospitals and other partners (research institutes, universities, companies) have joined forces.

▶   https://www.medizininformatik-initiative.de/de

For example, ten universities and university hospitals, two universities and one industrial partner are working together in the MIRACUM consortium (Medical Informatics in Research and Care in University Medicine) to establish an IT infrastructure for data from research and patient care (data integration centres) and to make it usable for research projects in the long term, for example for the development of predictive models and precision medicine.

▶   https://www.miracum.org/ (consortium leader Medical Informatics FAU Erlangen-Nürnberg, Kunz is a partner).

## 1.2   Protein Analysis Is Easy with the Right Tool

An important special case is the analysis of proteins. Many experiments in molecular biology focus on this particularly important type of molecule. Typically, general properties are first determined by experiments, such as certain binding sites, the weight of the protein, appearance, cofactors or catalytic properties. This is followed by detailed biochemical analyses. The Swiss Bioinformatics Institute has compiled a detailed software package for these numerous ways of analysing proteins. The site is again in English because such analyses are carried out here from all over the world, namely with regard to the properties of the protein sequence (secondary structure, amino acid composition and properties, antigenicity, etc.) as well as the protein structure, including the properties of the independent folding units in the protein, the protein domains.

**Analysis with BLAST**
A good first step is the already mentioned BLAST. This allows a protein sequence (blastp) to be compared for similar entries in a database, and also identifies conserved domains and motifs, such as catalytic and active sites.
In addition, there are more precise and specific tools, which are presented below.

**Entry Page on the Web: ExPASy (https://www.expasy.org)**
The Swiss Bioinformatics Institute had initially (1990s) built up the Swiss-Prot database under the direction of Amos Bairoch. It was particularly carefully maintained and still has a very high degree of correctness and correction of entries, even though it has now essentially been absorbed into the UniProt Knowledge base (UniProt KB):

▶      https://web.expasy.org/docs/swiss-prot_guideline.html

takes the interested person to this link. As explained on the page, there are also detailed comments on the sequence here. These so-called "header entries" provide a wealth of information about protein sequences, followed by the actual sequence.

**How Do I Quickly Analyze Protein Data?**

The ExPASy site brings expert help to get started with protein analysis. *"Proteomics"* means the analysis of large amounts (*"omics"*) *of* protein data.

▶      https://www.expasy.org/proteomics

In addition to various databases, you can also find a lot of bioinformatics information here:

| | |
|---|---|
| *Proteomics* | Large-scale analyses of proteins |
| *Protein sequences and identification* | Identification of proteins by sequence |
| *Mass spectrometry and 2-DE data* | Identification of peptides found in mass spectroscopy or protein spots found in 2D gel. Evaluation software and databases for these steps |
| *Protein characterisation and function* | Domain analyses in particular |
| *Families, patterns and profiles* | Proteins with the same function form a family. In particular, always the same ("conserved") amino acids, patterns and position-specific frequencies of amino acids for these families are summarized here |
| *Post-translational modification* | After production at the ribosome, proteins are further modified, these are the post-translational modifications |
| *Protein structure* | Finding or calculating the three-dimensional protein structure. A fast homology prediction via the SWISS-MODEL server is also offered here |
| *Protein-protein interaction* | Predicting which protein interacts with which other protein |
| *Similarity search/ alignment* | There are also a number of alternatives to BLAST here. Multiple protein sequences can also be compared |
| *Genomics* | How are the associated genes related to the proteins they encode? |
| *Structural bioinformatics* | In particular, the properties of protein structures are determined, for example globular proteins are particularly soluble |
| *Systems biology* | A nice introductory page on system effects of proteins, for example protein signalling cascades and phosphatases to switch off such signals |
| *Phylogeny/evolution* | Proteins develop according to specific patterns; in particular, building units, the protein domains, are assembled to form new proteins |
| *Population genetics* | How are important proteins and protein properties distributed in a population? What are the different types? |
| *Transcriptomics* | How are protein and its coding mRNA related? |

(*continued*)

(continued)

| Biophysics | What are the biophysical properties (solubility, stability, helix content, etc.) of my protein? |
|---|---|
| Imaging | How can proteins be visualized and images analyzed? |
| IT infrastructure | Computer infrastructure, service |
| Drug design | Helping to create new drugs to specifically target a protein |
| Glycomics | How sugar residues further modify proteins. In particular, this is how cells recognise their cell neighbours, bacteria cling to glycoproteins. Sugar-binding proteins are called lectins |

**How Do I Identify Important Amino Acids for Protein Function?**
The PROSITE page is particularly helpful for this.

▶   https://prosite.expasy.org

This examines an entered protein sequence to determine whether or not certain sequence motifs are preserved, for example signatures (hand-curated) or profiles (automatically calculated, consensus sequences, taking different sequences into account) that indicate a particular enzyme function.

This allows me to check whether my protein sequence is really an active enzyme (then all amino acids for catalysis are complete) or whether it only looks like one. If this happens in a genome sequence, this is termed a "pseudogene", a "false" gene regarding the enzyme function because important catalytic amino acids are missing and the enzyme therefore cannot function.

In addition, the independent folding units in the protein, the protein domains, are also examined to see whether they are present in the protein, e.g. whether all parts, i.e. domains, are present for a functional enzyme: at least one catalytic domain (50–150 amino acids) that carries out the enzymatic reaction. This is then often joined by numerous other types, e.g. DNA interaction if it is a transcription factor. Examples are:

- cofactor-binding domains (if the enzyme binds a cofactor),
- regulatory domains (for switching the enzyme on and off),
- interaction domains (with other proteins or to form dimers of two identical protein units for the enzyme, e.g. glutathione reductase only functions as a dimer, so needs an interaction domain for its function),
- structural domains (e.g., if it is a structural protein, like collagen).

**How Can I Estimate the Protein Structure?**
Structure prediction with homology modelling, for example by SWISS-MODEL, is helpful for this.

▶   https://swissmodel.expasy.org

SWISS-MODEL offers the possibility to predict the three-dimensional structure of the protein based on the sequence.

This is a relatively quick prediction, and the three-dimensional coordinates are then available for the user to download. However, it requires a protein with a known three-dimensional structure as a template in order to calculate how much the user's sequence differs from this in its three-dimensional structure. Whether a template can be found is determined by a special sequence comparison with the proteins in the SWISS-MODEL database.

SWISS-MODEL is a very solid, fast and often confirmed approach to determine a three-dimensional structure according to protein template. However, there are many other, often much more complex ways of calculating the protein structure (e.g. homology modelling with MODELLER):

▶    https://salilab.org/modeller/tutorial/

Since structures are not always available that can serve as a template, so-called *ab initio* and optimization algorithms calculate an approximate solution for the structure determination based on the sequence and the minimization of the free enthalpy. Prominent representatives here are neural networks, evolutionary algorithm or Monte Carlo simulation. One example is the QUARK server from the Zhang lab:

▶    https://zhanglab.ccmb.med.umich.edu/QUARK/

**Marking of the Known Structural Parts in the Protein Sequence**
For independent verification, we offer at the chair a labeling of the known three-dimensional structural domains to any sequence (the technical language says domain annotation, that is why our tool is called "AnDom"). This is a slightly different procedure and works for any sequence. It just looks to see if at least a small piece of the sequence is not similar to a known three-dimensional protein structure. Thus, it is completely independent of the ExPASy predictions and can check them. In general, independent databases and softwares from different authors and methods check each other. This allows to significantly increase the quality of the predictions, e.g. to collect all structure predictions (broad search) or to accept only those found by both websites (particularly validated predictions).

This then sometimes makes the predictions a bit tight. This happens when only short parts of the sequence have sufficient similarity to the structural databases that AnDom has. It can also happen that the protein structure is new, i.e. not similar enough to any known structure to allow prediction. Just as when using BLAST, very small random expectation values (1 in one million and lower probabilities) mean that the assignment using AnDom has been very successful in revealing a structure similarity. In contrast, a random similarity can be recognized by a high random hit rate (higher than 1 in 1000). It may even happen that such a small similarity is found several times even by a random sequence. In this case, the expected value is e.g. 4, if on average a random sequence would find four such hits in the AnDom structure database.

▶ https://andom.bioapps.biozentrum.uni-wuerzburg.de/index_new.html

Again, the HI virus from Fig. 1.1 will serve as an example here (Fig. 1.4). AnDom finds a protease domain in the protein sequence (top: b.50.1.1 according to the SCOP classification). The alignment is also shown (bottom), which once again shows the high degree of agreement between the search sequence (query) and the protease domain found (Sbjt = *subject*) (93% identical). Please also use our tutorial for further information.

---

### Conclusion

- In this first chapter, you have already quite actively learned and practiced the most important technique in bioinformatics, namely sequence analysis, especially of protein sequences. Modern molecular biology generates sequences in abundance. The steady increase of databases (NCBI, GenBank, Swiss-Prot) allows one to quickly find out which previous sequences are close to this new sequence by sequence similarity (BLAST tool). Domain databases and analyses allow to dissect a protein into its folding units, each of which carries a specific molecular function. RNA and DNA sequences are also quickly assigned a function through sequence comparisons.

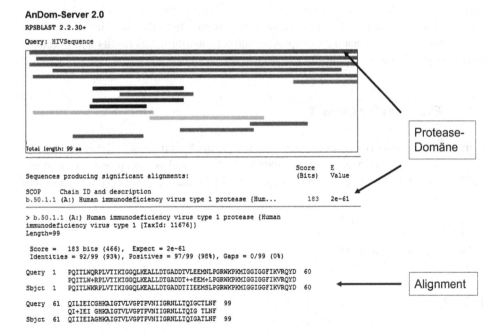

**Fig. 1.4** Search with the AnDom software for protein domains for the HI virus (for details see text). The result shows a high similarity (E-Value 2e-61, 93% *identities*) with the human HIV-1 protease domain (SCOP-ID b.50.1.1) and the corresponding alignment (see text and tutorial)

- Undeniably, sequence analysis is currently the field of bioinformatics that is growing the fastest, producing the quickest results, and allowing initial insights into biology. Hence, in the later chapters, there is sequence analysis software that allows us to quickly trace partial results. It is crucial to be able to learn about this software on the web and practice the different setting options.
- The tutorials and exercises encourage you to do so. Results from different software programs check each other. If they all examine the same sequence, it is always about the same biology, and contradictions then indicate that something was overlooked in the function assignment and must be checked. Sound biological knowledge should critique the results, experiments or further data then corroborate the bioinformatic results. Programming sequence comparison software and databases is useful if this enables a better analysis of the biological question - in all other cases, it is better to use the numerous software that is already available. The internet is only a mouse click away. ◄

**Outlook**

In addition to protein sequence analysis (Chap. 1), RNA (Chap. 2) and DNA sequences (Chap. 3) are important for rapid bioinformatics analysis and description of important molecules of the cell. Next, one would like to understand how these important molecules of the living cell (DNA, RNA, and proteins) interact in networks. These bioinformatic analyses happen either in metabolic networks (Chap. 4) or signaling networks (Chap. 5). Since these are already the most important analysis techniques of current bioinformatics, we then offer an in-depth look at basic strategies of bioinformatics working methods in Part II and look at fascinating examples of current bioinformatics results and developments in Part III.

## 1.3    Exercises for Chap. 1

In the exercises, important parts of the book will be dealt with in more detail in order to consolidate and practise what you have learned. Tasks marked as examples serve as application tasks in which you are to work independently with the computer in order to become more familiar with bioinformatics. In addition, we have provided numerous tutorials in the appendix, which also support the material of the textbook and the exercises and should contribute to a better understanding.

We recommend that you briefly review the material from Chap. 1 at Chap. 6 using the exercises.

**Task 1.1**

(a) What is and does bioinformatics do (feel free to explain with an example)?
(b) There are three areas of bioinformatics, informatically speaking: Databases, Programs/Software, and Modeling/Simulations. Describe important differences between these areas.

**Task 1.2**

An important task of bioinformatics is the collection and management of data and the provision of helpful tools. Name and describe two databases containing information on, for example, genes and gene expression datasets.

**Task 1.3**

Example:

The MEDLINE database (also known as PubMed) is a large, worldwide open library about medicine and biology. Here you can find publications and sequences as well as a lot of other information and links. So PubMed is a good first entry site to use when starting a search. Familiarize yourself with the PubMed database (https://www.ncbi.nlm.nih.gov/pubmed) and find out about the *artificial* sequence for the "TAR *protein*". Hint: Search with "*synthetic*", all searches are in English after all; the search is only limited enough by keywords if only one sequence is found by the query. Only then can you clearly answer the following questions.

1.  Which of the following statements about sequence length (amino acid = aa) is correct?
    A.  The protein sequence is 267 aa long.
    B.  The protein sequence is 367 aa long.
    C.  The protein sequence is 276 aa long.
    D.  The protein sequence is 376 aa long.
2.  Which of the following statements about the title is correct?
    A.  The sequence was filed under the title "Cloning of human full-length CDS in Creator (TM) recombinational vector system" in PubMed.
    B.  The sequence has been filed under the title "Uploading of human full-length CDS" in PubMed.
    C.  The sequence has been filed under the title "Uploading of recombinational vector system" in PubMed.
    D.  The sequence has been filed under the title "Cloning of recombinational vector system" in PubMed.
3.  Which of the following statements is correct?
    A.  Hines et al. submitted the sequence to the journal *Biological Chemistry* and Molecular Pharmacology, Harvard Institute of Proteomics on 05-JAN-2015.
    B.  Darwin et al. submitted the sequence to the journal *Biological Chemistry* and Molecular Pharmacology, Harvard Institute of Proteomics on 05-JAN-2005.
    C.  Hines et al. submitted the sequence to the journal *Biological Chemistry* and Molecular Pharmacology, Harvard Institute of Proteomics on 05-MAR-2005.
    D.  Hines et al. submitted the sequence to the journal *Biological Chemistry* and Molecular Pharmacology, Harvard Institute of Proteomics on 05-JAN-2005.

**Task 1.4**

Bioinformatics has taken off since the mid-1990s, when the first genome projects were successfully completed, because of its rapid sequence analyses. Sequence comparison (for example with the BLAST software) is thus a particularly frequently used and popular bioinformatics method for identifying genes or proteins in the genome.

Explain the BLAST algorithm (hint: it is sufficient to describe how the algorithm can become so fast). Also describe its usefulness for biology. If both are still unclear, simply refer to the chapter again.

**Task 1.5**

Develop a simple program that examines a sequence for possible sequence similarities in a database (hint: enumerate what parts this program would consist of).

**Task 1.6**

Which of the following statements about BLAST is correct (multiple answers possible)?

A. BLAST = *Basic Local Alignment Search Tool.*
B. BLAST = *Basic Low Alignment Search Tool.*
C. BLAST is an algorithm for finding locally similar sequence segments in a database.
D. BLAST uses a heuristic search and here the two-hit *method* (2-hit method).

**Task 1.7**

Example: The sequencing of a diseased person has revealed the following protein sequence:
>unknownsequence 1.7

```
PQITLWQRPLVTIKIGGQLKEALLDTGADDTVLEEMNLPGRWKPKMIGGIGGFIKVRQYDQIL
IEICGHKAIGTVLVGPTPVNIIGRNLLTQIGCTLNF
```

Which BLAST algorithm would you choose for your patient sequence?

A. blastn.
B. blastp.
C. blastx or tblastx.
D. tblastn.

**Task 1.8**

You now want to know exactly which virus the person has contracted. Perform a BLAST search yourself using the protein sequence (https://blast.ncbi.nlm.nih.gov/Blast.cgi).

Which of the following statements is correct (multiple answers possible)?

A. The sequence is almost certainly the pol protein and protease of the HIV-1 virus.
B. The unknown sequence shows low similarity to the pol protein and protease of the HIV-1 virus.
C. When searching for a sequence that is as similar/identical as possible, a match should always have as large an *E-value* as possible and a low identity.
D. The *E-Value* (expected value) shows how likely it is that the hit will be found again in the database with a similar or better score.

**Task 1.9**

What is a dot plot and what can I use it for (hint: look up this software on the internet)?

**Task 1.10**

Example: Dotplot by hand.

1. By hand, perform a dot plot of the word BIOINFORMATICS to compare the word with itself.
2. Use software (e.g., Dotter [https://sonnhammer.sbc.su.se/Dotter.html], JDotter [https://athena.bioc.uvic.ca/virology-ca-tools/jdotter/], or Cheetah [https://mips. gsf.de/services/analysis/gepard]) and perform a dot plot of the following sequence with yourself:

>unknownsequence 1.10

PQITLWQRPLVTIKIGGQLKEALLDTGADDTVLEEMNLPGRWKPKMIGGIGGFIKVRQYDQI
LIEICGHKAIGTVLVGPTPVNIIGRNLLTQIGCTLNF

| Useful Tools and Web Links | |
|---|---|
| Perl | https://www.perl.org/ |
| Java | https://www.oracle.com/technetwork/java/index.html |
| Python | https://www.python.org/ |
| C++ | https://www.cplusplus.com/ |
| R | https://www.r-project.org/ |
| BLAST | https://blast.ncbi.nlm.nih.gov/Blast.cgi |
| PDB | https://www.rcsb.org/pdb/home/home.do |
| BioNumbers | https://bionumbers.hms.harvard.edu/ |
| PubMed | https://www.ncbi.nlm.nih.gov/pubmed/ |
| EBI | https://www.ebi.ac.uk/services |
| DDBJ | https://www.ddbj.nig.ac.jp |
| ExPASy | https://www.expasy.org |
| PROSITE | https://prosite.expasy.org |
| SWISS-MODEL | https://swissmodel.expasy.org |
| MODELLER | https://salilab.org/modeller/tutorial/ |
| QUARK | https://zhanglab.ccmb.med.umich.edu/QUARK/ |
| AnDom | https://andom.bioapps.biozentrum.uni-wuerzburg.de/index_new.html |

## Literature

Altschul SF, Gish W, Miller W et al (1990) Basic local alignment search tool. J Mol Biol 215(3):403–410

Antiretroviral Therapy Cohort Collaboration (2008) Life expectancy of individuals on combination antiretroviral therapy in high-income countries: a collaborative analysis of 14 cohort studies. Lancet 372(9635):293–299. https://doi.org/10.1016/S0140-6736(08)61113-7

Hoog R, Lima V, Sterne JA et al (2008) Life expectancy of individuals on combination antiretroviral therapy in high-income countries: a collaborative analysis of 14 cohort studies. Lancet
372(9635):293–299. https://doi.org/10.1016/S0140-6736(08)61113-7

## Further Reading

Altschul SF, Madden TL, Schäffer AA et al (1997) Gapped BLAST and PSI-BLAST: a new generation of protein database search programs. Nucleic Acids Res 25(17):3389–3402 (Review.
PubMed PMID: 9254694 *The classic 1990 paper on the best-known sequence comparison
algorithm, the Basic Alignment Sequence Research Tool [BLAST]. Surprisingly, it was not until
1997 that gaps were considered ["gapped BLAST"], and it became possible to repeat the search
with multiple sequences if the function was not yet clear ["position specific iterative" BLAST
or psi-BLAST])
Bienert S, Waterhouse A, Beer TA de et al (2017) The SWISS-MODEL Repository-new features
and functionality. Nucleic Acids Res 45(D1):D313–D319. https://doi.org/10.1093/nar/gkw1132
(PubMed PMID: 27899672; PubMed Central PMCID: PMC5210589 *This is the latest version of
the homology program Swiss-Model, a very convenient program that predicts from the sequence
of a protein its three-dimensional structure, just send by e-mail the sequence to the server)
Gaudermann P, Vogl I, Zientz E et al (2006) Analysis of and function predictions for previously conserved hypothetical or putative proteins in Blochmannia floridanus. BMC Microbiol 2006(6):1
(*This paper provides a good introduction to how one can still determine the function of a protein
with sequence and structural analyses, even if BLAST initially finds no evidence of function)
Gupta SK, Bencurova E, Srivastava M et al (2016) Improving re-annotation of annotated eukaryotic
genomes. In: Wong K-C (Herausgeber) Big data analytics in genomics. Springer, S171–195.
https://link.springer.com/chapter/10.1007%2F978-3-319-41279-5_5 (*In this work, we explain
how to improve annotation [labeling] in a higher [eukaryotic] genome)
NCBI Resource Coordinators (2017) Database resources of the National Center for Biotechnology
Information. Nucleic Acids Res 45(D1):D12–D17. doi:https://doi.org/10.1093/nar/gkw1071
(PubMed PMID: 27899561; PubMed Central PMCID: PMC5210554 *This explains the bioinformatics opportunities at NCBI, the world's premier bioinformatics entry site)
SIB Swiss Institute of Bioinformatics Members (2016) The SIB Swiss Institute of Bioinformatics'
resources: focus on curated databases. Nucleic Acids Res 44(D1):D27–D37. doi:https://doi.
org/10.1093/nar/gkv1310 (*Hier werden die Bioinformatik-Möglichkeiten am Schweizer
Bioinformatik-Institut erklärt)
Srivastava M, Malviya N, Dandekar T (2015) Application of biotechnology and bioinformatics tools
in plant-fungus interactions. In: Bahadur B, Rajam MV, Sahijram L, Krishnamurthy KV (Hrsg)
Plant Biol Biotechnol. Springer India, S 49–64 (*Here we explain how to bioinformatically study
protein interactions)

# Magic RNA

**2**

**Abstract**

About half of the human genome is actively transcribed as RNA, new regulatory and non-protein-coding RNA types such as miRNAs and lncRNAs in higher cells and the CRISPR/Cas9 system from bacteria underline the importance of RNA for molecular biology. Typically, one analyzes RNA sequence, structure, and folding energy orientationally first using RNAAnalyzer software, Rfam database, and RNAfold server. GEO and GeneVestigator databases show gene expression differences that can be analyzed in more depth using R and Bioconductor as scripting language and program framework. Both are important tools, but they have to be learned like a language in order to be able to write instructions for biostatistical analysis (so-called "scripts"). Non-coding RNA is also important for diseases, and bioinformatics helps to uncover this, e.g. chast-lncRNA in heart failure.

## 2.1 RNA Sequences Are Biologically Active

What does magic mean? It means that words are immediately translated into action! For example, you mutter an incantation of the air spirit, and the medicine man uses it to set the air in motion. In everyday life, you can't do that, or only if you have a lot of money. Then with this "wishing machine", the money, one can also put every purchasable wish into action.

So in our everyday world, the thought (easy) and the deed (sweaty, grueling, tiring) are well separated. But in the molecular world this is not so, in particular RNA has even magical properties in this sense.

We can form single words especially with RNA building blocks ("nucleotides"), but at the same time this chain of RNA building blocks then already has active properties, can accelerate biochemical reactions or even make them possible in the first place - in a word: magic!

This is due, on the one hand, to the smallness of the dimensions on which we are moving here, namely a few angstroms (Å, i.e. ten billionths of a metre), as well as to the special properties of RNA. It is not as stable as deoxyribonucleic acid, i.e. DNA, which is therefore very suitable as a long-term storage medium. RNA stores for shorter times, after which it can either be digested with its additional OH group or otherwise continue to react. And this is also the reason for its "magic" activity, it can accelerate or advance a reaction at the same time.

This also makes it clear what existed before today's division of labor between genetic information (DNA) and enzymatic action (proteins): namely, the RNA world. That was more than 3 billion years ago. The first cells were just coming into being, and it was there that RNA nucleotides of varying lengths both stored information and accelerated reactions. The oldest molecule was an RNA polymerase made of RNA, which catalytically transcribed its description, faster than it was destroyed by environmental stresses. If you still want to know what was before RNA: metabolism on surfaces that held certain molecules and obtained energy from sulfur compounds until the first membranes and first nucleotides accumulated more and more on these surfaces (Scheidler et al. 2016).

Since that time, RNA has been essential for all life. The protein factories (ribosomes) of the body consist of RNA in their central parts. All peptide bonds in the ribosome are made by catalytic ribosomal RNA (rRNA), and many vitamins and excipients in our enzymes are still made of nucleotides (especially adenine, e.g. FAD, NAD, NADH, NADP, NADPH, cAMP, ATP, etc.).

But that's not all: RNA can not only build proteins (with the help of tRNA and rRNA), whereby the genes are transcribed via mRNA (messenger RNA), but there are also numerous regulatory functions of RNA. As microRNA (miRNA), it degrades messenger RNA more quickly (and one small molecule directs many, sometimes hundreds of messenger RNAs), as long non-coding RNA (lncRNA) it even switches off entire chromosomes, as smallRNA (sRNA) in bacteria it switches off or on promoters or individual genes, as a riboswitch (e.g. riboswitch finder [https://riboswitch.bioapps.biozentrum.uni-wuerzburg.de/]) it allows or rejects the translation of genes.

It can be seen that an important part of bioinformatics is trying to identify and describe the function and hidden signals in RNA molecules. The basic question is: Where is the signal in the RNA molecule? First, in the order of its building blocks, i.e. in the nucleotides (the so-called sequence), but then also in the folding of the RNA, the secondary structure, how the RNA forms. In addition, one can also look at how stable the folding of the RNA is, the so-called folding energy.

So, with these three characteristics, I can check a wide range of RNA molecules if I know what sequence, secondary structure, and energy the RNA molecule must have for a particular property. For example, one can check all three characteristics for a number of molecules using the RNAAnalyzer program or look up exciting RNA types in the Rfam database.

If you want to write such an RNA detection program yourself, you first need access to very many RNA sequences under which RNA molecules with a certain property (a "regulatory element") are hidden. Then you have to check which RNA molecules do not show these properties by chance (false-positive alarm), but actually possess them. In doing so, one must also not hastily discard molecules that may have this property after all. In practice, it takes a lot of trial and error to look more and more closely at the sequence, secondary structure and energy until regulatory RNA elements can be accurately identified. One example is *iron-responsive elements* (IREs) in messenger RNA. A protein, the IRE binding protein (IRE-BP), binds to these when the iron level is low. This then prevents further reading of the mRNA (always read from the beginning, the 5′-end, to the end, the 3′-end). The reading frame, i.e. the protein building instructions in the mRNA, is located downstream of the IRE. If the iron level is higher, the protein binds to iron, and the messenger RNA containing the iron-sensitive element is translated. In practice, it also helps to look at the biological function of the messenger RNA, because that must have something to do with iron metabolism if you suspect such an RNA element in that messenger RNA. So that's an important way to test this and come up with meaningful results. Interestingly, for a structure in RNA that mediates regulation, that is a so-called RNA element, both secondary structure and primary sequence and folding energy play important roles. In an IRE, for example, one finds the consensus sequence CAGUGN and a C alone without G as a partner in the opposite strand ("*bulged*"), a loop *stem-loop* structure consisting of two stems on top of each other (in between is the unpaired C), and a folding energy between −2.1 to −6.7 kcal/mol (Fig. 2.1).

RNA is therefore at the root of life and is a particularly active intermediate carrier of information. Just recently, much faster sequencing techniques than in the past have made it possible to read virtually all RNA molecules in the cell. Because bioinformatics can classify this large amount of sequenced RNA quickly enough (Chang et al. 2013), we are only now beginning to recognize the many functions that RNA mediates. Examples of such newly recognized RNA molecules are the regulatory miRNAs and lncRNAs that have been newly described for the past 5–10 years (Kunz et al. 2015, 2016, 2017; Fiedler et al. 2015). These play essential roles in various diseases, and bioinformatics can make an important contribution to uncovering this. To this end, we have developed various methods and analytical tools for integrative analysis of RNAs (Kunz et al. 2018, 2020; Stojanović et al. 2020; Fuchs et al. 2020). For example, our bioinformatics work could help to uncover the function of Chast-lncRNA in heart failure (Viereck et al. 2016) or molecular mechanisms of miRNA-21 in cardiac fibrosis (Fuchs et al. 2020).

## 2.2 Analysis of RNA Sequence, Structure and Function

A number of options are available for analysing RNA, e.g. databases such as Rfam, software such as the RNAAnalyzer and RNAfold. In the following, we would like to introduce these.

```
Iron-resp egg..:
 Position:      44
 Sequence:      cugugucuugcuucaacaguguuugaacggaacagaccc
 Structure:     ((((...(((.(((((......)))))))))).))))....    -5.900000 kcal/mol Quality: good
 Structure:     ((((.((.((.(((((.....))))))))))))))....      -6.200000 kcal/mol Quality: good
 Structure:     ((((.(((...(((((......)))))).)))))))....     -6.100000 kcal/mol Quality: good
 Structure:     ((((.(((.(.(((((......)))))))))))))))....    -6.200000 kcal/mol Quality: good
```

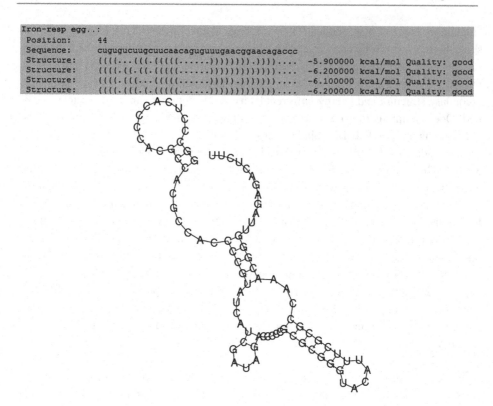

**Fig. 2.1** Bioinformatic analysis of a regulatory RNA element. Representation of the RNAAnalyzer of an IRE with associated sequence, structure and folding energy

### Rfam: All Known Families of RNA in One Database

▶     https://rfam.xfam.org/

One possibility is to collect all RNA molecules in a database. For this purpose, Rfam, for example, provides an overview of all RNA molecules (Gardner et al. 2011) that have been characterized in more detail to date. In particular, RNA molecules can be grouped into families. This means that a certain structure with which the RNA performs its function was retained in evolution and is then found in quite a large number of organisms.

IREs are one example. If such an RNA structure is present, the subsequent further mRNA sequence is only read and a protein that uses or utilises iron in some form is then produced by the ribosome if the iron level is sufficiently high.

▶     https://rfam.xfam.org/search/keyword?query=IRE

Rfam searches with the keyword "IRE", the correct entry is then:

▶ https://rfam.xfam.org/family/RF00037

Those who want to read more in-depth information about techniques and RNA functions in context can check out our books on regulatory RNA at Google-Books (Dandekar and Bengert 2002; Dandekar and Sharma 1998).

**RNAAnalyzer: A Quick Analysis for Each RNA Molecule**

▶ https://rnaanalyzer.bioapps.biozentrum.uni-wuerzburg.de

Another way to understand RNA and regulatory elements is to analyze the secondary structure and sequence motifs through a program. In our program developed for this purpose, the RNAAnalyzer, you can enter any RNA sequence, which is then searched for regulatory elements. The result is a list of regulatory element hits and important further descriptions, such as whether there is a lot of secondary structure, whether proteins can bind to the RNA or whether the RNA molecule is perhaps an mRNA, but also numerous other pieces of information (Bengert and Dandekar 2003).

One way to further check or supplement these results is to use the AnDom software (cf. Chap. 1, Protein analyses). For regulatory RNA, another alternative is the RegRNA server from Taiwan (https://regrna2.mbc.nctu.edu.tw/), which also offers a rapid analysis for RNA using related methods independently.

**RNAfold and mFold Show RNA Structure**
Another important method to analyze the RNA structure is to check the RNA folding with the pairing scheme: A always pairs with U (two hydrogen bonds), G with C (three hydrogen bonds). With the help of these rules and other rules (G pairs with U, only one hydrogen bond; thermodynamic parameters such as the Tinocco parameters), it is possible to systematically try out with the computer which structural folding of the RNA will lead to the highest number of base pairings and, in particular, hydrogen bonds and energy. This is also known as dynamic programming (Eddy 2004), because the sequence is broken down into small substrings and the optimal RNA structure is calculated iteratively (for longer RNA molecules, more and more memory is allocated dynamically for the base pairings).

Simple approaches such as the Nussinov algorithm are based on the optimal base pairing of the RNA, whereas extensions additionally consider the folding energy. The best known is the prediction algorithm of Zuker and Stiegler (1981), e.g. mFold server (https://unafold.rna.albany.edu/?q=mfold; out of operation since November 1, 2020.) or its further developments such as the RNAfold server (https://rna.tbi.univie.ac.at/cgi-bin/RNAWebSuite/RNAfold.cgi). The Sankoff algorithm takes phylogeny into account in addition to alignment and folding energy (e.g. LocARNA tool; https://www.bioinf.uni-freiburg.de/Software/LocARNA/). However, other software for RNA folding is also available (e.g., ViennaRNA package; https://www.tbi.univie.ac.at/RNA/; Freiburg RNA tools; https://rna.informatik.uni-freiburg.de/). By looking at several folding types (i.e., still the second- and third-best structure), I can see what remains conserved. These are usually also the structural regions actually present in the cell. In parallel with experiments, this gives a precise idea of what the RNA structure looks like in the living cell.

**Conclusion**

- RNA is an important level of information processing. About half of the human genome is actively transcribed and new RNAs such as miRNA and lncRNAs highlight the importance of deciphering the information encoded in RNA. In this chapter, we have therefore focused on the analysis of RNA sequence, structure and folding energy.
- RNA and regulatory RNA elements can initially be analysed with the RNAAnalyzer software, the Rfam database and the RNAfold server. For those who want to learn more, the tutorials show further steps (practice is important here, the tutorials offer a first introduction) to systematically analyze the transcriptome of a cell (e.g. GEO and GeneVestigator databases). For more in-depth statistical analysis of gene expression differences, R and Bioconductor are available. Both are important tools and have to be learned like a language in order to be able to write instructions for biostatistical analysis (so-called "scripts", both are scripting languages).
- In the field of computational analysis of RNA, new surprises and insights can be expected in the coming years, e.g. strong genetic engineering and matching software through the CRISPR/Cas9 system and the pathophysiology of newly discovered small RNAs in many bacteria and infectious agents (sRNAs). Non-coding RNA is also important in disease and bioinformatics is helping to uncover this, e.g. chast-lncRNA in heart failure (Viereck et al. 2016). ◄

## 2.3     Exercises for Chap. 2

In the exercises, important parts of the book will be dealt with in more detail in order to consolidate and practise what you have learned. Tasks that are marked as examples serve as application tasks in which you are to work independently with the computer in order to become more familiar with bioinformatics. In addition, we have provided numerous tutorials in the appendix, which also support the material of the textbook and the exercises and should contribute to a better understanding.

We recommend that you briefly review the material from Chap. 2 in Chap. 3 using the exercises.

**Task 2.1**

Example: As a result of transcription, a complete RNA sequence (mRNA, but also non-coding miRNA, lncRNAetc.) is formed, i.e. a copy of the DNA, whereby the nucleotides of the DNA (A, T, G and C) are translated into the nucleotides of the RNA (A, U, G and C) and the deoxyribose is exchanged for ribose. An RNA can form a secondary structure (alpha-helix and beta-sheet), which can be predicted bioinformatically.

Perform RNA folding with RNAfold using the following sequence (https://rna.tbi.uni-vie.ac.at/cgi-bin/RNAWebSuite/RNAfold.cgi):

> RNAsecondary structure

```
ATAAGAGACCACAAGCGACCCGCAGGGCCAGACGTTCTTCGCCGAGAGTCGTCGGG
GTTTCCTGCTTCAACAGTGCTTGGACGGAACCCGGCGCTCGTTCCCCACCCCGGC
CGGCCGCCCATAGCCAGCCCTCCGTCACCTCTTCACCGCACCCTCGGACTGCCCCAA
GGCCCCCGCCGCCGCTCCAGCGCCGCGCAGCCACCGCCGCCGCCGCCGCCTCTC
CTTAGTCGCCGCCATGACGACCGCGTCCACCTCGCAGGTGCGCCAGAACTACCACC
AGGACTCAGAGGCCGCCATCAACCGCCAGATCAACCTGGAGCTCTACGCCTCCTACG
TTTACCTGTCCATGTCTTACTACTTTGACCGCGATGATGTGGCTTTGAAGAACTTTGC
CAAATACTTTCTTCACCAATCTCATGAGGAGAGGGAACATGCTGAGAAACTGAT
GAAGCTGCAGAACCAACGAGGTGGCCGAATCTTCCTTCAGGATATCAAGAAACCAG
ACTGTGATGACTGGGAGAGCGGGCTGAATGCAATGGAGTGTGCATTACA
TTTGGAAAAAAATGTGAATCAGTCACTACTGGAACTGCACAAACTGGCCACTGA
CAAAAATGACCCCCATTTGTGTGACTTCATTGAGACACATTACCTGAATGAG
CAGGTGAAAGCCATCAAAGAATTGGGTGACCACGTGACCAACTTGCGCAAGATGGGAGC
GCCCGAATCTGGCTTGGCGGAATATCTCTTTGACAAGCACACCCTGGGAGACAGTGATAA
TGAAAGCTAAGCCTCGGGCTAATTTCCCCATAGCCGTGGGGTGACTTCCCTGGTCACCAAGGC
AGTGCATGCATGTTGGGGTTTCCTTTACCTTTTCTATAAGTTGTACCAAAACAT
CCACTTAAGTTCTTTGATTTGTACCATTCCTTCAAATAAAGAAATTTGGTACCCAGG
TGTTGTCTTTGAGGTCTTGGGATGAATCAGAAATCTATCCAGGCTATCTTCCAGATTCCTT
AAGTGCCGTTGT
```

1. Which of the following statements about RNA folding is correct (multiple answers possible)?
   (A)   An RNA secondary structure should always have a very high folding energy, then it is most stable.
   (B)   RNAfold does not find a possible secondary structure for the exercise example.
   (C)   For the exercise example, RNAfold calculates a *minimum free energy* (folding energy) of $-360.20$ kcal/mol.
   (D)   RNA folding (also bioinformatically predicted) is, from this point of view, a very simple process, since there is only one linear structure.
   (E)   RNA folding (also bioinformatically predicted) is a complex process from this point of view, since there are, for example, several secondary structural forms (e.g. *stem-* and *hairpin-loop*).

2. Create a short random RNA sequence (approx. 20–25 nucleotides) and let RNAfold fold it. Subsequently, double the sequence length and fold it again.

How do the amounts of the released energies of the short and long sequence relate to each other?

**Task 2.2**

There are also certain RNA motifs, also called regulatory RNA elements, that perform a specific function. Name and explain different RNA elements, such as IREs or riboswitches.

**Task 2.3**

What criteria do I have to consider in order to analyze an RNA (RNA molecule) for RNA motifs?

**Task 2.4**

Explain how to identify regulatory RNA elements.

**Task 2.5**

Which of the following statements about regulatory RNA elements is correct (multiple answers possible)?

(A) Regulatory RNA elements are not found in humans.

(B) For RNA motif searches, it's enough if I just look at the sequence.

(C) I can find many RNA families in the Rfam database.

(D) IRE and riboswitches are examples of regulatory RNA elements.

(E) It is best to combine several criteria (sequence, structure and energy) for an RNA motif search.

**Task 2.6**

Example: Find by hand (Ctrl + F or grep or Perl script) a typical conserved IRE motif (CAGUGN or CAGTGN) in the following sequence:
> RNAanalyzer.

```
ATAAGAGACCACAAGCGACCCGCAGGGCCAGACGTTCTTCGCCGAGAGTCG
TCGGGGTTTCCTGCTTCAACAGTGCTTGGACGGAACCCGGCGCTCGTTCCCCACCCC
GGCCGGCCGCCCATAGCCAGCCCTCCGTCACCTCTTCACCGCACCCTCGGACTGCCCC
AAGGCCCCCGCCGCCGCTCCAGCGCCGCGCAGCCACCGCCGCCGCCGCCGCCTC
TCCTTAGTCGCCGCCATGACGACCGCGTCCACCTCGCAGGTGCGCCAGAACTACC
ACCAGGACTCAGAGGCCGCCATCAACCGCCAGATCAACCTGGAGCTCTACGCC
TCCTACGTTTACCTGTCCATGTCTTACTACTTTGACCGCGATGATGTGGCTTT
GAAGAACTTTGCCAAATACTTTCTTCACCAATCTCATGAGGAGAGGGAACAT
GCTGAGAAACTGATGAAGCTGCAGAACCAACGAGGTGGCCGAATCTTCCTT
CAGGATATCAAGAAACCAGACTGTGATGACTGGGAGAGCGGGCTGAATGCA
ATGGAGTGTGCATTACATTTGGAAAAAAATGTGAATCAGTCACTACTGGAACTGCA
CAAACTGGCCACTGACAAAAATGACCCCCATTTGTGTGACTTCATTGAGACACAT
TACCTGAATGAGCAGGTGAAAGCCATCAAAGAATTGGGTGACCACGTGACCAACTTGCGCAA
GATGGGAGCGCCCGAATCTGGCTTGGCGGAATATCTCTTTGACAAGCACACCCTGGGAG
ACAGTGATAATGAAAGCTAAGCCTCGGGCTAATTTCCCCATAGCCGTGGGGTGACT
TCCCTGGTCACCAAGGCAGTGCATGCATGTTGGGGTTTCCTTTACCTTTTCTATAAG
TTGTACCAAAACATCCACTTAAGTTCTTTGATTTGTACCATTCCTTCAAATAAAG
AAATTTGGTACCCAGGTGTTGTCTTTGAGGTCTTGGGATGAATCAGAAATCTATCCAGGC
TATCTTCCAGATTCCTTAAGTGCCGTTGT
```

1. Can a potential IRE motif be found?
2. Now use the RNAAnalyzer (https://rnaanalyzer.bioapps.biozentrum.uni-wuerzburg.de/) for this examination.

Which of the following statements is correct (multiple answers possible)?

(A) The exercise example is an IRE.
(B) Besides the IRE, the RNAAnalyzer does not find any other elements for the exercise example, e.g. no "*Catalytic RNA*".
(C) The RNAAnalyzer finds an IRE at position 71 for the exercise example.
(D) One IRE is the consensus sequence "CAGUGN", the RNAAnalyzer also found this in the exercise example.

**Task 2.7**
Example: Perform a search with the Riboswitch Finder (https://riboswitch.bioapps.biozentrum.uni-wuerzburg.de/) using the following sequence (please just use the sequence example from the Riboswitch Finder page): Streptococcus pyogenes STPY1 (https://riboswitch.bioapps.biozentrum.uni-wuerzburg.de/examples.html).

Which of the following statements is correct (multiple answers possible)?

(A) The Riboswitch Finder finds three possible hits for a riboswitch for the example sequence, they are all on the minus strand.
(B) All hits found for the example sequence are of poor quality (sequence, structure, energy), thus indicating no possible riboswitches.
(C) The Riboswitch Finder finds three possible riboswitches on the plus strand at position 1288 for the example sequence.
(D) The hits found for the example sequence have, among others, three *stem-loops* in their secondary structure.
(E) Riboswitches are the only regulatory RNA elements in prokaryotes.

**Task 2.8**
Example: Analyze the 18 S-rRNA gene from Cordulegaster boltonii (GenBank ID: FN356072.1) for a possible ITS2 secondary structure using the ITS2 database (https://its2.bioapps.biozentrum.uni-wuerzburg.de/).

**Task 2.9**
Example:
(a) Familiarize yourself with non-coding RNAs (e.g. miRNAs and lncRNAs). Use e.g. https://www.microrna.org, https://www.mirbase.org, https://lncipedia.org/ and https://www.targetscan.org, but also our two articles (Kunz M et al. Bioinformatics of cardiovascular miRNA biology. J Mol Cell Cardiol 2015 Dec;89(Pt A):3–10. https://doi.org/10.1016/j.yjmcc.2014.11.027; Kunz M et al. Non-Coding RNAs in Lung Cancer: Contribution of Bioinformatics Analysis to the Development of

Non-Invasive Diagnostic Tools. Genes (Basel) 2016 Dec 26;8(1). pii: E8. https://doi.org/10.3390/genes8010008).

(b) Become familiar with different target prediction algorithms and their different parameters (e.g., TargetScan, miRanda, and PITA).

**Useful Tools and Web Links**

| | |
|---|---|
| **Rfam** | https://rfam.xfam.org/ |
| **RNAAnalyzer** | https://rnaanalyzer.bioapps.biozentrum.uni-wuerzburg.de/ |
| **mFold web server** | https://unafold.rna.albany.edu/?q=mfold (out of service since November 1, 2020) |
| **RNAfold web server** | https://rna.tbi.univie.ac.at/cgi-bin/RNAWebSuite/RNAfold.cgi (replaces mfold since 1.11.20) |
| **ViennaRNA Package** | https://www.tbi.univie.ac.at/RNA/ |
| **Freiburg RNA Tools** | https://rna.informatik.uni-freiburg.de/ |
| **regRNA** | https://regrna2.mbc.nctu.edu.tw/ |
| **Riboswitch Finder** | https://riboswitch.bioapps.biozentrum.uni-wuerzburg.de/ |

Dandekar and Bengert (2002) RNA Motifs and Regulatory Elements. Springer Verlag, 2002 (https://books.google.de/books?id=hOLtCAAAQBAJ&hl=de)
Dandekar and Sharma (1998) Regulatory RNA. Springer Verlag, 1998 (https://books.google.de/books?id=j7LoCAAAQBAJ&hl=de)

(c) Look for miRNAs that indicate a possible interaction with Brca1 (e.g. https://www.microrna.org and https://www.targetscan.org – do miRNAs find each other?).

## Literature

Bengert P, Dandekar T (2003) A software tool-box for analysis of regulatory RNA elements. Nucl Acids Res 31:3441–3445

Chang TH, Huang HY, Hsu JB et al (2013) An enhanced computational platform for investigating the roles of regulatory RNA and for identifying functional RNA motifs. BMC Bioinformatics 14(2):4

Dandekar T, Bengert P (2002) RNA motifs and regulatory elements. Springer. https://books.google.de/books?id=hOLtCAAAQBAJ&hl=de

Dandekar T, Sharma K (1998) Regulatory RNA. Springer. https://books.google.de/books?id=j7LoCAAAQBAJ&hl=de

Eddy SR (2004) How do RNA folding algorithms work? Nat Biotechnol 22:1457–1458

Gardner PP, Daub J, Tate J et al (2011) Rfam: Wikipedia, clans and the "decimal" release. Nucleic Acids Res 39(Database issue):D141–D145. https://doi.org/10.1093/nar/gkq1129

Scheidler C, Sobotta J, Eisenreich W et al (2016) Unsaturated C3,5,7,9-monocarboxylic acids by aqueous, one-pot carbon fixation: possible relevance for the origin of life. Sci Rep 6:27595. https://doi.org/10.1038/srep27595 (PubMed PMID: 27283227; PubMed Central PMCID: PMC4901337)

Zuker M, Stiegler P (1981) Optimal computer folding of large RNA sequences using thermodynamic and auxiliary information. Nucl Acid Res 9:133–148

## Own RNA Analysis Examples Together with Method Protocols

Fiedler J, Breckwoldt K, Remmele CW et al (2015) Development of long noncoding RNA-based strategies to modulate tissue vascularization. J Am Coll Cardiol 66(18):2005–2015. https://doi.org/10.1016/j.jacc.2015.07.081

Fuchs M, Kreutzer FP, Kapsner LA et al (2020) Integrative bioinformatic analyses of global transcriptome data decipher novel molecular insights into cardiac anti-fibrotic therapies. Int J Mol Sci 21(13):4727. https://doi.org/10.3390/ijms21134727

Kunz M, Xiao K, Liang C et al (2015) Bioinformatics of cardiovascular biology. J Mol Cell Cardiol 89(Pt A):3–10. https://doi.org/10.1016/j.yjmcc.2014.11.027

Kunz M, Wolf B, Schulze H et al (2016) Non-coding RNAs in lung cancer: contribution of bioinformatics analysis to the development of non-invasive diagnostic tools. Genes (Basel) 8(1):pii:E8. https://doi.org/10.3390/genes8010008

Kunz M, Göttlich C, Walles T et al (2017) MicroRNA-21 versus microRNA-34: lung cancer promoting and inhibitory microRNAs analyzed in silico and in vitro and their clinical impact. Tumour Biol 39(7). https://doi.org/10.1177/1010428317706430

Kunz M, Pittroff A, Dandekar T (2018) Systems biology analysis to understand regulatory miRNA networks in lung cancer. Methods Mol Biol 1819:235–247. https://doi.org/10.1007/978-1-4939-8618-7_11

Kunz M, Wolf B, Fuchs M et al (2020) A comprehensive method protocol for annotation and integrated functional understanding of lncRNAs. Brief Bioinform 21(4):1391–1396. https://doi.org/10.1093/bib/bbz066

Stojanović SD, Fuchs M, Fiedler J et al (2020) Comprehensive bioinformatics identifies key microrna players in ATG7-deficient lung fibroblasts. Int J Mol Sci 21(11):4126. https://doi.org/10.3390/ijms21114126

Viereck J, Kumarswamy R, Foinquinos A et al (2016) Long noncoding RNA chast promotes cardiac remodeling. Sci Transl Med 8(326):326ra22. https://doi.org/10.1126/scitranslmed.aaf1475

# Genomes: Molecular Maps of Living Organisms

**3**

**Abstract**

Based on sequence comparisons, special algorithms assemble the sequence fragments of modern sequencing techniques. After bacterial genomes and the yeast cell genome were completely sequenced and bioinformatically analysed in the 1990s, human genomes and numerous other eukaryotic (cells with a cell nucleus) genomes followed from 2001. The function of individual genes is identified by sequence comparisons: Protein function analysis (see Chap. 1), but also annotation of regulatory genome elements (ENCODE consortium) are main tasks of genome analysis. The genome sequence is available for almost all known organisms. It is thus possible to successfully predict the essential molecular components of these organisms.

## 3.1    Sequencing Genomes: Spelling Genomes

In the previous chapter we dealt with RNA as a "magic" molecule. But what about the permanent storage of information in the cell, the totality of DNA, the genome?

DNA means deoxyribonucleic acid, abbreviated to DNA in English, and is an excellent storage medium for information that living organisms have been using for almost 3 billion years. As is the case with our modern storage media, the read-in and read-out technology is quite important, because mostly only transcripts are produced, via RNA (see previous chapter). If, on the other hand, a unicellular organism reproduces or a multicellular organism grows, the cells of the body divide. And before they split into two halves, the genetic information in the cells has to be duplicated. There is an enzyme for this, the polymerase, and with it, adenine, guanine, cytosine and thymidine pair up as a new DNA strand to the opposite strand. With many nucleotides per second, an exact copy is thus produced. This process was first used by Frederick Sanger to read genetic information. He marked the

© Springer-Verlag GmbH Germany, part of Springer Nature 2023

T. Dandekar, M. Kunz, *Bioinformatics*,

https://doi.org/10.1007/978-3-662-65036-3_3

newly produced DNA radioactively, but also mixed dideoxy adenine triphosphate with the normal deoxy adenine triphosphate, so that the enzyme always stutters at the adenine and breaks off with about 1% probability at each adenine. This way, you can then visualize all the adenines in the sequence after sorting the radiolabeled fragments by size and putting on a film. If I use other dideoxy nucleotides, I also read the other nucleotides. I can also replace the radioactivity with nucleotides of different luminosity and use a laser to determine the nucleotides online. All this led to the fact that one could determine the DNA sequences ever faster, in order to store the sequence flood finally in large computer data bases. After the sequencing reaction and the separation of the fragments had been miniaturised further and further, the sequencing speed increased further and further so that it is now possible to read many millions of nucleotides per track and process many tracks simultaneously. In order to determine the genome sequence, the DNA of an organism is first chopped up ("shotgun" method) and then all these small pieces are sequenced simultaneously at lightning speed. However, this makes another task more and more difficult, namely to put the many sequence snippets together in the right way, i.e. to determine the genome sequence correctly from the snippets found by putting them together ("mapping" and "assembly" of the genome sequence). In particular, regions in which sequences are repeated again and again (repeat regions) are difficult to represent correctly in terms of their length and number of repeats.

Then we can begin to read the finished genome sequence, i.e. to understand its content (cf. Fig. 3.1). Many parts can be understood by sequence comparison, for example with the program BLAST. If this sequence section resembles an already labeled piece of DNA from another organism, I assume that this is also the function of this gene section in the newly sequenced organism. However, since similarities can be weak, labeling the genome sequence at the dissimilar sites can cause problems (technical term annotation; checking an existing label is called reannotation). As a simple rule of thumb, one adopts only those BLAST results that have an expected chance (E-value) of less than 1 in 1 million.

For the other parts of the genome sequence, which do not reveal their function so easily by high similarity, one has to analyse them in more detail. Here, machine learning and artificial intelligence methods (Chap. 14) help to understand the sequence. For example,

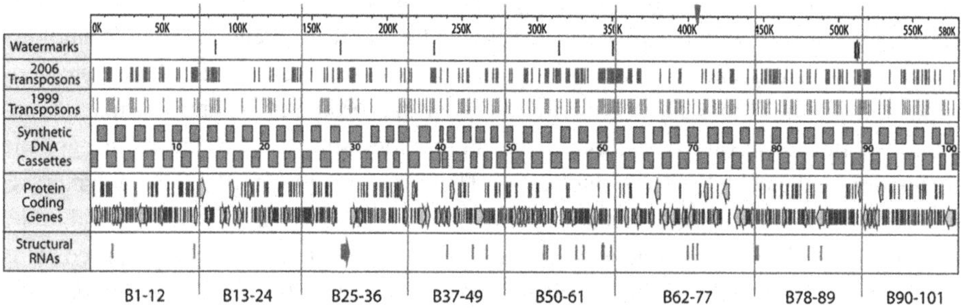

**Fig. 3.1** The figure shows a finished piece of the genome sequence. (Figure from Gibson et al. 2008)

stochastic models such as hidden Markov models (Sean R Eddy 2004) allow hidden system states (e.g. exon, intron) to be predicted from a sequence (observations, e.g. ATCCCTG ...) using a Markov chain (Bayesian network; supervised machine learning). Hidden Markov models are widely used for genome annotation (exon-intron region; e.g. GenScan program), but also for protein domain prediction (e.g. Pfam, SMART, HMMER, InterPro databases) and network regulation (e.g. signal peptides; SignalP, TMHMM programs).

In addition, there are numerous special software that detect RNA sequences (e.g. Rfam, tRNAscan), viral sequences, repeat regions (e.g. Repeat Masker) and other sites in the genome (e.g. enhancers, miRNAs, lncRNAs) and label them accordingly.

An important step is also to take a closer look at the promoter. Transcription factors bind to DNA sequence motifs (Patrik D'haeseleer 2006) in the promoter (so-called transcription factor binding sites, TFBS) and thus regulate gene expression (transcription). These conserved DNA patterns, usually consisting of 8–20 nucleotides, can be recognized by computers using binding site pattern recognition algorithms based on experimental data, such as chromatin immunoprecipitation DNA sequencing (Chip-Seq). A distinction is made between probabilistic (binding site; position weight matrix), discriminant (sites + non-functional sites) and energy (site + binding free energy) TFBS models (Stormo 2010, 2013). Databases such as Transfac and JASPAR contain the TFBS matrices for different organisms. These can be used, for example, to search a sequence for TFBS to understand gene expression (e.g. MotifMap, Alggen Promo, TESS, etc. programs), but also to find possible regulation via modular TFBS (TF modules) (e.g. using the Genomatix program). Besides, *ab initio approaches* (e.g. MEME Suite and iRegulon) try to find recurrent sequence patterns in multiple sequences via multiple alignment, which are then compared to known TFBS motifs for similarity. For example, we showed in one paper that heart failure-associated Chast-lncRNA is regulated by promoter binding of Nfat4 (Viereck et al. 2016).

In this way, from 1995 onwards (with *E. coli* and the yeast cell), the first genomes began to be completely labelled and published. This was followed by the genomes of eukaryotes (cells with a cell nucleus), which were about a thousand times larger, in particular that of humans (2001) and many other higher organisms (fly, mosquito, mouse, rat, chimpanzee, chicken, fish, etc.).

Another aspect is then to assemble the encoded proteins, RNAs and elements into higher networks. For example, a single enzyme does not stand alone, but forms metabolic networks (see next chapter). In the same way, a transcription factor that binds to the promoter of a gene does not stand alone, but is part of the overall regulation (so-called regulatory networks, see next but one chapter). The precise description of individual genes often requires not only DNA but also RNA ("transcriptome"), in particular in order to precisely determine the beginnings and ends of the segments overwritten in RNA. An integrative analysis yields the most accurate results here, even in the case of viruses with their compact genome (Whisnant et al. 2020).

One organism that has a fairly compact genome and yet is a fully viable self-contained cell is *Mycoplasma genitalium* (just over 580,000 nucleotides in size). In three exciting papers from 2009, Luis Serrano (experiments) and Peer Bork (bioinformatics) nicely illustrated these different levels of understanding the genome sequence, understanding the transcriptome and proteins, and understanding metabolism and regulation (Güell et al. 2009; Yus et al. 2009; Kühner et al. 2009). Figure 3.1 illustrates a completed piece of the genome sequence. We show here the *"origin of replication"* from Gibson et al. 2008, because in the case of bacteria in particular, this is where one starts to number the genes in their genome.

## 3.2    Deciphering the Human Genome

The deciphering of the human genome was a milestone in research. The sequencing techniques of the 1990s (capillary gel electrophoresis, automatic reading with a laser) were used systematically and intensively. Craig Venter, in particular, decided to go ahead in an industrial way and to finish much faster with the help of the first sequencing robots (only 3 years after 1998; Venter et al. 2001) than the group of typical university scientists and professors who had been working on the project for more than 10 years.

This race has certainly greatly accelerated the sequencing of the human genome, but also the development of the sequence analyses of bioinformatics that are necessary with it in order to put everything together "correctly". On the other hand, Craig Venter cannot be said to have "won". On the one hand, both working groups finished at about the same rate, but on the other hand, it has been the case that the map (i.e., collecting genetic markers, restriction sites, positional cloning of genes, etc.) of the public consortium under Erik Lander has been instrumental in enabling Venter to put his sequences together so quickly in the first place. Then in 2001, both consortia, the private company consortium and the public research consortium, published a first *"draft"* sequence of the genome (Lander et al. 2001; review in Lander 2011) – a rough map, but not only of the genes, but precisely of all the nucleotides that encode each gene.

This was the first time that the human genome had been "spelled out". However, the groundbreaking work of the ENCODE consortium (2012), for example, showed that after spelling, reading only really begins with a hundredfold better genome and, above all, transcriptome coverage, and one begins to understand the content and the subtleties of the human genome.

These results, which have continued to grow over the years, are now available on various entry pages.

For example, one can also seek out these results at NCBI for questions and analysis, e.g., via the link https://www.ncbi.nlm.nih.gov/geo/info/ENCODE.html.

### Entry Page of the Human Genome Project
A particularly good general access to human genome analysis and its history is provided by the entry page of the Human Genome Project.

▶       https://www.genome.gov/human-genome-project

The result is explained on *"All About The Human Genome Project (HGP)"*.

▶       https://www.genome.gov/10001772

An alternative view has the entry page of the *"Department of Energy"*. Here, many large-scale projects in physics were managed, which is why this page also highlights the *"Big Data"* aspect.

▶       https://genomics.energy.gov

A detailed review of all data is available in the archive of the Human Genome Project.

▶       https://web.ornl.gov/sci/techresources/Human_Genome/index.shtml

## 3.3   A Profile of the Human Genome

So what does our own genome look like? It is important to know that the human genome comprises about 3.2 billion base pairs (haploid, a complete set, for example in a sex cell) and is distributed in all body cells as a diploid total stock on 46 chromosomes: 44 autosomes, one pair of each chromosome (1 to 22) as well as two sex chromosomes, XX (woman) or XY (man). There are about 23,700 genes coding for proteins in the human genome (current status to be looked up at https://www.ensembl.org/Homo_sapiens/Info/ Index). There are also many thousands of RNA genes.

Since only 2–3% of the genome is needed for protein reading frames and only about 10% of the genome for the additional regulatory signals in mRNA, RNA precursors and finally genes with promoter sequences, the genome was initially seen to be loaded with up to 90% ballast. In particular, with selfish DNA distributed throughout the genome as short (SINE, *small interspersed elements*) and long elements (LINE, *large interspersed elements*, e.g. ALU sequences). Other such elements are transposons and former retroviral sequences. Other repetitive regions characterize promoters (GC regions). Stabilizing, structural DNA (around centromeres, at chromosome ends e.g. telomeres etc.) also occupies some space in the chromosome.

Nevertheless, after closer analysis, much more meaningful information is available in the human genome. First of all, there are the many splice variants from the protein genes, which increase the variance of the proteins in the different organs and life stages (especially in the embryo). There are numerous other genes, especially for the 22-nucleotide miRNAs that are excised from precursors, and the long *non-coding RNAs* (Liu et al. 2017). Therefore, the total amount of encoded genetic information is even higher. In total (genetic estimates), about 100,000 genetic traits are passed from generation to generation through the genome. Figure 3.2 makes this graphically clear.

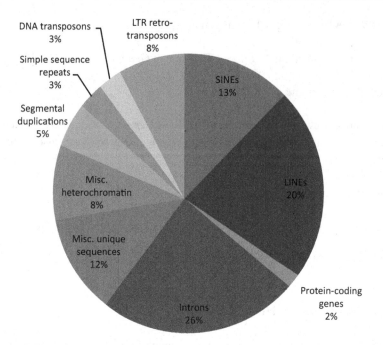

**Components of the Human Genome**

**Fig. 3.2** What is in our genome? If we look at the figure, it becomes clear that we only consist of about 25% genes, of which only 2–3% code for proteins (the majority are *selfish DNA*, LINE and SINE). (Image from https://upload.wikimedia.org/wikipedia/commons/6/64/Components_of_the_Human_Genome.jpg)

Interestingly, the publication of the human genome in 2001 was more of a race than an exhaustive analysis. This very elaborate detailed analysis is the goal of the ENCODE project. We should therefore also take a look at the subsequent, detailed analyses of the ENCODE consortium:

►     https://www.encodeproject.org

This consortium.

►     also: https://www.genome.ucsc.edu/ENCODE/

has further investigated the human genome in great detail after its first sequencing, intensively re-sequenced all areas, but also investigated contained DNA elements and created an encyclopaedia:

▶ https://www.genome.gov/10005107, ENCyclopedia Of DNA Elements.

In particular, the ENCODE consortium was able to show that at least half of the genome is transcribed at least some of the time, in addition to protein genes, especially various RNAs.

Numerous ENCODE publications (ENCODE 2011) continue to reveal new details of regulation in the human genome:

▶ https://www.nature.com/encode/#/threads

For example, in the regulation of histones, RNA, the transcriptome and promoters. In the meantime, the human genome has been sequenced many times, among other things to determine individual differences. A current project is even examining 10,000 human genomes (Telenti et al. 2016).

### Conclusion

- Based on sequence comparisons, special algorithms assemble the sequence fragments of modern sequencing techniques (see tutorials). After bacterial genomes and the yeast cell genome were completely sequenced and bioinformatically analysed in the 1990s, human genomes and numerous other eukaryotic (cells with a cell nucleus) genomes followed from 2001. The function of individual genes is identified by sequence comparisons. Protein function analysis (see Chap. 1), but also annotation of regulatory genome elements (ENCODE consortium) are main tasks of genome analysis.
- Eukaryotic genomes are billions of nucleotides in size, bacterial genomes only a few million. This means that there is room for long introns in the eukaryotic genome. Half of the human genome is transcribed, but there is also plenty of room for short (SINE) and long (LINE) repetitive elements and transposons.
- This combination of genome sequencing and bioinformatics means that the genome sequence is available for almost all known organisms. Bioinformatics can thus successfully predict the essential molecular components of these organisms: we live in the age of post-genomics (whenever the genome sequence is known). ◀

## 3.4 Exercises for Chap. 3

It is a good idea to briefly review the exercises for Chap. 2. You should also briefly look at the exercises for Chap. 3 later for repetition at Chaps. 5 and 7.

**Task 3.1**
Describe how the human genome is constructed.

**Task 3.2**

Describe what is meant by annotation or reannotation of a genome.

**Task 3.3**

Name and describe different sequencing techniques.

**Task 3.4**

Describe important steps in genome sequencing (also address bioinformatics challenges or important hurdles). Do you know any important pioneers of genome sequencing?

**Task 3.5**

Think about the bioinformatics requirements/challenges of ever-improving sequencing techniques, but also associated hurdles/limitations.

**Task 3.6**

Name and describe databases where you can find information about the genome.

**Task 3.7**

Explain how to bioinformatically screen a promoter sequence for transcription factor binding sites (name and briefly describe software/databases). What are the advantages of comparing e.g. several software/databases?

**Task 3.8**

Develop a simple program that reads in a promoter sequence, examines it for transcription factor binding sites, and outputs the result back. What parts would this program consist of? Also consider what challenges and sources of error this program would have to deal with.

**Task 3.9**

Analyze a sample RNA sequence:

Using the following sequence, perform a promoter search for possible transcription factor binding sites using ALGGEN PROMO software (https://alggen.lsi.upc.es/cgi-bin/promo_v3/promo/promoinit.cgi?dirDB=TF_8.3):

```
>FP018429 BRCA1_1
TTCCAAGGAACAGTGTGGCCAAGGCCTTTCGTTCCGCAATGCATGTTGGAAATAGTAGTTCTT
TCCCTCCACCTCCCAACAATCCTTTTATTTACCTAAACTGGAGACCTCCA
TTAGGGCGGAAAGAGTGGGGTAATGGGACCTCTTCTTAAGACTGCTTTGGACAC
TATCTTACGCTGATATTCAGGCCTCAGGTGGCGATTCTGACCTTGGTACAGC
AATTACTGTGACGTAATAAGCCGCAACTGGAAGCGTAGAGGCGAGAGGGCG
GGCGCTTTACGGCGAACTCAGGTAGAATTCTTCCTTTTCCGTCTCTTTCTTTTTATGTCACCAGG
GGAGGACTGGGTGGCCAACCCAGAGCCCCGAGAGATGCTAGGCTCTTTCTGTCCC
GCCCTTCCTCTGACTGTGTCTTGATTTCCTATTCTGAGAGGCTATTGCTCAGC
GGTTTCCGTGGCAACAGTAAAGCGTGGGAATTACAGATAAATTAAAACTGTGGAA
```

CCCCTTTCCTCGGCTGCCGCCAAGGTGTTCGGTCCTTCCGAGGAAGCTAAGGCCGCGTTGGG
GTGAGACCCTCACTTCATCCGGTGAGTAGCACCGCGTCCG

Which of the following statements are correct (multiple answers possible)?

(A)  ALGGEN PROMO does not find transcription factor binding sites in the DNA sequence.

(B)  All hits found are also actual transcription factor binding sites, so in each case it is sufficient to predict them bioinformatically only.

(C)  Transcription factors bind to DNA motifs, I can predict these bioinformatically.

(D)  ALGGEN PROMO finds a transcription factor binding site for NF-AT2 in the promoter (with the *"matrix dissimilarity rate"* setting of 15). I can then use this information for further experimental studies, such as whether NF-AT2 has an influence on transcription, in this case of BRCA1.

**Task 3.10**

Explain a Hidden Markov Model (feel free to use an example).

| Useful Tools and Web Links | |
|---|---|
| **Pfam** | https://pfam.xfam.org/ |
| **Rfam** | https://rfam.xfam.org/ |
| **SMART** | https://smart.embl-heidelberg.de/ |
| **ProDom** | https://prodom.prabi.fr/prodom/current/html/home.php |
| **UniProt** | https://www.uniprot.org/ |
| **GenScan** | https://genes.mit.edu/GENSCAN.html |
| **HMMER** | https://hmmer.org/ |
| **SignalP** | https://www.cbs.dtu.dk/services/SignalP/ |
| **TMHMM** | https://www.cbs.dtu.dk/services/TMHMM/ |
| **Transfac** | https://www.gene-regulation.com/pub/databases.html |
| **TESS** | https://www.cbil.upenn.edu/tess/ |
| **MotifMap** | https://motifmap.ics.uci.edu/ |
| **Alggen Promo** | https://alggen.lsi.upc.es/cgi-bin/promo_v3/promo/promoinit.cgi?dirDB=TF_8.3 |
| **Genomatix** | https://www.genomatix.de/ |
| **MEME Suite** | https://meme-suite.org/ |
| **iRegulon** | https://iregulon.aertslab.org/ |
| **tRNAscan** | https://lowelab.ucsc.edu/tRNAscan-SE/ |
| **Repeat Masker** | https://www.repeatmasker.org/ |
| **ENCODE** | https://www.encodeproject.org |
| **NIH** | https://www.genome.gov |
| **Genomic Science program** | https://genomics.energy.gov |
| **Human Genome Project** | https://web.ornl.gov/sci/techresources/Human_Genome/index.shtml |
| **Ensembl** | https://www.ensembl.org/Homo_sapiens/Info/Index |

## Literature

Eddy SR (2004) What is a hidden Markov model? Nat Biotechnol 22:1315–1316. https://doi.org/10.1038/nbt1004-1315

Gibson DG, Benders GA, Andrews-Pfannkoch C et al (2008) Complete chemical synthesis, assembly, and cloning of a mycoplasma genitalium genome. Science 319(5867):1215–1220. https://doi.org/10.1126/science.1151721

Güell M, van Noort V, Yus E et al (2009) Transcriptome complexity in a genome-reduced bacterium. Science 326(5957):1268–1271. https://doi.org/10.1126/science.1176951 (PubMed PMID: 19965477)

Kühner S, van Noort V, Betts MJ et al (2009) Proteome organization in a genome-reduced bacterium. Science 326(5957):1235–1240. https://doi.org/10.1126/science.1176343 (PubMed PMID: 19965468 *Here, genome and proteome of the small bacterial organism M. pneumoniae is explained in an exemplary manner)

Lander ES (2011) Initial impact of the sequencing of the human genome. Nature 470(7333):187–197. https://doi.org/10.1038/nature09792 (*Here, Eric Lander describes what followed from his first human genome sequence ten years later)

Lander ES, Linton M, Birren B et al (2001) Initial sequencing and analysis of the human genome. Nature 409(6822):860–921. https://doi.org/10.1038/35057062 (*The landmark paper about the first description of the human genome)

Liu SJ, Horlbeck MA, Cho SW et al (2017) CRISPRi-based genome-scale identification of functional long noncoding RNA loci in human cells. Science 355(6320). pii: aah7111. https://doi.org/10.1126/science.aah7111 (*This recent work describes that there are thousands of human lncRNAs [over 200 nucleotides long], and 16401 lncRNA loci after they have been studied in seven Cell lines studied in more detail. 499 lncRNAs were identified as essential for cell growth, with 89% being cell type specific. Presumably, there are also thousands of miRNA loci; the ENCODE consortium had evidence for many miRNAs).

Patrik D'haeseleer (2006) What are DNA sequence motifs? Nat Biotechnol 24:423–425. https://doi.org/10.1038/nbt0406-423

Stormo G (2010) Zhao Y (2010) Determining the specificity of protein–DNA interactions. Nat Rev Genet 11:751–760. https://doi.org/10.1038/nrg2845

Stormo GD (2013) Modeling the specificity of protein-DNA interactions. Quant Biol 1(2):115–130. https://doi.org/10.1007/s40484-013-0012-4

Telenti A, Pierce LC, Biggs WH et al (2016) Deep sequencing of 10,000 human genomes. Proc Natl Acad Sci U S A 113(42):11901–11906 (PubMed PMID: 27702888; PubMed Central PMCID: PMC5081584 *This paper shows the current state of human genome sequencing: In the meantime, even 10000 genomes can be compared on an industrial scale, for instance for conserved single nucleotide polymorphisms. https://www.ncbi.nlm.nih.gov/pubmed/27702888)

The ENCODE Project Consortium (2012) An integrated encyclopedia of DNA elements in the human genome. Nature 489:57–74. https://doi.org/10.1038/nature11247 (*The ENCODE consortium has created an encyclopedia of all DNA elements in the human genome and is about 100 times more accurate than the original initial sequencing. It also showed that about half of the human genome is actively transcribed, much more than the protein genes [30% of the genome; coding regions only 3%])

Venter JC, Adams MD, Myers EW et al (2001). The sequence of the human genome. Science 291(5507):1304–1351. Erratum in: Science 292(5523):1838 (PubMed PMID: 11181995 *This is the famous human genome sequencing paper that J. Craig Venter and his little Armada of sequencing robots accomplished in just three years)

Viereck J, Kumarswamy R, Foinquinos A et al (2016) Long noncoding RNA chast promotes cardiac remodeling. Sci Transl Med 326:326ra22. https://doi.org/10.1126/scitranslmed.aaf1475

Whisnant AW, Jürges CS, Hennig T et al (2020) Integrative functional genomics decodes herpes simplex virus 1. Nat Commun 11(1):2038. https://doi.org/10.1038/s41467-020-15992-5

Yus E, Maier T, Michalodimitrakis K et al (2009) Impact of genome reduction on bacterial metabolism and its regulation. Science 326(5957):1263–1268. https://doi.org/10.1126/science.1177263 (PubMed PMID:19965476 *This article from the Serrano group and Bork group describes how the genome and metabolism, as well as its regulation, of Mycoplasma pneumoniae has adapted)

# Modeling Metabolism and Finding New Antibiotics

**4**

**Abstract**

Metabolic modelling allows metabolism to be analysed in detail. Biochemical knowledge and databases such as KEGG determine the set of all enzymes involved. It is then possible to calculate which metabolic pathways and enzyme chains keep the metabolites in a network in equilibrium (flux balance analysis), which of these are also no longer decomposable (elementary mode analysis) and which of these are sufficient to represent all real metabolic situations by combining a few pure flux modes (extreme pathway analysis). To calculate the flux strength, one needs further data, e.g. gene expression data and software (e.g. YANA programs). Further analyses look at metabolic control (metabolic control theory) and describe the rates (kinetics) of the enzymes involved more precisely. This allows a better description and understanding of metabolism, prediction of essential genes and resulting antibiotics as well as metabolic responses, for example in tumour growth.

The genome sequence allows bioinformatics to gain a much better overview of the organism. In particular, this allows us to determine much better than before which enzymes and metabolic pathways occur in an organism.

Is it possible for the bioinformatician to calculate, for a given set of enzymes, what metabolism might come out of it?

The surprising answer is "yes"; so-called metabolic modeling (Mavrovouniotis et al. 1990; Schuster and Schuster 1993) can indeed answer this question.

© Springer-Verlag GmbH Germany, part of Springer Nature 2023
T. Dandekar, M. Kunz, *Bioinformatics*,
https://doi.org/10.1007/978-3-662-65036-3_4

## 4.1     How Can I Model Metabolism Bioinformatically?

The reasoning is as follows: All metabolic sources ("sources") serve, after all, to nourish and supply the organism, and in the same way there are excretions which dispose of the unnecessary metabolic products ("sinks"). But for all other metabolites ("internal metabolites"), the body and each cell of the body must ensure that they are supplied and degraded in the same amount over time, so that this internal metabolite is neither permanently missing nor increasing. This can be translated into a mathematical calculation (algorithm) to calculate which enzyme chains balance all the internal metabolites involved (programs we have developed for this purpose include YANA and Metatool; an overview of numerous other programs for this purpose is provided by Dandekar et al. 2014). Interestingly, this initially somewhat abstract result (all "elementary flux modes") is an accurate description of all metabolic possibilities for this organism with these enzymes. Figure 4.1 provides a general overview of metabolic modeling, and Fig. 4.2 of elemental mode analysis.

We have investigated this in more detail, e.g. for the metabolic network of glycolysis and the pentose phosphate pathway, and were able to show (Schuster et al. 2000) that by exact mathematical calculation one can also find additional alternatives from these two metabolic pathways, e.g. other enzyme combinations that nevertheless balance all the internal metabolites used. These allow the organism to adapt to completely different metabolic conditions, e.g. to produce a lot of NADPH, energy or nucleotides (Fig. 4.3).

However, apart from marvelling at the numerous metabolic possibilities that even simple bacteria have as well as higher cells, these flow analyses can also be used for various applications.

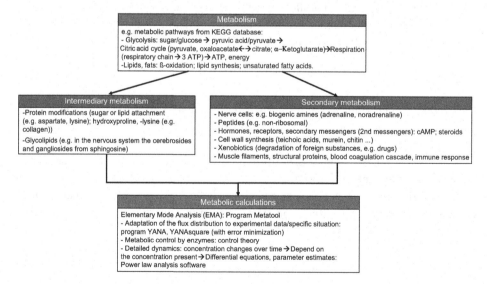

**Fig. 4.1** Overview of metabolic modelling

**Fig. 4.2** Overview of
elemental mode analysis

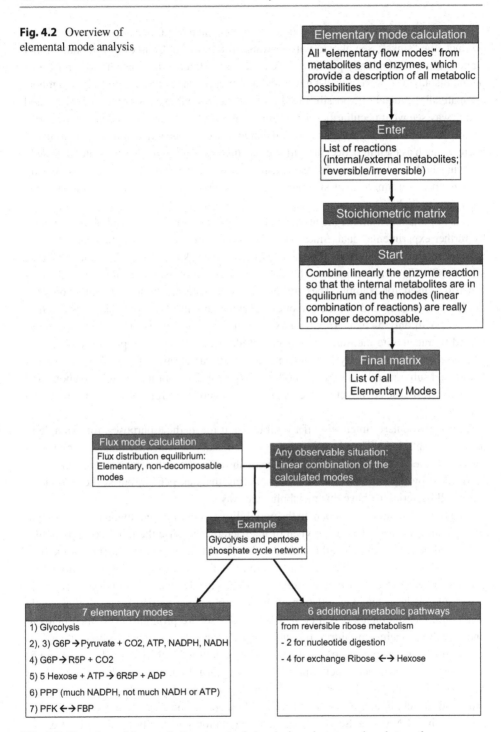

**Fig. 4.3** Overview of the metabolic network of glycolysis and pentose phosphate pathway

In particular, I can use it to investigate how I can achieve the best possible yield of a product (the "sinks", see above) with starting products (the "sources"), for example if I want to biotechnologically produce citric acid for the kitchen or nanocellulose for transparent displays – to give a well-known and a very modern example. Similarly, I can now compare all the metabolic possibilities for one organism with various other organisms and in this way see what peculiarities are present or even what diversions and alternatives one organism has and the other does not. You can also see in this way that different strains of bacteria, such as meningococci, use different pathways to achieve the same rate of growth, allowing both a pathogenic, disease-causing lifestyle and a more benign lifestyle with greater effort on amino acid synthesis but less aggressiveness against the human host (Ampattu et al. 2017).

It is important to validate the modelled (and thus only predicted) metabolic differences by further experimental data. Since individual errors are corrected by the metabolic flux network model (in a metabolic flux, all enzymes must work together at the same rate), data such as RT-PCR measurements on the mRNA expression of metabolically active enzymes can also be used, for example. These mRNA measurements are "indirect" because only the mRNA is measured and not the protein or enzyme activity; however, this works well in practice, with only 5–10% error for fluxes from a network of 30–100 enzymes, as confirmed by metabolite measurements (Cecil et al. 2011, 2015). Examples of applications include the changing lifestyle of chlamydiae (bacteria) during infection (as elementary bodies and subsequently as reticular bodies; Yang et al. 2019) or the mutual metabolic and regulatory responses to infection events in fungal infections of fungus and host (Srivastava et al. 2019).

This is particularly interesting if I want to use it for medical purposes, for example to develop an antibiotic. Then I am interested in the metabolic pathways that as many bacteria as possible have in common, but which are absent in the sick person and can therefore be blocked by the antibiotic without endangering the sick person, but at the same time killing all bacteria that have this metabolic pathway.

The flux calculations also open up the possibility of identifying individual enzymes that are particularly critical for the survival of the bacteria (because the failure of a particular enzyme affects, for example, all flux modes that provide an essential cofactor for the bacterium and not just a few). This may also help in finding new drugs against insidious fungal infections. One can also re-examine the detailed effects of an antibiotic with gene expression analyses and a calculation of the resulting metabolite fluxes as well as single metabolite measurements (Cecil et al. 2011; YANAsquare program). This then helps to find new drugs against multidrug-resistant staphylococci, for example (Cecil et al. 2015).

At present, we also want to link the different modelling levels (Chaps. 1, 2, 3, 4 and 5) more intensively in order to better protect plants against drought stress and infections, for example by identifying key enzymes that have an alternative regulatory function (e.g. aconitase, which, in addition to its metabolic function in the citric acid cycle, also regulates IRE in mRNAs, see Sect. 2.2) and alter regulation favourably for drought stress or resistance to infection.

## 4.2    Useful Tools for Metabolic Modelling

As we have already learned, all metabolic sources serve to nourish and supply the organism ("sources"). In addition, there are excretions ("sinks"), but also internal metabolites, which must be supplied and degraded to the same extent. In this context, mathematical algorithms can calculate the existing metabolic fluxes for all enzymes and reactions involved and are a helpful tool for metabolic modelling. Table 4.1 presents a number of applications of metabolic modelling.

Table 4.2 shows next how the modeling is then technically performed, whereby this is again only a selection of useful programs.

**Metatool** (Table 4.2) has been in use since 2005 (von Kamp and Schuster 2006). It allows the calculation of the stable metabolic pathways available to the metabolism for a given set of enzymes. It is constantly being further developed. In addition to the integer version 4.9, which runs stably on Windows, there is a new version 5.1 and variants for Linux and Windows.

The **YANA programs,** which are programmed in Java and can therefore be used flexibly on any computer, can be used to calculate not only the different flux possibilities for metabolism, but also how strong the flux is in a particular situation, especially through a single enzyme. The programs can analyze larger and larger networks faster and faster; a genome-wide network is described in the 2011 paper (Schwarz et al. 2005, 2007; Cecil et al. 2011).

**CellNetAnalyzer (CNA)** is a MATLAB toolbox. Via a graphical user interface, various computer methods and algorithms are offered for the analysis of the structure of metabolic networks as well as for the analysis of signaling networks and regulatory networks.

Metabolic networks are modeled using a stoichiometric matrix and boundary conditions. Thus, CNA uses very similar principles of flux balancing as Metatool does with elementary mode analysis. Their calculation is also offered as well as *"minimal cut sets"* (how do I safely cut a metabolic pathway?). The different algorithms are also offered for the construction of strains as well as for metabolic engineering (e.g. optimal yields in biotechnology).

**Table 4.1**  Applications of metabolic modelling. (Dandekar et al. 2014)

| | |
|---|---|
| Metabolic fluxes ("modes") that occur only in bacteria but not in humans | Antibiotics |
| Yield (final product) for given starting product(s) | Biotechnology |
| Growth equation | Calculation of growth in plants, bacteria, pathogens |
| Metabolic overview | Characterization of microbes and organisms |
| (Often from genome sequence) | Characterization of the adaptation potential, identification of organisms |
| Calculation of robustness | Prediction of essential genes |

**Table 4.2** Programs for metabolic modeling

| Metatool | https://pinguin.biologie.uni-jena.de/bioinformatik/networks/metatool/ |
|---|---|
| YANA | https://www.bioinfo.biozentrum.uni-wuerzburg.de/computing/ yanasquare/ |
| CellNetAnalyzer | https://www2.mpi-magdeburg.mpg.de/projects/cna/cna.html |
| COPASI | https://copasi.org/ |
| *Flux Balance Analysis* | https://systemsbiology.ucsd.edu/Downloads/FluxBalanceAnalysis |
| COBRA Toolbox | https://opencobra.github.io/ |

CNA also uses Boolean networks as well as multi-digit logic and interaction graphs and can thus also model signal networks and regulation. The stable system states are determined and the dynamics are simulated with differential equations (via a so-called *plugin*, an additional program that uses the software ODEfy). Finally, one can also consider network properties such as the signal network length and any feedback loops that may be present.

The **COPASI** *"Biochemical System Simulator"* allows to analyze biochemical networks in their structure and dynamics (Kühnel et al. 2008; Kent et al. 2012; Bergmann et al. 2016). It is also possible to read in models (in SBML format) and model the network using differential equations ("ODEs") or stochastic (*"Gillespie's stochastic simulation"*), so that random events (e.g. nutrient supply) can be simulated well.

*Flux Balance Analysis* (FBA) is the software of the world-famous old master of metabolic simulations, Bernhard Palsson. You can also model metabolic and, with extensions, signal networks.

The **COBRA toolbox** (Kent et al. 2012) is useful for metabolic modeling and signaling cascades. A detailed tutorial, including the starting metabolic model for *E. coli,* is available and a whole community of users and developers. Orth et al. (2010) introduce an instructive *E.* coli metabolism model in a separate paper.

## Conclusion

Metabolism is fundamental to the nutrition, growth and reproduction of all living beings. Metabolic modelling allows us to look at this in detail. Bioinformatics first uses biochemical knowledge and databases such as KEGG to determine the set of all enzymes involved. It is then possible to calculate (see exercises and tutorials) which metabolic pathways and enzyme chains keep the metabolites in a network in equilibrium (flux balance analysis), which of these are also no longer decomposable (elementary mode analysis) and which of these are sufficient to represent all real metabolic situations by combining a few pure flux modes (*extreme pathway analysis*).

In order to calculate the flux strength, one needs further data, e.g. gene expression data and software (e.g. YANA programs). Further analyses look at metabolic control (metabolic control theory) and describe the rates (kinetics) of the enzymes involved in more detail. This is mathematically complex, but leads to deeper insights into their regulation and function.

Bioinformatics thus makes it possible to better describe and understand metabolism, to predict essential genes and resulting antibiotics as well as metabolic responses, for example in tumour growth or for bacterial cell wall synthesis. ◄

## 4.3 Exercises for Chap. 4

As an introduction, it is advisable to work through the exercises in Chap. 11 (Sect. 11.1, 11.2, 11.3, 11.4, 11.5, and 11.6).

**Modelling metabolic networks:**

**Task 4.1**
Describe how metabolic pathways can be calculated bioinformatically. Also state possible problems with metabolic modelling.

**Task 4.2**
Name a computational program for metabolic pathways.

**Task 4.3**
Explain how to compile all the enzymes of glycolysis for a metabolic pathway. What advantages do you have in each case when you compare several databases?

**Task 4.4**
Explain what is meant by elemental mode analysis.

**Task 4.5**
You want to develop a new antibiotic. Which enzymes in your metabolic pathway could be interesting antibiotic targets?

**Task 4.6**
Perform elementary mode analysis on the citrate cycle/citric acid cycle in *E. coli*. First download Metatool (https://www.bioinfo.biozentrum.uni-wuerzburg.de/computing/metatool_4_5/). Then create the Metatool file for the citrate cycle/citric acid cycle in *E. coli* *yourself* and carry out an analysis.

Look at the metabolic network in Metatool and answer the following questions:

1. How many modes do I get?
2. How do I interpret my found modes in terms of finding drugs/targets against bacteria?

To better understand an elemental mode analysis, you should also answer the following questions:

3.  What happens to the number of modes when I change a metabolite from internal to external? Why does this happen?
4.  What happens to the number of modes if I set all metabolites from external to internal? Why does this happen?
5.  What happens to the number of modes when I change an enzyme from irreversible to reversible? Why does this happen?
6.  What happens to the number of modes if I change all the enzymes from reversible to irreversible? Why does this happen?

**Task 4.7**

Perform elementary mode analysis for pyrimidine metabolism. In doing so, compare the metabolism between humans and *S. aureus*. Proceed according to Example 4.6 and answer the following questions:

1.  How many modes do I get in humans and *S. aureus?*
2.  Are there differences in pyrimidine metabolism between humans and *S. aureus?*
3.  How do I interpret my found fashions in terms of finding drugs/points of attack against diseases?

| **Useful Tools and Web Links** | |
| --- | --- |
| **Database** | Information on metabolism can be found, for example, in the KEGG database (https://www.genome.jp/kegg/), Roche Biochemical Pathways (https://www.roche.com/sustainability/what_we_do/for_communities_and_environment/philanthropy/science_education/pathways.htm) and EcoCyc (https://ecocyc.org/). Since 2020, KEGG now has a new small grey box "change pathway type" in the upper left corner, which shows the selection of available enzymes for an organism (green boxes), missing ones are shown in white |
| **Software** | A tutorial about Metatool can be found at: https://pinguin.biologie.uni-jena.de/bioinformatik/networks/metatool/metatool5.0/metatool5.0.html. Also important are YANA (https://www.bioinfo.biozentrum.uni-wuerzburg.de/computing/yanasquare/), YANAsquare (https://www.bioinfo.biozentrum.uni-wuerzburg.de/computing/yanasquare/), COPASI (https://copasi.org/) and CellNetAnalyzer (https://www2.mpi-magdeburg.mpg.de/projects/cna/cna.html) |

## Literature

Ampattu BJ, Hagmann L, Liang C et al (2017) Transcriptomic buffering of cryptic genetic variation contributes to meningococcal virulence. BMC Genomics 18(1):282

Bergmann FT, Sahle S, Zimmer C (2016) Piecewise parameter estimation for stochastic models in COPASI. Bioinformatics 32(10):1586–1588. https://doi.org/10.1093/bioinformatics/btv759 (PubMed PMID: 26787664)

Cecil A, Rikanović C, Ohlsen K et al (2011) Modelling antibiotic and cytotoxic effects of the dimeric isoquinoline IQ-143 on metabolism and its regulation in *Staphylococcus aureus*, *Staphylococcus epidermidis* and human cells. Genome Biol 12(3):R24

Cecil A, Ohlsen K, Menzel T et al (2015) Modelling antibiotic and cytotoxic isoquinoline effects in *Staphylococcus aureus*, *Staphylococcus epidermidis* and mammalian cells. Int J Med Microbiol 305(1):96–109

Dandekar T, Fieselmann A, Majeed S et al (2014) Software applications toward quantitative metabolic flux analysis and modeling. Brief Bioinform 15(1):91–107. https://doi.org/10.1093/bib/bbs065

Kent E, Hoops S, Mendes P (2012) Condor-COPASI: high-throughput computing for biochemical networks. BMC Syst Biol 6:91. https://doi.org/10.1186/1752-0509-6-91

Kühnel M, Mayorga LS, Dandekar T et al (2008) Modelling phagosomal lipid networks that regulate actin assembly. BMC Syst Biol 2:107. https://doi.org/10.1186/1752-0509-2-107

Mavrovouniotis ML, Stephanopoulos G, Stephanopoulos G (1990) Computer-aided synthesis of biochemical pathways. Biotechnol Bioeng 36:1119–1132

Orth JD, Fleming RM, Palsson BØ (2010) Reconstruction and use of microbial metabolic networks: the core *Escherichia coli* metabolic model as an educational guide. EcoSal Plus 4(1). https://doi.org/10.1128/ecosalplus.10.2.1

Schuster R, Schuster S (1993) Refined algorithm and computer program for calculating all non-negative fluxes admissible in steady states of biochemical reaction systems with or without some flux rates fixed. Comput Appl Biosci 9(1):79–85

Schuster S, Fell DA, Dandekar T (2000) A general definition of metabolic pathways useful for systematic organization and analysis of complex metabolic networks. Nat Biotechnol 18(3):326–332

Schwarz R, Musch P, von Kamp A et al (2005) YANA – a software tool for analyzing flux modes, gene-expression and enzyme activities. BMC Bioinformatics 6:135. (PubMed PMID: 15929789; PubMed Central PMCID: PMC1175843)

Schwarz R, Liang C, Kaleta C et al (2007) Integrated network reconstruction, visualization and analysis using YANAsquare. BMC Bioinformatics 8:313. (PubMed PMID: 17725829; PubMed Central PMCID: PMC2020486)

Srivastava M, Bencurova E, Gupta SK et al (2019) *Aspergillus fumigatus* challenged by human dendritic cells: metabolic and regulatory pathway responses testify a tight battle. Front Cell Infect Microbiol 9:168. https://doi.org/10.3389/fcimb.2019.00168. PMID: 31192161; PMCID: PMC6540932

von Kamp A, Schuster S (2006) Metatool 5.0: fast and flexible elementary modes analysis. Bioinformatics 22(15):1930–1931

Yang M, Rajeeve K, Rudel T et al (2019) Comprehensive flux modeling of *Chlamydia trachomatis* proteome and qRT-PCR data indicate biphasic metabolic differences between elementary bodies and reticulate bodies during infection. Front Microbiol 10:2350. https://doi.org/10.3389/fmicb.2019.02350. PMID: 31681215; PMCID: PMC6803457

# Systems Biology Helps to Discover Causes of Disease

**5**

**Abstract**

The systems biology modelling of signalling cascades and protein networks allows deeper insights into the function of the proteins involved and thus helps to understand the causes of diseases, to better describe infection processes and immune responses, or to elucidate complex processes in biology such as cell differentiation and neurobiology. Stronger mathematical models describe signalling networks precisely in terms of changes over time and their speed using differential equations. This explains the process exactly, but spends additional time e.g. determining the velocities (kinetics; time series analysis). Boolean models, on the other hand, only require information about which proteins are involved in the network and which protein interacts with which other proteins in what way (activating or inhibiting). Simulations based on a Boolean model (e.g. with SQUAD or Jimena) must be checked iteratively in many cycles to see whether the behaviour in the computer model also matches the actual outcome observed in the experiment, at least qualitatively. The computer model is thus adapted to the data step by step (data-driven modeling).

Let us now turn to systems biology in application. Bioinformatics models also allow us to gain new insights into system effects, and in particular to understand how a signalling cascade functions as a whole. The easiest way to understand this is to think of a disease, such as stroke or heart attack. Not only the heart is "broken", but the whole person is affected. Often his/her life is in danger, and only decisive and the best modern medicine can still save people with heart attacks. But is it not hopeless to model and even understand such a complex system as a whole? Well, this question always arises when I want to look at a system in its entirety. For example, all living things, including humans, are part of an environment. And only when I also model this, do I understand everything, which in turn

© Springer-Verlag GmbH Germany, part of Springer Nature 2023
T. Dandekar, M. Kunz, *Bioinformatics*,
https://doi.org/10.1007/978-3-662-65036-3_5

is not so easy, because the region, the country, indeed the whole world are again part of an even greater whole. That is why half of the work is actually done when we succeed in defining an interesting section that we can model sufficiently to gain new insights from it. For example, how does a heart attack happen? Often it happens in a sudden flash. But similarly often a heart failure, the so-called cardiac insufficiency, announces itself, gets worse and worse, and only then does the heart attack occur. Can we perhaps stop this process, can modern medicine and bioinformatics help here?

## 5.1    Application Example: How Does Phosphorylation Cause Heart Failure?

This is a section of the overall context around heart failure and heart attack that can be modeled quite well. New research shows that in heart failure, an important growth stimulus comes from a growth signal, a phosphorylation, on the Erk protein. Normally, the Erk protein is only doubly phosphorylated (Thr183 and Tyr185). But when the heart is overloaded, the protein gets a third phosphorylation (Thr188), migrates into the nucleus and leads to genes being read ("transcription factor"; together with other transcription factors, e.g. NFATc4 and GATA4), which now cause the cardiac muscle cell and thus the cardiac muscle to grow ("hypertrophy"; Fig. 5.1). If we want to simulate this process in the model, we first need the partner proteins of the Erk signaling cascade and the most important alternative signaling pathways (Figs. 5.1 and 5.2). Only then can we see whether we can change something in this fateful cascade by administering a drug (receptors on the cell membrane, above in Fig. 5.1). To do this, we first need to compile the signalling cascade (knowledge, literature, experiments, databases) and then translate it into a network using a machine-readable drawing program. This can be done by programs such as Cytoscape or CellDesigner. Just as with a drawing program, one draws in proteins and their connections ("interactions"), noting whether these are inhibitory (as a "truncated line") or activating (as an "arrow") connections. But the advantage of the above-mentioned programs is that they save the illustration e.g. inXML (*Extensible Markup Language*) or SBML (*Systems Biology Markup Language*) format, so that now the computer also understands the drawing ("machine-readable"). This then opens up numerous further evaluation possibilities through already well-established software. For example, it is possible to display which different *pathways* are involved in a given set of proteins ("*Gene Ontology*" analysis, see Appendix).

But what is so fatal in our cascade? Is it not good if the heart reacts to stress with growth and thus becomes stronger? Well, we have identified Erk kinase and its third phosphorylation as an important switch in our model. We could simulate in the model whether we could perhaps strengthen or weaken Erk phosphorylation by stimulating the receptors of the cardiac muscle cell (see Figs. 5.1 and 5.2) (see legend to the figure). It is interesting to note that an enlarged heart goes through a detrimental circle ("*circulus vitiosus*"), and a poorer supply of blood (due to arterial calcification, for example) leads to less oxygen in

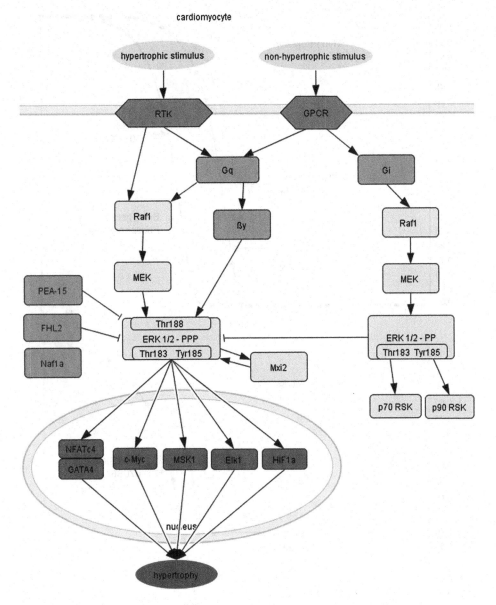

**Fig. 5.1** Representation and simulation of the signalling cascade in heart failure. Using various databases, a signalling network can be reconstructed in which the individual interactions of the proteins/molecules involved represent, in simplified form, important molecular relationships leading to heart failure (activation as →, inhibition as -|; here by the CellDesigner software)

the heart muscle. But this is a stimulus for the third Erk phosphorylation, so that the muscle grows. But since the normal heart is already optimally adapted, the enlargement of the heart muscle cell results in an even poorer oxygen supply and so on and so forth. So the question arises: What is the best way to intervene in this system?

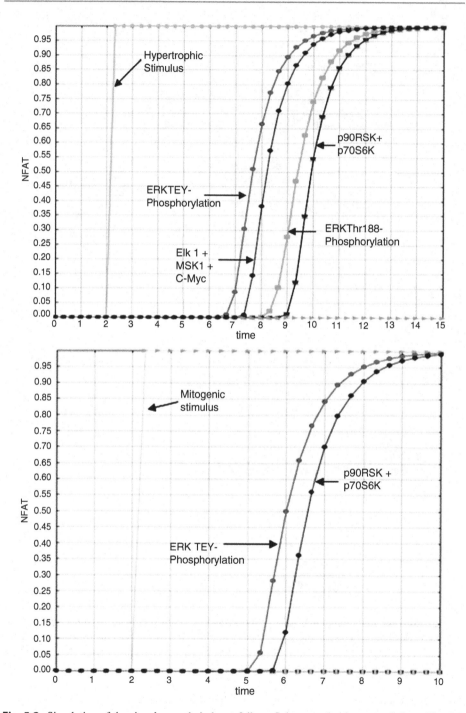

**Fig. 5.2** Simulation of the signal cascade in heart failure. Subsequently, the network from Fig. 5.1 can be used for dynamic simulation, whereby the logical interconnection of the interactions is

(continued)

For this purpose, our network, which is already available in machine-readable form, is next dynamically simulated by software. In this process, the logical interconnection is translated into dynamics, since various differential equations, e.g. exponential function, simulate the rise and fall of the respective signal strength in the network. Here, first the receptor is stimulated, then a protein directly below it, finally the Erk protein is phosphorylated and so on. This is done, for example, by the software SQUAD (*Standardized Qualitative Dynamical Modeling Suite*) or, alternatively, by the software CellNetAnalyzer (CNA), if we provide the network in a machine-readable form in each case. Now we can compare different signals on the cell membrane, such as the effect when I stimulate the beta receptors (e.g. an adrenaline rush going through our body, the heart then beats faster and we fight or run away). Then I get phosphorylation at the Erk protein, which then makes the heart muscle stronger (positive inotropic effect), it can then beat faster. The third phosphorylation is added during overexertion. Our network section shows us that we achieve this via two other signalling pathways, each of which is stimulated by different signals (Fig. 5.2).

Once this is understood through bioinformatics simulations, it also becomes clear how we need to control our system: It is important to prevent the Erk switch from turning on further genes in the nucleus of the cardiac muscle cell, which in the situation of cardiac insufficiency under consideration lead to an increase in stress and oxygen deficiency in the cardiac muscle cell. This prevents further unfavourable growth and overstretching of the heart muscle. One possibility that is currently being tested in pharmacology (although only in cell culture for the time being) is suitable peptides that prevent one Erk protein from hitting another Erk protein ("dimerisation inhibitor"). Without its peer ("dimerisation"), however, it cannot enter the cell nucleus at all and activate genes there. In this way, one could achieve that, on the one hand, the two "good" phosphorylations of the Erk protein are still supported with heart-strengthening drugs ("positively inotropic", as the medical profession says, i.e. the path on the far right in the model of Fig. 5.1), but, on the other hand, the "bad" third phosphorylation of the Erk protein and the further, here harmful, enlargement of the heart muscle are prevented.

---

translated into mathematical functions (e.g. exponential functions) in order to reproduce the respective signal strength, i.e. the activation state of the individual proteins/molecules. In this way, it is possible to simulate, for example, the activation of the RTK receptor by a hypertrophic stimulus, which activates the Raf1 protein and subsequently phosphorylates the Erk protein, which then switches on genes in the cell nucleus (with further transcription factors, such as NFATc4 and GATA4) that cause the heart muscle cell to grow. Shown is the change in the network in response to a hypertrophic stimulus (time 2, Top) and a non-hypertrophic (mitogenic) stimulus (also at time 2, Bottom). It can be seen that different transcription factors are turned on over time in response to a hypertrophic stimulus (e.g., c-Myc and Elk1). It is also clear that ERK TEY phosphorylation (Thr183 and Tyr185) is active first, followed by Thr-188 phosphorylation (simulated here by SQUAD software). With the help of such systems biology descriptions, one can effectively model a biological system and is thus able to understand the system behavior, such as how heart failure occurs or how one can intervene in the cascade (e.g. possible therapy)

The diagram once again illustrates the system behaviour of these signalling cascades. On the one hand, bioinformatic simulation enables us to understand how a cell with its signalling cascades reacts in a healthy or pathological way to external signals. On the other hand, it is also possible to test individual strategies in detail and determine which signals are stronger or weaker and what then prevails in each case. Of course, one could also find this out directly with many experiments, but this is much more time-consuming and also requires many, many experiments.

The model description we have just used is "semi-quantitative", i.e. we explain exactly which is stronger or weaker, which signal is important first, second and last. But we are not yet exactly quantitative, so that one already has exact quantities/concentrations. Of course, there are such exact quantitative models in bioinformatics. However, the disadvantage of such models is that they need much more additional information, in particular how fast the most important processes change with time, or how strong the signals are at the beginning and at least at four further time points. Then I can calculate with which function I describe the change with time, i.e. I can set up the so-called differential equation of this property. This makes sense, for example, if I want to thin the blood and I do not want to make the platelets too weak or too strong for this purpose. That's why we set up a fairly accurate model with differential equations for this (and collected a lot of experimental data before-hand). In many other cases, however, one does not have the time to measure everything so precisely experimentally, and a semi-quantitative model is then already very good for describing the corresponding system effects, for example when we want to protect plants against pests or heat stress, to give a completely different example. To do this, we then looked more closely at the effect of plant hormones ("cytokinins"), with which you would spray the plant in the event of a bacterial infestation, for example, and then you have a completely biological and readily degradable pest control agent (Naseem et al. 2012a). In order to find the right cytokinin, the complex further effects were simulated more precisely in a systems biology model, as shown above for the heart (cytokinins also control many other processes, for example in the growth of plants).

In summary, then, systems biology descriptions are an important area of bioinformatics today for better understanding systems behavior and signal processing in cells and organisms. Often, relatively few data are sufficient for this purpose, because even a rather small semi-quantitative model answers the questions about the best or most interesting system effect, as in the case of heart failure, blood thinning, plant pests or, another exciting topic, for example, cancer and cancer drugs (antibodies, cytostatics). These drugs need to be optimally combined and correctly dosed – ideally even individually and patient-specific. This is precisely where bioinformatics can calculate the best strategy for the patient.

There is often also more informatic preliminary work to the semiquantitative model. This applies to biological systems and their system effects, which are hidden in large amounts of data (e.g. gene expression data, genome sequences, metabolites, pharmaceutical levels, etc.) and where the decisive system components must first be filtered out using statistics or even complex sequence analysis programs. This is also an important and laborious task for bioinformatics. But nevertheless, one would subsequently set up again a semi-quantitative or also an exactly quantitative model of the system in order to describe it more precisely, exactly as just illustrated.

## 5.2   Generalization: How to Build a Systems Biology Model?

Can we generalize our approach? Yes, it is one way to describe the systems biological relationships,

1. First gather all the components (consider the "topology" of the model, that is, its structure).
2. Then I take this structure and now simulate the system behavior in the computer ("simulation").

### First: Gather All Components of the System (Its Network or Topology)

Each of the steps has its own challenges. For step 1 to work, I need to write my network with software so that the output is an image that a machine can read. For example, the CellDesigner and Cytoscape software allow such network descriptions. Their output format, an XML format (i.e. all parts of the image are marked in a computer-readable way), can be read by a program. The two softwares differ only in minor details. In any case, however, they work by connecting proteins ("nodes") with each other ("edges"). This creates a network. In order for this network to predict what will actually happen in the cell, Boolean logic is important in linking, i.e. which protein is linked to another protein and how (activating or inhibiting). In addition, it is important to pay attention to "And", "Or" and "Not", i.e. whether, for example, activation only occurs when two proteins jointly activate a third (an "And"), whether one of the two is sufficient for this (then corresponds to an "Or" linkage) or whether one must not be there ("Not" as well as with SQUAD, "Nor", "And not").

In practice, this requires taking into account many sources about the biological system as well as collecting missing information from databases or determining it from one's own domain and sequence analyses (consider phosphorylation sites, function of the proteins involved, existing interaction domains or known substrate-enzyme relations).

In addition, it is important to be clear about how to decide which article or finding is most likely in the case of conflicting sources (see Sect. 6.2). It is best to store this information in a separate table so that you can later prove on the basis of which data the model was arrived at.

### Second: Simulate the Network and Its Dynamics

It is now possible to follow Boolean logic and thus make statements about signal chains. One way to do this is to construct a Petri net (Li et al. 2011; Schlatter et al. 2012) from the network using appropriate software and thus reproduce the signal cascade in a first form.

However, a so-called semiquantitative model reproduces the processes somewhat more accurately. Starting from the Boolean network, differential equations, e.g. exponential functions, are linked in such a way that they reproduce this logic, but by means of the mathematical transformation between the completely switched-on or switched-off state, they lay a compensation curve ("interpolate"). In order to reproduce the logic in the network correctly, the software SQUAD, for example, creates chained exponential terms (uses exponential function), which also take into account the "and", "or" and "not". It reads networks written with CellDesigner e.g. as SBML format and requires a Windows XP or Linux operating system. These limitations no longer apply to the Jimena software (Karl and Dandekar 2013a). It runs platform-independently using Java and can read YeD files, among others, but also various versions of CellDesigner. Surprisingly, this way I also get all order relations in the model correctly, i.e. which receptor is excited before which one and which link in a signal chain is activated earlier or later. In most cases, the molecules close to the receptor are excited first, followed by the later, mediating proteins. If the topology (structure) of the model provides for a feedback *loop*, this can then return the signal to the beginning of the signal chain, either inhibiting (negative feedback) or activating (positive feedback, sometimes also called "*feedforward loop*").

This brings us to another important point. The software can only simulate correctly what is also reproduced correctly in the network. This means that a period of constant testing and trial and error begins until the simulation reproduces the correct sequence of events in this signal network as faithfully as possible.

Since this is a semi-quantitative model, the next step is to normalize the different units of the model according to the experimental data. This means that the typical times of the signal cascade, receptor excitation, phosphorylation of kinases, etc. are determined (so-called *data-driven modeling*). Hundreds of biological problems have already been simulated in this way in recent years. The Boolean semiquantitative model is therefore quite popular in biology, because one can begin to describe the biological system with relatively little information, and then step by step learn more and more about the model through simulations and experiments.

If so much data is put into the model, one can of course wonder what new insights the model can bring out. But it is the case that a few experiments are sufficient to normalize the model and to qualitatively confirm the correctness of all links (correct stimulus response and sequence). With the model, I can now predict the outcome for all times and all signal and switching sequences that are possible in the network.

For example, we used this approach to simulate the behavior of lung carcinoma (Stratmann et al. 2014; Göttlich et al. 2016a) and colon carcinoma cells (Baur et al. 2019) and then tested through new combinations and options for therapies in addition to standard therapies.

With regard to the Erk signaling network, the interesting thing was that through the bioinformatic model we can mimic new approaches to treating heart failure (Brietz et al. 2016a), such as the negative *feedback loop* through Rkip or the approach of using dimerization inhibitors against Erk dimerization – both ways to prevent heart failure at the molecular level. Furthermore, the bioinformatic model also clarifies the downstream targets (i.e. target proteins) of heart failure, which can also be pharmacologically influenced to prevent or favourably influence heart failure.

**Alternatives to Semiquantitative Modeling**

If further data is known, especially about the velocity and stimulus strength in the signal chain, the data-driven modeling can be taken even further and the exact velocities, affinities and chemical equilibria can be calculated more precisely. With this, enough information is then available to represent this process with exact equations, so-called differential equations, which thus have the change of a quantity on the left side and describe this change on the right side via the quantity itself and further determining factors. If I know all the influencing factors, I know the constants and kinetic properties of the signal cascade (in mathematical terms, this is called the "parameters" of the differential equation), and I can then use them to model the system accurately and precisely. An example is the inhibitory cAMP and cGMP signalling pathways in the platelet, which thus dampen the platelet in its activation. Here we had enough information from experiments that we patiently repeated over and over again for several years to set up such a model (Wangorsch et al. 2011). This area of accurate modeling is also being pursued by many systems biology groups. A simple approach to set up such models oneself is the software PLAS (*Power law analysis and simulation*; https://enzymology.fc.ul.pt/software/plas/), which also introduces one to all the steps for this more accurate simulation via tutorials. However, as a beginner you have to make many decisions about the parameters. But if there are too many "free" parameters, one runs very easily the risk that the system is described incorrectly, because one can always choose the free parameters in the equations in such a way that the system seems to fit the little available data, but then very easily misses the mark with new experiments or data. This is easily prevented in the semiquantitative models. This is because these are coarser, but have fewer free parameters and therefore are not as quick to be wrong in their predictions as the much more accurate quantitative models. Finally, it is worth mentioning that one can also stop at step 1 and also just examine the structure of the model in detail. This works for signaling cascades as well as for metabolism. For the latter, glycolysis or the citric acid cycle, for example, are very illustrative textbook examples, which are followed by further insights from, for example, the linear metabolic pathway of glycolysis and from the cyclic pathway of the citric acid cycle for metabolism. An overview of these different systems biology methods and approaches is provided in the English textbook by Klipp et al. (2016).

Finally, we have collected an introductory selection of our own work on Boolean models and semiquantitative modeling based on them (see below), which should give an overview of the basics, but also various examples of applications, and help the interested reader to continue learning.

**Biological Examples for Boolean Modeling**
**Basics:**

In a review paper, we systematically compared different approaches to Boolean modeling and dynamic modeling, e.g. SQUAD, ODEFY, and CellNetAnalyzer (Schlatter et al. 2012). Another good starting publication is Di Cara et al. (2007) on SQUAD. Our software Jimena is a nice further development (Karl and Dandekar 2013a). Jimena also offers to distinguish between direct and dynamic network control quantitatively and qualitatively in networks (Karl and Dandekar 2015).

**Specific models for different cells and processes:**

- heart: Brietz et al. (2016a) and Breitenbach et al. (2019a, b),
- liver: Philippi et al. (2009),
- immune cells: Czakai et al. (2016),
- tumours: Stratmann et al. (2014), Göttlich et al. (2016a), Baur et al. (2020), and Kunz et al. (2020),
- plants (hormones and infections): Naseem et al. (2012, 2013a, b), and Kunz et al. (2017),
- bacteria: Audretsch et al. (2013),
- platelets: Mischnik et al. (2013a, b).

**Extension of such semi-quantitative models to fully dynamic models:**

Two papers on dynamic modeling via platelets are helpful here for comparison:

Mischnik et al. (2014) describe the function of the signal molecule Src, but now with differential equations and estimates of the velocities of all processes ("kinetic parameters"). It is crucial to switch between active and inactive platelets. In the process, the mathematical description was also verified in detail experimentally.

Wangorsch et al. (2011) again describe the function of inhibitory cyclic nucleotides in the platelet using differential equations that take into account the different rates of the processes involved and the absolute signal strength. In particular, I can cause the platelet to become inactive by increasing the level of cAMP. This can be used medically, for example, to prevent a new blood clot in the case of strokes. The behaviour for different active substances and their combination is described in detail in the paper.

In both works, this was used to accurately estimate the kinetic parameters through experimental data and then develop corresponding optimal fitting differential equations (ODEs). One can also calculate in general what the optimal pharmacological intervention should be (Breitenbach et al. 2019a, b).

In addition to this selection of one's own work on the topic, there are of course also large repositories of models, so that one can compare models from many authors or search for the optimal one for a question, which one can then possibly adapt to one's own question, for example:

https://systems-biology.org/resources/model-repositories/ (from the journal "Systems Biology and Applications").

celldesigner.org/models.html (from the software CellDesigner, very nicely linked to the Panther Pathway database).

https://www.ebi.ac.uk/biomodels/ (The "Biomodels Database" of the EBI, with many mathematical, pharmacological and physiological dynamic models collected from the literature).

The examples above show that semiquantitative models can be used to cover the entire range of systems biology regulation and biological signalling networks. The particular advantage of the method lies in the fact that it is possible to model processes without precise data on the speed ("kinetics"). If, on the other hand, one wants to model a dynamic process, in particular a signal cascade, in more detail, one must determine these data about the velocity. This is done by methods of time series analysis: If one has measured the process (for example the phosphorylation of a kinase that transmits a signal in the cell) for five or more time points, there is enough data to estimate how fast this process proceeds. It is therefore possible to describe the speed (kinetics) precisely in mathematical terms using a parameter (in the example: the speed). There are a number of bioinformatics tools for estimating parameters. Easy to learn and good to use for this parameter estimation is the software Potters Wheel (https://www.potterswheel.de/Pages/; Maiwald and Timmer 2008) and its successor Data2Dynamics (Steiert et al., 2019).

This software can also be used to investigate which parameters need to be accurately estimated and which do not (*sensitivity analysis*). It also allows to see which of the parameters can be well estimated from the data (*identifiability analysis*) and which cannot (either because the data are not sufficient or because the network is wired in such a way that, for example, the parameter always depends on another one that cannot be estimated either or because the parameter is simply not determined by this data at all).

---

### Conclusion

- Systems biology modelling of signalling cascades and protein networks allows deeper insights into the function of the proteins involved and thus helps to understand the causes of diseases, to better describe infection processes and immune responses, or to elucidate complex processes in biology, such as cell differentiation and neurobiology. Stronger mathematical models describe signalling networks precisely in terms of changes over time and their speed using differential equations. This explains the process exactly, but additional time is needed, e.g. with the determination of the velocities (kinetics, *data driven modeling*, time series analysis).
- Boolean models only require information about which proteins are involved in the network and which protein interacts with which other proteins and how (activating or inhibiting). Therefore, they are well suited for an introduction. If you want to reproduce one of the presented examples yourself, it is easy (use the same components and links and software). However, if you want to create your own new model, many cycles are necessary, because you have to check again and again in simulations based on the Boolean model (e.g. with SQUAD or Jimena) whether the behaviour in the computer model also matches the outcome actually observed in the experiment, at least qualitatively, and thus adapt the computer model to the data step by step.
- Conversely, the model then allows to describe all situations that have not yet been measured or reproduced in the experiment. In particular, the effect of drugs and their combinations, the activity of all proteins involved, the effect of signals, mutations or even immune substances (e.g. cytokines). Systems biology modeling can be described as the central, current field of bioinformatics. It is also called network analysis, dynamic modelling or interactomics in order to emphasize these aspects more strongly. ◄

## 5.3    Exercises for Chap. 5

It is useful to repeat these tasks in Chap. 7.

This part of the exercise will focus on bioinformatic models in order to better understand possible system effects and the organism as a whole. A bioinformatic model can provide various information about the network topology, e.g. a Boolean model about the logical interconnection of the signaling components (e.g. activation, inhibition, *feedback/feedforward regulation*) or a kinetic model about metabolic pathways, but also predict the resulting network behavior. Bioinformatics models can answer any number of questions. Usually, the function of the signaling cascade and how it can be used therapeutically is of particular interest.

### Task 5.1

The basis for a bioinformatics model includes interactions, such as protein-protein interactions. You can find these in various databases and thus generate a corresponding signal network.

Which of the following statements are correct (multiple answers possible)?

A.  The STRING database provides little information on protein-protein interactions.
B.  One of the things I find in the PlateletWeb database is protein-protein interactions in platelets.
C.  In the KEGG, iHOP and HPRD database I find protein-protein interactions.
D.  Signaling cascades are, in this sense, a type of protein-protein interaction.
E.  Proteins can interact with each other directly or as complexes.

### Task 5.2

Name and describe databases/software where you can get information about interaction partners, e.g. of proteins.

### Task 5.3

Explain how to bioinformatically screen a protein for potential interaction partners (name and briefly describe two software/databases). What do they output in each case? What are the advantages of comparing several software/databases with each other?

### Task 5.4

An interaction database is the STRING server. What is the difference to other databases, such as PlateletWeb or HPRD?

### Task 5.5

Explain how to create a protein-protein interaction network.

**Task 5.6**

Example:

We want to take a closer look at protein-protein interactions and turn to network analysis and modelling of regulatory networks. You now have the opportunity to generate a network and then examine it for its biological function, for example, to detect well-connected proteins in a network, so-called hub proteins. Please note that for training purposes we have only chosen a small network around BRCA1 (BReast CAncer 1, also known as breast cancer gene 1). Normally, however, the network to be investigated is always much larger and more complex, which makes a comprehensive network analysis necessary.

Now search for all human interaction partners of BRCA1 from the string database (https://string-db.org/). Which of the following statements are correct (multiple answers possible)?

A. The interaction of BRCA1 and ESR1 was found experimentally and has a very low score (close to 0).
B. For the interaction of BRCA1 and ESR1 I get a high score (>0.99).
C. Each indicated interaction for BRCA1 was simultaneously found and predicted experimentally.
D. All indicated interactions with a score >0.99 were found experimentally.

**Task 5.7**

Describe a simple method for creating an interaction network and analyzing it for function.

**Task 5.8**

Example:

Now download all human interaction partners of BRCA1 from the string database (https://string-db.org/). Please use the parameter "*Experiments*" (i.e. only all experimental interactions) and a "*confidence score*" of 0.9. Save the network (under "*save*" as text summary; TXT – *simple tab delimited flatfile*).

Now that you have downloaded all experimentally determined interaction partners, you can visualize and further analyze your small network. For this purpose, please inform yourself about the software Cytoscape (https://cytoscape.org/) and download the free version 2.8.3. To make sure that the interactions match, please compare your network with the one in the solution section and please adjust it accordingly. Please save your network (currently still as .txt) also as .sif (*simple interaction file*), because Cytoscape needs this format. Now you just have to import your created interaction file BRCA1.sif into Cytoscape via File → Import Network (*Multiple File Type*) and you are ready to go.

Which of the following statements will you see after loading the BRCA1.sif file?

    A. Network contains 11 nodes and 18 edges.
    B. Network contains 18 nodes and 11 edges.
    C. Network contains 1 nodes and 8 edges.
    D. Network contains 8 nodes and 1 edges.
    E. Network contains 111 nodes and 181 edges.

**Task 5.9**

Example:

Now your actual network analysis begins. For this, Cytoscape has numerous plugins to choose from, such as the Biological Networks Gene Ontology Tool (BiNGO). Please briefly inform yourself about Cytoscape (https://cytoscape.org/) or Plugin → Manage Plugins – Search BiNGO (you can also download the current BiNGO version here). Now perform a BiNGO search for all proteins of the network (Plugins → Start BiNGO 2.44; please use the preset default parameters, but use Homo sapiens as organism).

Which of the following statements are correct (multiple answers possible)?

    A. The BiNGO analysis identifies relatively few biological processes (less than 20).
    B. In addition to the functions, the BiNGO analysis also shows me the *p-value* and which genes are involved.
    C. For example, the BiNGO analysis shows me the biological process *cell cycle checkpoint* (GO-ID 75) with a BRCA1 involvement.
    D. The BiNGO analysis identifies the biological process *induction of apoptosis* (GO-ID 6917) as significant (*p-value* < 0.05), but also that all proteins of the network are involved.

**Task 5.10**

Describe what a *Gene Ontology* is and how the GO terms are organized.

**Task 5.11**

Example:

Now take a look at the network topology. Please use the NetworkAnalyzer plugin for this. Please also inform yourself about this beforehand via Cytoscape (https://cytoscape.org/) or via Plugin → Manage Plugins – Search NetworkAnalyzer. Here you can also download the current NetworkAnalyzer version. Now perform an analysis (Plugins → NetworkAnalysis → AnalyzeNetwork → Treat the network as undirected) and familiarize yourself with the various parameters and plots (then use, for example, "*Parameter average number of neighbors*" and "*Plot Node Degree Distribution*").

Which of the following statements are correct (multiple answers possible)?

A. Among other things, I can use the NetworkAnalyzer to identify important hub proteins, i.e. strongly networked nodes.
B. The NetworkAnalyzer identifies an *"average number of neighbors" of* less than 1.
C. The NetworkAnalyzer identifies an *"average number of neighbors" of* more than 3.
D. Looking at the *Node Degree Distribution* plot, I see five nodes with three interactions and ten nodes with five interactions.
E. Looking at the *Node Degree Distribution* plot, I see one node with five interactions – this represents a hub node given the *average number of neighbors* parameter.

**Mathematical modeling of regulatory networks:**

**Task 5.12**
Name and describe software for mathematical modeling of biological networks.

**Task 5.13**
Describe three different approaches to mathematical modeling of biological networks (Boolean, quantitative, and semiquantitative).

**Task 5.14**
State advantages and disadvantages of mathematical modeling of biological networks.

**Task 5.15**
Describe how one would bioinformatically model a biological network, e.g., the cAMP pathway (briefly describe: what data, what steps, what possible software).

**Task 5.16**
Which statements about the mathematical modeling of regulatory networks are correct (multiple answers possible)?

A. Boolean, quantitative, and semiquantitative modeling are three mathematical modeling methods.
B. Boolean modeling always considers the on/off (0/1) state of a system.
C. Quantitative modeling is not able to consider the system state in the interval between 0 and 1 and thus cannot model kinetic data, e.g. via Michaelis-Menten kinetics.
D. Semiquantitative modeling is a combination of Boolean and quantitative modeling, where I do not necessarily need information about kinetics.

**Task 5.17**
What is meant by a *"steady state"* condition of a network?

| Useful Tools and Web Links | |
|---|---|
| **SQUAD** | https://www.vital-it.ch/software/SQUAD |
| **Jimena** | https://www.bioinfo.biozentrum.uni-wuerzburg.de/computing/jimena_c/ |
| **CellNetAnalyzer** | https://www2.mpi-magdeburg.mpg.de/projects/cna/cna.html |
| **PLAS** | https://enzymology.fc.ul.pt/software/plas/ |
| **Odefy** | https://www.helmholtz-muenchen.de/icb/software/odefy/index.html |
| **Cytoscape** | https://www.cytoscape.org/ |
| **CellDesigner** | https://www.celldesigner.org/ |
| **PottersWheel** | https://www.potterswheel.de/Pages/ |

# Literature

Audretsch C, Lopez D, Srivastava M et al (2013) A semi-quantitative model of quorum-sensing in staphylococcus aureus, approved by microarray meta-analyses and tested by mutation studies. Mol BioSyst 9(11):2665–2680. https://doi.org/10.1039/c3mb70117d (PubMed PMID: 23959234)

Brietz A, Schuch KV, Wangorsch G et al (2016) Analyzing ERK 1/2 signalling and targets. Mol BioSyst 12(8):2436–2446. https://doi.org/10.1039/c6mb00255b

Czakai K, Dittrich M, Kaltdorf M et al (2016) Influence of platelet-rich plasma on the immune response of human monocyte-derived dendritic cells and macrophages stimulated with aspergillus fumigatus. Int J Med Microbiol pii:1438–4221(16)30199-0. https://doi.org/10.1016/j.ijmm.2016.11.010 ([Epub ahead of print] PubMed PMID: 27965080)

Di Cara A, Garg A, De Micheli G et al (2007) Dynamic simulation of regulatory networks using SQUAD. BMC Bioinformatics 8:462. https://doi.org/10.1186/1471-2105-8-462 (*Nice original paper about SQUAD with good examples to practice)

Göttlich C, Müller LC, Kunz M et al (2016) A combined 3D tissue engineered in vitro/in silico lung tumor model for predicting drug effectiveness in specific mutational backgrounds. J Vis Exp 110:e53885. https://doi.org/10.3791/53885 (*Cancer is modeled here both *in vitro* and *in silico*, the two complementing each other)

Karl S, Dandekar T (2013) Jimena: efficient computing and system state identification for genetic regulatory networks. BMC Bioinformatics 14:306. https://doi.org/10.1186/1471-2105-14-306 (*Explains analysis using Jimena software, useful for all systems biology modeling)

Karl S, Dandekar T (2015) Convergence behaviour and control in non-linear biological networks. Sci Rep 5:9746. https://doi.org/10.1038/srep09746 (PubMed PMID: 26068060; PubMed Central PMCID: PMC4464179 *This work explores the interesting aspect of which proteins direct and determine the network. The Jimena determines both direct control and dynamic [via network effects] for each switch in the network ["network nodes", mostly proteins]. This can then be used to determine exactly which receptors, kinases, etc. are the main clocks in the network and whether they do it directly or via network effects)

Li C, Nagasaki M, Koh CH et al (2011) Online model checking approach based parameter estimation to a neuronal fate decision simulation model in *Caenorhabditis elegans* with hybrid functional Petri net with extension. Mol BioSyst 7(5):1576–1592. https://doi.org/10.1039/c0mb00253d

Maiwald T, Timmer J (2008) Dynamical modeling and multi-experiment fitting with PottersWheel. Bioinformatics 24(18):2037–2043. https://doi.org/10.1093/bioinformatics/btn350 (PubMed PMID: 18614583; PubMed Central PMCID: PMC2530888)

Mischnik M, Boyanova D, Hubertus K et al (2013a) A Boolean view separates platelet activatory and inhibitory signalling as verified by phosphorylation monitoring including threshold behaviour and integrin modulation. Mol Biosyst 9(6):1326–1339. https://doi.org/10.1039/c3mb25597b (PubMed PMID: 23463387 *This work uses platelets as an example to show how systems biology regulation controls the fragile balance between blood coagulation and blood flow to prevent thrombosis or bleeding. Inhibitory and activating pathways are modeled in detail)

Mischnik M, Hubertus K, Geiger J et al (2013b) Dynamical modelling of prostaglandin signalling in platelets reveals individual receptor contributions and feedback properties. Mol BioSyst 9(10):2520–2529. https://doi.org/10.1039/c3mb70142e (PubMed PMID: 23903629)

Mischnik M, Gambaryan S, Subramanian H et al (2014) A comparative analysis of the bistability switch for platelet aggregation by logic ODE based dynamical modeling. Mol BioSyst 10(8):2082–2089. https://doi.org/10.1039/c4mb00170b (PubMed PMID: 24852796)

Naseem M, Philippi N, Hussain A et al (2012) Integrated systems view on networking by hormones in Arabidopsis immunity reveals multiple crosstalk for cytokinin. Plant Cell 24(5):1793–1814. https://doi.org/10.1105/tpc.112.098335 (*This work shows how experiment and modeling interact in bioinformatics to elucidate a complex plant hormone network here)

Naseem M, Kaltdorf M, Hussain A et al (2013a) The impact of cytokinin on jasmonate-salicylate antagonism in Arabidopsis immunity against infection with *Pst DC3000*. Plant Signal Behav 8(10). https://doi.org/10.4161/psb.26791 (PubMed PMID: 24494231)

Naseem M, Kunz M, Ahmed N et al (2013b) Integration of Boolean models on hormonal interactions and prospects of cytokinin-auxin crosstalk in plant immunity. Plant Signal Behav 8(4):e23890. https://doi.org/10.4161/psb.23890 (PubMed PMID: 23425857)

Philippi N, Walter D, Schlatter R et al (2009) Modeling system states in liver cells: survival, apoptosis and their modifications in response to viral infection. BMC Syst Biol 3:97. https://doi.org/1 0.1186/1752-0509-3-97 (PubMed PMID: 19772631; PubMed Central PMCID: PMC2760522)

Schlatter R, Philippi N, Wangorsch G et al (2012) Integration of Boolean models exemplified on hepatocyte signal transduction. Brief Bioinform 13(3):365–376. https://doi.org/10.1093/bib/bbr065 (*Detailed overview of Boolean network models and how to model them comparatively)

Steiert B, Kreutz C, Raue A, Timmer J. Recipes for Analysis of Molecular Networks Using the Data2Dynamics Modeling Environment. Methods Mol Biol.2019;1945:341–362. https://doi.org/10.1007/978-1-4939-9102-0_16.

Stratmann AT, Fecher D, Wangorsch G et al (2014) Establishment of a human 3D lung cancer model based on a biological tissue matrix combined with a Boolean *in silico* model. Mol Oncol 8(2):351–365. https://doi.org/10.1016/j.molonc.2013.11.009 (Epub2013Dec18)

Wangorsch G, Butt E, Mark R et al (2011) Time-resolved in silico modeling of fine-tuned cAMP signaling in platelets: feedback loops, titrated phosphorylations and pharmacological modulation. BMC Syst Biol 5:178. https://doi.org/10.1186/1752-0509-5-178 (*Shows detailed modeling with differential equations and time series analysis)

## Application Examples

Baur F, Nietzer SL, Kunz M et al (2019) Connecting cancer pathways to tumor engines: a stratification tool for colorectal cancer combining human in vitro tissue models with Boolean in silico models. Cancers (Basel) 12(1):28. https://doi.org/10.3390/cancers12010028

Baur F, Nietzer S, Kunz M et al (2020) Connecting cancer pathways to tumor engines: astratification tool for colorectal cancer combining human *in vitro* tissue models with Boolean *in silico* models. Cancers (Basel) 12(1), 28. pii: E1761. https://doi.org/10.3390/cancers12010028

Breitenbach T, Liang C, Beyersdorf N et al (2019a) Analyzing pharmacological intervention points: a method to calculate external stimuli to switch between steady states in regulatory networks. PLoS Comput Biol 15(7):e1007075. https://doi.org/10.1371/journal.pcbi.1007075

Breitenbach T, Lorenz K, Dandekar T (2019b) How to steer and control ERK and the ERK signaling cascade exemplified by looking at cardiac insufficiency. Int J Mol Sci 20(9):2179. https://doi.org/10.3390/ijms20092179

Brietz A, Schuch KV, Wangorsch G et al (2016) Analyzing ERK 1/2 signalling and targets. Mol BioSyst 12(8):2436–2446. https://doi.org/10.1039/c6mb00255b

Göttlich C, Müller LC, Kunz M et al (2016) A combined 3D tissue engineered in vitro/in silico lung tumor model for predicting drug effectiveness in specific mutational backgrounds. J Vis Exp 110:e53885. https://doi.org/10.3791/53885 (*Cancer is modeled here both *in vitro* and *in silico*, the two complementing each other)

Karl S, Dandekar T (2013) Jimena: efficient computing and system state identification for genetic regulatory networks. BMC Bioinformatics 14:306. https://doi.org/10.1186/1471-2105-14-306 (*Explains analysis with Jimena software, useful for all systems biology modeling)

Klipp E, Liebermeister W, Wierling C et al (2016) Edda klipp systems biology: a textbook, 2nd edn. isbn: 978-3-527-33636-4

Kunz M, Dandekar T, Naseem M (2017) A systems biology methodology combining transcriptome and interactome datasets to assess the implications of cytokinin signaling for plant immune networks. Methods Mol Biol 1569:165–173. https://doi.org/10.1007/978-1-4939-6831-2_14

Kunz M, Jeromin J, Fuchs M et al (2020) In silico signaling modeling to understand cancer pathways and treatment responses. Brief Bioinform 21(3):1115–1117. https://doi.org/10.1093/bib/bbz033

Naseem M, Philippi N, Hussain A et al. (2012) Integrated systems view on networking by hormones in Arabidopsis immunity reveals multiple crosstalk for cytokinin. Plant Cell 24(5):1793–1814. https://doi.org/10.1105/tpc.112.098335 (*This work shows how experiment and modeling in bioinformatics work together to elucidate a complex plant hormone network here)

# How Do I Understand Bioinformatics?

After we have become acquainted with the sequence analysis of proteins, RNA molecules and DNA as basic techniques of bioinformatics and have already looked at their interaction in the form of metabolic and regulatory networks (Part I), we will now provide an in-depth insight into basic strategies of bioinformatic working methods in Part II.

## From a Computer-Technical (Informatics) Point of View, Three Points Are Interesting

1. In order to cope with the large amounts of data, good databases are important, in which one can search particularly efficiently and accurately (e.g., database indexing). Likewise, because of the abundance of data, search capabilities that are as fast as possible are crucial to shorten exact, lengthy computations as efficiently as possible (heuristic searches, Chap. 6, e.g., BLAST).
2. Actually, bioinformatics is always about cracking codes to understand biological processes. How does one measure (according to Shannon) the amount of information hidden in biological messages? And how do you crack the codes as efficiently as possible (Chap. 7, e.g. with sequence analyses)?
3. How long do computers actually need for a calculation? Problems become especially difficult with built-in combinatorics. In this case, the computer needs a multiple of computing time for just one unit more (NP problems). We get to know typical problems of this kind from bioinformatics, how they are solved and when only a larger computer helps (Chap. 8).

## From a Biological Point of View, the Following Points Are Generally Important

1. Bioinformatics analyzes biological systems. However, these all behave similarly in principle. What are the principles? How and with which software do I get the system behavior out? It is very important that a biological system adapts to the environment as optimally as possible and actively controls itself. These capabilities do not reside in individual components, but only emerge through the interaction of all parts (emergence). Pioneers of systems biology have summarized these principles (Chap. 9).
2. Every living being today is the result of millions of years of evolution of the population that produced it. Therefore, a good bioinformatics strategy is also to look at the evolution of a protein sequence, a protein structure, an organism. We will learn basic techniques for this (Chap. 10).
3. Finally, we can also look at the concrete implementation of design principles in a cell to efficiently address bioinformatics problems, i.e., in particular, to understand which molecular component we are looking at and how it acts in the cell. For this, we look at the flow of genetic information from genome to RNA to protein, as well as the control of genetic information and gene expression data. We look at how proteins find their place in the cell, how the cell moves, organizes metabolism and differentiates. Again, the information that is important for each of these can be quickly analyzed and recognized using bioinformatics algorithms (e.g., localization signals, enzyme network lookup in biochemical metabolism database, and even use it to make custom proteins; Chap. 11).

This lays the foundation for Part III, which explores fascinating results and current developments in bioinformatics.

# Extremely Fast Sequence Comparisons Identify All the Molecules That Are Present in the Cell

**Abstract**

With the BLAST server at NCBI (National Center of Biotechnology Information), you can get an answer in seconds to a few minutes. This is made possible by fast, but not entirely accurate, searches. Almost all of the fast bioinformatics programs on the net use such heuristics. In BLAST, for example, two short but perfect match alignments are first pretested in a database entry before an exact alignment with the database entry is performed, thus saving a lot of computing time: indexing the database (after all, you also look up this book via the table of contents much faster than by browsing through it). Besides speed, sensitivity (do I recognize all relevant entries?) and specificity (do I not get too many irrelevant entries?) are also important for a good heuristic search.

How and why do bioinformatic analyses actually work? A very basic step towards understanding is to understand which biomolecule you have in front of you. For this purpose, bioinformatics uses the analysis of the molecular sequence. It is important to remember that we first need the experimentally determined sequence. However, this sequence does not tell us which molecule is present. However, this can be solved by comparing the respective molecular sequence with all entries in a database (cf. Chap. 1). The interesting thing is that bioinformatics has developed very fast computational recipes (algorithms) for this task. This was necessary because the sequences have grown so quickly that we are now dealing with many millions of stored sequences and many billions of stored letters. How do you speed up bioinformatics algorithms so that they can cope with these large amounts of data?

## 6.1    Fast Search: BLAST as an Example for a Heuristic Search

The following accelerations are used for sequence comparisons (technical term: sequence alignment):

A so-called indexing first considers whether the database entry contains single short words (3 letters for protein sequences or 11 nucleotides for nucleic acid sequences) that are similar to the sequence of the molecule. If this is the case (a first "hit" or "*hit*" is found), the system immediately searches whether there is another hit not too far away (predefined window length). Only when this second hit is found does the BLAST algorithm start checking whether the remaining sequence letters of this database entry match the search sequence. This exact comparison of the two letter sequences ("*alignment*") is also accelerated by "dynamic programming", so that step by step more memory is available for the comparison of search sequence and database entry.

Thus, we see two principles of bioinformatics: Since all important biomolecules (DNA, RNA, proteins, but also, for example, carbohydrates and lipids) are built from recurring building units, most biomolecules can be recognized by the sequence of these building units, i.e. by their letter sequence (with each class of molecules using its own alphabet).

In the meantime, however, so much information about biomolecules has been stored in large databases that a major part of the informatics work in bioinformatics consists of using fast computational rules (algorithms) and conveniently constructed databases to cope with this flood of information so well that the correct biomolecule can be identified as quickly as possible (see Mount et al., 2004).

If you use BLAST on the NCBI website (https://blast.ncbi.nlm.nih.gov/Blast.cgi), for example, I get a result very quickly (a few seconds, sometimes one to 2 minutes). In that time, BLAST actually screens several billion nucleotides and many millions of sequence entries. This is an amazing speed-up. We now want to understand how to speed up bioinformatics searches in general so that you get a result quickly. This usually happens by foregoing the perfect search and taking a program that uses shortcuts to get a near-perfect solution.

When searching for a similar sequence, one way to do an exact search would be to compare letter by letter to determine exactly where a local match with high similarity is. Local similarity is therefore a popular choice for protein function searches, because you can then move on from a subsequence whose similarity was found in the database to the next best similarity. After I have recognized that a partial sequence, usually a protein domain, has a certain function, I shorten my protein by this domain and now search for a hit in the database with the remaining sequence, which then not so rarely assigns the next piece of the sequence, often a whole domain again, with a suspected function, and so on.

On the other hand, if I want to save time, I forgo the exact but slow calculation and do a less exact but fast search instead. This is exactly what a heuristic is. Figure 6.1 summarizes how BLAST speeds up the search using an index search followed by an exact local *alignment* between two hits of the upstream heuristic search (Hansen 2013; see tutorial for more information).

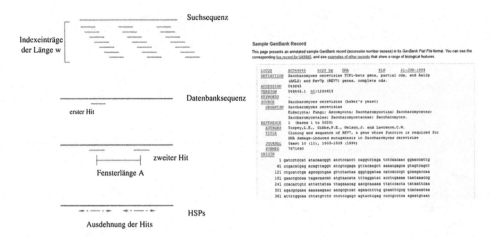

**Fig. 6.1** Two-hit method of BLAST and GenBank example entry. The left side shows the index search of BLAST (Figure from Hansen 2013) and the right side shows an example database entry for *S. cerevisiae* with name, label and start of sequence from GenBank (BLAST would search against this database entry)

## 6.2   Maintenance of Databases and Acceleration of Programs

For database searches, I need good bioinformatics databases in any case. Figure 6.2 explains an exemplary well-maintained database for this purpose, the UniProt database. This database carries the older Swiss-Prot database. Even earlier, this database was the personal project of Amos Bairoch. He looked at protein families and made notes on which amino acid residues were typical for zinc finger proteins, for example, which deviations occurred and whether an entire protein family could be described by a certain pattern. For example, zinc fingers can be described by two cysteines at a distance of three amino acids from each other, i.e.

Cysteine - - Cysteine ... Cysteine - - Cysteine,

in the single-letter code then finally.

CXXC [3..5 X] CXXC.

He then compiled such signatures into the PROSITE signature database, but the precisely labelled protein sequences (according to their family membership, structure in domains, sequence properties) as entries in the Swiss-Prot database. After some time, the work became too much for one person, and so the PROSITE database was gradually created. Around the turn of the millennium, it was concluded that protein labelling could no longer be a single task for one country because of the ever-increasing number of sequences. Together with the EBI in Cambridge and American scientists, the UniProt database was founded.

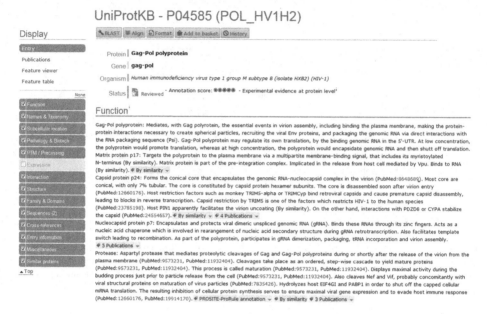

**Fig. 6.2** The example shows the header lines and the actual data part (only the function is shown here). The header shows whether the entry is "trusted", i.e. manually reviewed (*Reviewed*), or computer-annotated (e.g. *DataMining; Unreviewed*), but also how much information is available (*annotation score*)

This historical overview also summarizes briefly the essential problems and tasks of databases: Ideally, each sequence is viewed by hand, analyzed with various bioinformatics programs, and then accurately labeled. This is a lot of work, typically referred to as database maintenance. Since data sets in bioinformatics usually grow very quickly, this database maintenance is a chronic problem, often exacerbated by the fact that new databases are usually created by a new project and then not maintained after the PhD thesis or postdoctoral project ends. Only a few large institutions, which are mentioned here and at other places in the book, have enough staff to nevertheless maintain really well-maintained data, in particular the NCBI, the EBI and the SBI (Swiss Bioinformatics Institute).

Other problems of databases are cross-linking to other data (this is also difficult due to the constant growth of data), maintenance of content (especially when new types of content are added), the number of errors or outdated entries.

For the protein databases UniProt and PDB (one of the oldest bioinformatics databases, since the 1960s of the last century), as for many other databases, the uniform formatting of entries is a problem. And of course it is not only difficult for BLAST to find entries quickly and accurately in constantly growing databases. There are the two problems of **recall** (sensitivity; how many of the hits are also stored in the database as real entries?) and **precision** (specificity; do I find exactly what I am looking for or does my program suspect that it could be half the database?).

In any case, it is advisable to always take a look at the quality parameters of the databases first in order to be able to assess the actual information content and the usability of the information provided for one's own scientific work and resulting statements.

Figure 6.2 explains again very nicely the high quality of a UniProt data entry, here for our well-known example of the HIV-1. Each entry is divided into a header and the actual data part (here only a section shown for the function). A look at the header already gives a first important hint about the information content of the database entry (status). It shows how "trustworthy" the information is. UniProt distinguishes between *Reviewed* (manual/Swiss-Prot-annotated) and *Unreviewed* (computer-annotated from TrEMBL). In our example the entries were manually annotated and reviewed by UniProt curators, so they are trustworthy in this case. In addition, an annotation *score* is calculated for the provided information (maximum score of 5), which also indicates how much information is available for the respective entry, i.e. how well it is annotated. For HIV-1 UniProt displays the maximum score.

Users should therefore always take a look at the quality parameters before using the information provided.

So with that, we understand how bioinformatics now works so quickly and soundly. There are fast and yet surprisingly accurate programs used (heuristics). And there are good, highly sophisticated databases where you can trust the entries and yet they are very well maintained.

Therefore, a few other notable heuristics should be mentioned here. Besides BLAST sequence search, BLAT search is another speedup, as is Mega-BLAST (the expert then knows what is more easily overlooked by these variants of BLAST).

Even 3-D structures are made faster and shorter by heuristic searches. In particular, many reasonably fast modeling programs use the homology modeling step, that is, using known structures to model the unknown structure if it is sufficiently similar. This heuristic is not an exact model and assumes that the new structure is too similar to something. The heuristic is even more stringent in *threading*. Here it is assumed that even an unknown 3-D structure can be predicted by combining and testing known 3-D structures. To do this, the unknown structure is threaded onto the known 3-D structures on the basis of the sequence. One then calculates which region is best covered by which known structure. Not exact, just a heuristic.

One can be surprised at the protein interaction database STRING (EMBL) how quickly the interactions are calculated. A trick is used that is also used by a number of other databases. Here, all interactions are calculated in many weeks with each update of the database. The single database query now only looks up where the best entry for the query is located in the database. If one or more sequences are entered, this is done via a sequence comparison (with BLAST), if a *keyword* is entered, this is done via a fast text search.

Metabolic models often make the heuristic assumption of steady-state equilibrium and then calculate the underlying enzyme chains for this equilibrium (*flux balance analysis*; the same principle used: elementary mode analyses). Even if, for example, YANAsquare calculates flux strengths, it makes the simplified assumption that gene expression data already correctly reflect the different activities of the metabolic pathways (which is only true on statistical average or for sufficiently large networks).

Finally, even the semiquantitative models for signal modeling use heuristics, in particular the kinetics is estimated only from the Boolean networks of the process to be modeled. This allows me to get started with such a model when little is known in detail about the speed and nature of the proteins, enzymes, kinases, etc. involved.

How can you now program a heuristic search yourself?

The **BioPerl** and **Biojava modules** (https://bioperl.org/, https://biojava.org/) at the EBI (European Bioinformatic Institute) are a good way to quickly program a heuristic search or even a simple program or a larger program composed of simple parts. They provide ready-written modules (program parts) for reading, output, but also for web servers or database searches for the user. The PERL Cookbook (Christiansen and Torkington 2003) offers a lot of tips for concrete implementation with the PERL programming language. Even more tips are found in further publications (Angly et al., 2014; Vos et al., 2011; Stajich et al., 2002; Tisdal et al., 2001).

For calculations, the book "*numerical recipies*" (https://numerical.recipes) is a real treasure trove. Originally a book (Press et al., 2007), it now explains online in a clear way how I can quickly and easily compute small calculations or even surprisingly complex ones, which, however, come up again and again in many problems. Similar to a cooking recipe, the principles are explained and codes are provided, for example to make a Matlab code run faster (tutorial: https://numerical.recipes/nr3_matlab.html) or to use a "C+ +" code for even faster calculations instead. Examples of applications for these numerical recipes, also in bioinformatics, are e.g. efficient matrix and vector calculations (calculate metabolic fluxes efficiently), but also routines for geometric tasks (calculate protein structures) or the generation of random numbers (for population simulations in ecology).

### Conclusion

In this chapter we have tried to look a little behind the façade of the fast bioinformatics programs on the net, such as the BLAST server at the NCBI (National Center of Biotechnology Information) in Washington. In most cases, you can get an answer in seconds to a few minutes. This is made possible by fast but not entirely accurate searches (heuristics), and we have seen some tricks for doing this. For example, in BLAST, the heuristic is to first find two short but perfect match alignments in the same database entry before I check over the whole sequence length to see what the similarity is to the question sequence.

It is equally important to make the database (e.g. GenBank, UniProt) quickly readable, for example by indexing it (after all, you look up this book much more quickly via the table of contents than by leafing through it). In addition to speed, sensitivity (do I recognise all relevant entries?) and specificity (do I not get too many non-relevant entries?) are also important for a good search.

In the tutorial in the appendix a short introduction to programming including installing BLAST or in general a web server is given. Generally speaking, web-based programs and good bioinformatics algorithms and scripts for bioinformatics analysis are still developing rapidly and there are many fascinating programming tasks. ◄

## 6.3   Exercises for Chap. 6

You are welcome to work on the tasks of Chap. 1 for this chapter as well.

**Task 6.1**
A simple illustration: How do you look something up in a book? Discuss different approaches.

**Task 6.2**
Comparison "fast" and "super fast": How do BLAST, FASTA and Psi-BLAST differ in terms of their search strategy?

**Task 6.3(a)**
What is BLAT (not a typo, bioinformatics question)?

**Task 6.3(b)**
What are the advantages of BLAST as sequence comparison tool?

**Task 6.4**
Which sequence comparison search is fastest? Give some examples and consider which is the very fastest. Compare the advantages and disadvantages.

**Task 6.5**
Which annotation is best? Compare: Annotations in GenBank, UCSC Genome Browser and Swiss-Prot/UniProt.

**Task 6.6**
In your opinion, how should an "ideal" database/server be constructed (what basic parts should the database/server consist of)?

**Task 6.7**
List ways in which ideally a database should be maintained and kept up to date.

| Useful Tools and Web Links | |
| --- | --- |
| BLAST | https://blast.ncbi.nlm.nih.gov/Blast.cgi |
| NCBI | https://www.ncbi.nlm.nih.gov/pubmed/ |
| EBI | https://www.ebi.ac.uk/services |
| SBI | https://www.sib.swiss/ |
| UniProt | https://www.uniprot.org/ |
| PDB | https://www.rcsb.org/pdb/home/home.do |
| STRING | https://string-db.org/ |
| YANAsquare | https://www.bioinfo.biozentrum.uni-wuerzburg.de/computing/yanasquare/ |
| BioPerl | https://bioperl.org/ |
| Biojava | https://biojava.org/ |
| Numerical recipies | https://numerical.recipes/ |

## Literature

Christiansen T, Torkington N (2003) Perl cookbook. Solutions & examples for Perl programmers. O'Reilly Media, Beijing (Final Release Date: August 2003, Pages: 968 *This book is simply very well written and provides a very good introduction to the Perl programming language)

Hansen A (2013) Bioinformatik: Ein Leitfaden für Naturwissenschaftler. Birkhaeuser, Basel. (Erstveröffentlichung 2004, isbn 3-7643-6253-7, Taschenbuchauflage – 4. Oktober 2013)

Press WH, Teukolsky SA, Vetterling WT, Flannery BP (2007) Numerical recipes, 3rd edn. The art of scientific computing (English) Bound edition. Cambridge University Press, New York. isbn 978-0-521-88068-8

Here are some more book suggestions for Perl and its programming that are not explicitly discussed in the chapter. For other programming languages please look in the tutorial (later chapter in the book)

Angly FE, Fields CJ, Tyson GW (2014) The bio-community Perl toolkit for microbial ecology. Bioinformatics 30(13):1926–1927. https://doi.org/10.1093/bioinformatics/btu130

Mount D (2004) Bioinformatics: sequence and genome analysis, 2. Aufl. Cold Spring Harbor Laboratory Press, Cold Spring Harbor, New York (© 2004 • 665pp., illus., appendices, index paperback, isbn 978-087969712-9 *David Mount also does a very good job of introducing the underlying algorithms of sequence analysis. However, it is aimed at somewhat advanced students. Also used as a textbook at LMU Munich)

Stajich JE, Block D, Boulez K et al (2002) The bioperl toolkit: Perl modules for the life sciences. Genome Res 12(10):1611–1618

Tisdall J (2001) Beginning Perl for bioinformatics an introduction to Perl for biologists. 1. Aufl. O'Reilly Media, Sebastopol, Kalifornien, USA (Final Release Date: October 2001, Pages 386)

Vos RA, Caravas J, Hartmann K et al (2011) BIO: phylo-phyloinformatic analysis using Perl. BMC Bioinformatics 12:63. https://doi.org/10.1186/1471-2105-12-63

# How to Better Understand Signal Cascades and Measure the Encoded Information

**Abstract**

Shannon has made it possible to measure how much information is contained in a message. It is calculated how many bits of information are contained in each part (word, nucleotide, etc.) of a message. Interestingly, this can identify any number of codes, languages, and encodings in the cell. Since living cells are not computers, but numerous biochemical reactions run simultaneously side by side and sometimes quite disorderly, causing a lot of commotion and disturbance, it is important to send this information as clearly as possible, for example to amplify signals by signal cascades. The more precisely the signal is understood and implemented in the cell, the better the cell survives. Therefore, survival pressure already ensured that the genetic information is well coded and well transferred into various other codes. These codes can again be "cracked" by bioinformatics for good predictions, for example for sequence analysis.

## 7.1 Coding with Bits

How much data have I actually collected in a specific case, how do I measure the abundance of data? In order to measure the cellular messages (e.g. messenger RNA between cell nucleus and cytoplasm or hormone between endocrine gland and other body cells), the Shannon entropy is a useful measure: one bit of information is the smallest unit of information, a "yes" or "no" decision. Shannon entropy now calculates (Fig. 7.1, left) for each piece of information transmitted how many "yes/no" decisions are hidden in it. A letter is, after all, one of 26 possibilities, so it contains about four and a half bits (because with 4 "yes/no" decisions you can distinguish between 16 possibilities, and with one more question you can even cover 32 possibilities [2 to the power of n, abbreviated written 2**n

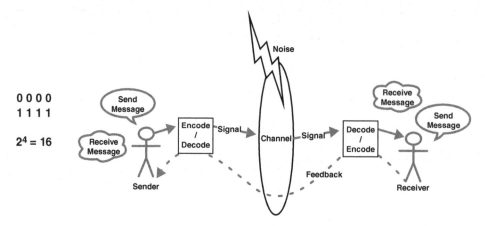

**Fig. 7.1** Schematic representation of the Shannon entropy. The Shannon entropy calculates the expected information value of a message. Typical units are binary bits, i.e. yes/no decisions that can be encoded and decoded. The figure shows the limits of the maximum information units that can be transported from the sender via channel to the receiver, which are subjected to signal noise

possibilities]). Shannon developed this system further, so that words and sentences are then assigned their information content according to their length.

Next, one can compare the quality of different signal sources: For example, the information value is very low if the same character is always sent, but very high if very different characters are always sent in a new sequence, such as a radio station.

After all, you have to take into account what it looks like inside living cells: Countless reactions take place, there is a lot of hustle and bustle. Therefore, biological signals are often amplified in signal cascades, so that one can still understand the signal despite the "noise" (all the other reactions and signals taking place). The quality of the signal depends on the ratio of signal to background noise (signal-to-noise). Shannon has set up a whole theory on how communication via communication channels can run as optimally as possible despite interference.

If the bioinformatician wants to model and better understand cell growth, differentiation or the death of cells, these theories are taken into account and the amplification, weakening and modelling of cellular signals in different signalling cascades is investigated, as well as the weakening of kinase cascades by phosphatases, for example, so that the cell stops growing again. At this complex level (function of the various signalling cascades in the cell), a deeper understanding of the processes surrounding cell growth and cell differentiation is then indeed possible.

## 7.2    The Different Levels of Coding

Now that we have discussed how to calculate information in principle and send it clearly enough to be understood despite the background noise (biological noise): In the forest, in the environment, in the nervous system or even in the cell, there are always disturbances and sources of noise), we can now start to look more closely at how the cell encodes information at different levels with an adapted code (Fig. 7.1, right).

The figure shows a selection. Each two bits corresponding to Shannon coding or Shannon entropy are represented by a nucleotide. If you look at proteins, there are 20 amino acids encoded with 64 codons, i.e. 6 bits (because 2 to the power of 6 or 2**6 is 64).

The three-dimensional protein structure code is much more complex. There are so many possibilities here that the information value of a defined protein structure is very high (to be calculated in a simplified way by the number of bits that a PDB structure file has when it is downloaded, which is already hundreds of thousands of bits). Informatically clever is the use of internal coordinates to encode protein structures with few bits: Only the path from one amino acid to the next is ever specified. This can be done with the angles phi and psi at the central carbon atom (alpha-C atom) of each amino acid (AlQuraishi 2019). If I then use four or eight standard conformations to merely represent the protein structure in a highly simplified way, I only need 2 or 3 bits for each amino acid position in a protein folding simulation (Saxena et al. 1997).

Finally, there are other codes, for example at the cell membrane (membrane lipids, but also specific membrane modifications), the RNA sequence structure code within the cell for regulatory RNA, metabolic regulation (e.g. iron) as well as localisation in the cell, and finally the sugar code at the cell surface, with which cells recognise each other and via which transplant rejection is also coded. Finally, there are phospholipids that, for example via gangliosides and cerebrosides (i.e. sugar-lipid structures), assign the wiring in the brain and different neuronal structures to each other in detail in order to ensure the plasticity of our brain during embryology.

All these codes are not only used and needed in the cell, but you can also decode them with bioinformatics, especially via sequence.

In this way, it is possible to translate the fairly universal genetic code (program "Translate" from the Expert Protein Analysis System, EXPASY, at the "Swiss Institute of Bioinformatics" https://web.expasy.org/translate/) and better understand its rarer variants for certain codons, for example in mitochondria, some bacteria and also protozoa (Heaphy et al. 2016) (https://www.ncbi.nlm.nih.gov/Taxonomy/Utils/wprintgc.cgi). Similarly, signals in regulatory RNA can be analyzed, for example with the RNA analyzer (https://rna-analyzer.bioapps.biozentrum.uni-wuerzburg.de/), but also, for example, sugar codes (https://www.functionalglycomics.org/; https://ncfg.hms.harvard.edu/) or code analyses in lipids, for example to assign lipids to the correct type after mass spectrometry (Ahmed et al. 2015).

## 7.3   Understanding Coding Better

So what can we take away as insights? It's a lot like a conversation in a busy pub. The signals of the cell are constantly fighting against the background noise. Apart from our own signalling cascade, which we are currently interested in, such as the Erk kinase cascade (see Chap. 5), all other signalling pathways are also active. The cell works with biochemical reactions and not like a digital silicon computer. Therefore, signals can only reach their destination if they are amplified in a cascade. Nice examples are the blood clotting cascade, so that the broken vessel is guaranteed to be closed again safely and quickly, and also the opposite blood clot dissolving cascade (plasminogen cascade). In the blood, for example, there is then also the complement cascade for the immune system and so on. So in general, biology has to come up with a lot of things to cope with the noise. One possibility to reach highest sensitivity is given for example by the photoreceptors of our eye, where three inhibitory mechanisms all together return to the resting state and the initial situation is a hyperpolarization.

A computer or even you yourself with the next transfer with IBAN number use check bits to be sure that nothing has been changed by mistake. This mechanism also exists. First of all, all kinds of sequence signals are used for this purpose, which you can find out with the ELM server, for example, and which ensure in a relatively error-tolerant way that every protein gets to the right place. However, the stability signals and signals that ensure that a "wrong" protein, for example one that is too short, is rapidly degraded (so-called "*non-sense mediated decay*", NMD, for stopping too early in the case of mRNA from eukaryotes) are also a kind of check bit for proteins. Similar check bits exist for RNA, such as various methylguanosine caps that mark different types of RNA as mature and regulate the nuclear or cytoplasmic transport of that RNA and its proteins. Another strategy to better understand the notoriously complex codes in biological systems is simplification (technical term: dimensionality reduction). The aim is to transform and visualise high-dimensional data in a new coordinate system (usually 2D). For this purpose, methods of multivariate statistics such as PCA (Principal Component Analysis; for examples in R see our web application [Fuchs et al. 2020] or https://rpubs.com/amos593/419546) are applied (explorative data analysis). Through dimensionality reduction, one wants to get an overview of the data and reduce its complexity by decomposing it into principal components. Through this structuring one wants to extract relevant variables (features) and groups, for example for the construction of predictive models (Chap. 14), but also to make visible possible batch effects in the data that may need to be corrected (especially in omics analyses). For example, the pattern of gene expression is determined by the interaction of many 1000 genes. To get an overview of the most important components involved, PCA can be used to calculate the two main components of the differences between datasets, giving a quick overview of which combination of important genes decisively determines the differences. The method is applicable to all complex datasets, e.g. cardiac fibrosis (Fuchs et al. 2020), but also in ecology, for example to quickly screen bacterial communities (Kim et al. 2020).

One can also look at the challenges of reliable signal transmission and coding in the cell in a mathematically exact way for signal cascades and the phosphatases that switch off the signal and thus better understand how these cellular signals are formed and transmitted (Heinrich et al. 2002). Phosphatases are important for the regulation of signal amplitude, signal frequency, and signal duration because the phosphatase must shut down fast enough. Kinases are important for the regulation of signal amplitude and signal height, because the kinase must amplify strongly enough.

Finally, however, the recognition and decoding of such signals with the aid of bioinformatics is also medically important. An important example, for example, was the third phosphorylation in Erk kinase, which supports heart failure (see Chap. 5). Many cancers arise from the fact that a mutation in a body cell causes a growth kinase to be constantly turned on. An important example is the B-Raf kinase. Unmutated, it allows skin cells to grow. In the mutated version, such as from too much UV radiation while tanning at the beach, it leads to melanoma, or black skin cancer. How good it is that sunburns cause the skin to exfoliate: These skin cells have all voluntarily perished (via the cell death or *apoptosis pathway*) so they don't harm us as cancer cells. This *apoptosis pathway* is another equivalent of check bits in a computer: in particular, the p53 protein makes sure that either DNA repair still works successfully and is carried out, or the cell goes into apoptosis. The miRNAs are also important regulators in cancer (Lujambio and Lowe 2012). Constant coding and decoding is vital to us, and it is exciting to trace this using bioinformatics (Richard et al. 2016).

---

**Conclusion**

- Shannon has made it possible to measure how much information is contained in a message. It is calculated how many bits of information are contained in each word of the message. For example, a nucleotide in DNA comes in four forms. To identify one, I have to answer two yes/no questions (is it a purine/pyrimidine? Which of the two purines/pyrimidines is it?), so a nucleotide carries two bits .

- Interestingly, one can thus identify any number of codes, languages and codings in the cell. Since living cells are not computers, but numerous biochemical reactions run simultaneously side by side and sometimes quite disorderly, thus causing a lot of commotion and disturbances, it is important to send this information as clearly as possible, for example to amplify signals through signal cascades. The more precisely the signal is understood and implemented in the cell, the better the cell survives. Therefore, survival pressure already ensured that the genetic information is well encoded and well transferred into various other codes.

- Bioinformatics only has to replicate this in the computer programs used and can then decipher and "crack" code after code in the cell quite accurately. A good starting point for using this are the numerous programs for sequence analysis, which are explained here from the first chapter onwards. Sequence analyses have therefore also become the most important basic tool in bioinformatics.

## 7.4     Exercises for Chap. 7

**Task 7.1**
Encoding signals: How many bits are in a message that encodes the number seven with dual numbers?

**Task 7.2**
How many bits do I need to represent the number one thousand (1000) as a bit?

**Task 7.3**
Of course you can do this with a nice webtool, can you find one?

**Task 7.4**
How many bits does a word have, e.g. the word "WORD"?

**Task 7.5**
How many bits does a biological word have?
    So we are already in the middle of biology and the tasks and problems in the cell:

**Task 7.6**
How does a signal reach its receiver safely despite the loud noise in the cell? Put together some examples that are biologically exciting (e.g. from this chapter/book).

**Task 7.7**
Calculation of the amplification of a signalling cascade: The Ras-Raf-Mek-Erk cascade amplifies the cellular signal by a factor of ten at a time. What happens when the receptor activates a Ras molecule?

**Task 7.8**
Importance of the Ras-Raf-Mek-Erk cascade: Give a biological example of what this signal is important for. Also tell what can go wrong in the process.

**Task 7.9**
Set up the differential equation of the Ras-Raf-Mek-Erk cascade.

**Task 7.10**
What mathematical models of protein kinase signal transduction do they know?

**Task 7.11**
How does a metabolic signal safely reach its location? Put together important factors for this.

## Task 7.12

What are *"moonlighting"* enzymes? Find an example.

## Task 7.13

Higher sensitivity of metabolic regulation due to simultaneous outward and reverse reaction:

(a) Consider for which biochemical processes it can actually make sense what is constantly going on in the cell (in the "bubbling soup"), namely that outward and reverse reactions can take place simultaneously.

(b) Carry out a calculation example for this.

**Useful Tools and Web Links**

| | |
|---|---|
| PDB | https://www.rcsb.org/pdb/home/home.do |
| RNAAnalyzer | https://rnaanalyzer.bioapps.biozentrum.uni-wuerzburg.de/ |
| Functional Glycomics | https://www.functionalglycomics.org/; https://ncfg.hms.harvard.edu/ |
| ENCODE | https://www.encodeproject.org (see Diehl and Boyle 2016). |

- This is an important link to the human genetic code, namely the famous *"Encyclopedia of DNA Elements"* of the human genome, which you can both look up and analyze here. There is also much original literature describing it.
  OMIM: https://www.omim.org
- *"Online Mendelian Inheritance in Man"* makes it very clear how a wrong letter (a genetic mutation) leads to disease.
  Lipid-Pro: https://www.neurogenetics.biozentrum.uni-wuerzburg.de/services/lipidpro/
- This is the software we developed that helps classify lipids and decode their code.
  Bionumbers: https://bionumbers.hms.harvard.edu
- Here, the number codes that play a role in numerous biological processes are explained nicely and engagingly (Milo et al. 2010).

## Literature

Ahmed Z, Mayr M, Zeeshan S et al (2015) Lipid-Pro: a computational lipid identification solution for untargeted lipidomics on data-independent acquisition tandem mass spectrometry platforms. Bioinformatics 31(7):1150–1153. https://doi.org/10.1093/bioinformatics/btu796. (* This is software we developed that helps classify lipids and decode their code.)

AlQuraishi M (2019) End-to-end differentiable learning of protein structure. Cell Syst. 8(4):292–301. e3. https://doi.org/10.1016/j.cels.2019.03.006

Diehl AG, Boyle AP (2016) Deciphering ENCODE. Trends Genet 32(4):238–249. https://doi.org/10.1016/j.tig.2016.02.002. (Review. PubMed PMID: 26962025 * A very nice overview of the results of ENCODE.)

Fuchs M, Kreutzer FP, Kapsner LA et al (2020) Integrative bioinformatic analyses of global transcriptome data decipher novel molecular insights into cardiac anti-fibrotic therapies. Int J Mol Sci 21(13):4727. https://doi.org/10.3390/ijms21134727

Heaphy SM, Mariotti M, Gladyshev VN et al (2016) Novel ciliate genetic code variants including the reassignment of all three stop codons to sense codons in *Condylostoma magnum*. Mol Biol Evol 33(11):2885–2889. https://doi.org/10.1093/molbev/msw166

Heinrich R, Neel BG, Rapoport TA (2002) Mathematical models of protein kinase signal transduction. Mol Cell 9(5):957–970. (*Describes how the cell asserts itself against the background noise.)

Kim M, Lee J, Yang D et al (2020) Seasonal dynamics of the bacterial communities associated with cyanobacterial blooms in the Han River. Environ Pollut 266(Pt 2):115198. https://doi.org/10.1016/j.envpol.2020.115198

Lujambio A, Lowe SW (2012) The microcosmos of cancer. Nature 482(7385):347–355. https://doi.org/10.1038/nature10888. (Review. PubMed PMID: 22337054; PubMed Central PMCID: PMC3509753 *Shows miRNA codes and how they are important for cancer.)

Milo R, Jorgensen P, Moran U et al (2010) BioNumbers – the database of key numbers in molecular and cell biology. Nucleic Acids Res 38(Database issue):D750–D753. https://doi.org/10.1093/nar/gkp889. (*Hier werden die Nummerncodes, die in zahlreichen biologischen Prozessen eine Rolle spielen, schön und ansprechend erklärt.)

Richard A, Boullu L, Herbach U et al (2016) Single-cell-based analysis highlights a surge in cell-to-cell molecular variability preceding irreversible commitment in a differentiation process. PLoS Biol 14(12):e1002585. https://doi.org/10.1371/journal.pbio.1002585. (*Uses advantageously Shannon entropy in a nice biological application example)

Saxena P, Whang I, Voziyanov Y et al (1997) Probing Flp: a new approach to analyze the structure of a DNA recognizing protein by combining the genetic algorithm, mutagenesis and non-canonical DNA target sites. Biochim Biophys Acta 1340(2):187–204. https://doi.org/10.1016/s0167-4838(97)00017-4

# When Does the Computer Stop Calculating? 8

**Abstract**

The question of when a bioinformatics problem will be completed is difficult to answer for problems with built-in combinatorics. Alan Turing generally modeled all computable problems using the Turing machine, an idealized abstract computer. All non-Turing computable problems cannot be solved by computers and remain tasks for humans. Many particularly interesting problems in bioinformatics are NP (nondeterministic polynomial complexity) problems, such as protein structure prediction and most network and signal computation or image processing. In general, more powerful computers, the bundling of many computer nodes (parallelisation) and application-specific chips can also directly increase computer performance, for example with omics data.

We remember that bioinformatics analyses biological data with programs (Sect. 2.1), collects them in databases (Sect. 2.2) and then maps the biological relationships in models. But how good are bioinformatic models? Well, bioinformatics tries to use computers to make "good" and comprehensible biology. One can have fundamental reservations about this. After all, life is a quality rather than a quantity. Experiences are not seldom simply indescribable, and also a bacterium or also your own mind and even the brain are not simply a kind of chip (bacterium) or supercomputer (we ourselves). We are infinitely much more, and who cannot understand this at all, should now go to a good theater play (no cinema effect, it is better to experience this "live") or talk for a few minutes with a patient in a psychiatric ward, then may be he will better fathom what we want to say.

© Springer-Verlag GmbH Germany, part of Springer Nature 2023    93
T. Dandekar, M. Kunz, *Bioinformatics*,
https://doi.org/10.1007/978-3-662-65036-3_8

## 8.1    When Does It Become a Challenge for the Computer?

However, the moment we recognise that this may be the unavoidable limitation of our computational approach and that we naturally systematically do not take these imponderable, qualitative aspects into account in our bioinformatic models, we are already a significant step further. So let's keep in mind: bioinformatics tries to describe biological reality in clear, transparent models, for example how a normal cell becomes a cancer cell. By using the computer and experimental data, I make myself blind in terms of experience and other direct interactions with nature, but I have the undeniable advantage of having quantitative statements about the biological process through numbers and measures (*"give numbers to the arrows" is what* Leroy Hood once called it). This alone, through these quantitative glasses, prevents being drowned in too much unprovable theory. For example, the model predicts that 80% of cancer cells will die from treatment. We can simply measure in experiments how far this is true. This also brings up another important implication for all bioinformatics analyses. For example, if we have found a related sequence to a sequence of which we know more about the function by sequence comparison, then we should continue this chain of reasoning (from sequence comparison to sequence comparison) until we have a clear experiment on the last sequence that biochemically or molecularly confirms the function of the protein associated with the sequence. Only then do we have solid ground within the framework of our model.

So much for the bioinformatic model, which should therefore always base its own calculations on solid, experimental data. Now a few sentences about the calculations: After all, it could be that these calculations take a very long time, and anyone who has ever "kicked" his computer with such a complicated, lengthy calculation knows the problem of wondering, "When will this limited computational box finally stop calculating?" The problems where this is unresolved are called NP problems (NP stands for nondeterministic polynomial time). There is no simple formula (a polynomial) that allows one to calculate how long the computer will compute based on the length of the input. Unfortunately, most biologically exciting problems are such NP problems. This is because biomolecules and all higher processes in the cell are usually modular, made up of similar or identical units (see Part 1). Thus, the addition of only one further unit leads to a multiple increase in computation time, and such combinatorial problems ("combinatorial explosion") therefore almost always occur in our biological modelling. This leads to corresponding uncertainties in the computation time. However, one can help oneself with fixed specifications, so-called *"stopping criteria"*, i.e. stopping specifications for the computer, e.g. "please stop after one hour of computing time". But more important is the fact that with a fixed calculation time it is not possible to estimate how good the solution found up to that point is in comparison to the best or optimal solution. But that's just life: Not so easy to grasp!

To conclude this chapter, it is therefore worth pointing out that the outstanding mathematician Alan Turing succeeded in defining the capabilities of a computer quite precisely (Hodges 2014). He devised an abstract machine, the so-called Turing machine (Fig. 8.1), which could only perform five basic operations. He was able to show that every exact

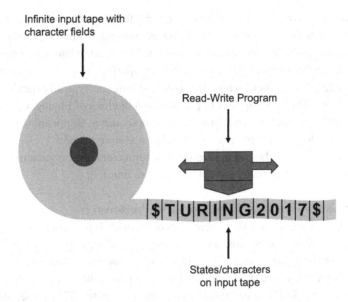

**Fig. 8.1** Simplified representation of a Turing machine. The Turing machine consists of an infinitely long input tape with separated character fields, a read-write program (can read/write in both directions) and the characters of the input tape. The read-write program reads the input tape field by field and can change the characters according to the program instruction (transfer function). This procedure can be used to determine which calculations are bioinformatically computable with the computer (Turing-computable, non-Turing-computable)

computation possible at all was also performed by (usually very many) concatenating the five operations of his abstract machine. This allows us to determine very precisely which computations in bioinformatics can be done at all with computers (no matter how modern they are or will become) ("Turing-computable") and which cannot ("not Turing-computable"). So much for the limits and restrictions of bioinformatics models and of modelling in general.

## 8.2   Complexity and Computing Time of Some Algorithms

Let's now turn to another problem: How much longer does my calculation take when the task becomes more difficult? This question is generally called the complexity of a computational problem.

**Polynomial Complexity**
In this case, everything is not too computationally intensive. A simple calculation expression, a so-called polynomial, gives the calculation time as a function of the length.

For example, if an RNA has a length of n nucleotides and is to be folded (i.e., the secondary structure is calculated), each nucleotide is typically juxtaposed with every other one along its entire length, and thus sampled for all possible pairs. So this computational task is quadratically complex, taking 100 time units for 10 nucleotides and 10,000 time units for 100 nucleotides. Therefore, RNA folds are only calculated for molecules that are not too large, and database searches are usually not fast for complete molecule folds.

Many computational tasks, e.g. the sequence comparison of protein sequences in the genome, i.e. again each protein with every other protein, are typically quadratic in their time requirements. The same applies to pairwise calculations of phylogenetic trees with phylogenetic software, such as when one calculates a sequence alignment with CLUSTAL and associated phylogenetic trees with the *Neighbor-Joining method*.

Many database searches here again require a quadratic or cubic amount of time (sifting through 10 times more data requires 1000 times more time).

### Non-deterministic Polynomial Complexity (NP-Problems)
The case is quite different when the problem becomes many times more difficult with each step. These are problems that grow exponentially, for example (complexity EXP). The complexity class NP is now the set of all problems solvable by nondeterministic Turing machines in polynomial time. Put simply: All problems solvable in polynomial time by a computer that can randomly select multiple computational paths. This subset of EXP contains a very large number of relevant problems. Since the problems from P can be solved non-deterministically in polynomial time if they must, P is a subset of NP. These NP-hard problems are very hard to estimate in computational time. It is true that if the solution is correct (given by a good fairy, for example), one can check it in polynomial time to see if it is correct. But from this one does not find it fast or at all without the good fairy.

The best-known problem is the *travelling salesman problem* ("TSP"), who wants to visit many cities with an optimally short route on his way. One can only be really sure after quite long calculations, but these become more than 100 times more complex with each additional city, for example, with the 200th city even more than 200 times more difficult with each additional city.

Actually, many problems of real interest in bioinformatics are NP-complete, i.e., equivalent to TSP, theorem of Cook (1971) and Levin (1973), respectively. The Theorem of Cook (1971) founded a new class of problems in terms of computation time, more generally, complexity theory. Cook showed that there exists a subset of the class NP to which all problems from NP can be reduced. Named after him, Cook's theorem states namely that the *satisfiability* problem of propositional logic, SAT, is NP-complete. Thus, the SAT problem is representative of NP-problems, and all problems that can be transformed into a SAT problem are equivalent to it (class of NP-complete problems). Levin (1973) showed this important insight, i.e., when a computer cannot find a solution and finish, quite independently and in its own way. An example is for instance the protein folding problem, i.e. the prediction of the protein structure, where each additional amino acid makes the computation of the coordinates of the three-dimensional structure many times harder. Homology modeling or the calculation of system states also belong to this class. Accordingly, each additional kinase or phosphatase makes the problem at least twice as difficult, and usually even more ambiguous. In any case, this should be taken into account

**Table 8.1**  Degree of difficulty of a P problem (sequence alignment) compared to an NP problem (protein folding)

| Algorithm | Runtime complexity (m, n = sequence length of a, b) |
|---|---|
| *Heuristic algorithms:* | |
| Blast | $O(n*m)$ |
| *Dynamic algorithms:* | |
| Needleman-Wunsch | Cubic: $O(n^3)$; e.g. at 5 = 125 |
| Smith-Watermann | Quadratic: $O(n^2)$; e.g. with 5 = 25 |
| *Protein folding:* | |
| For x possible folds | Exponential: xn (e.g. for 2 convolutions: 2n; for 7 convolutions: 7n) |

in bioinformatic considerations. Table 8.1 shows the degree of difficulty of a P-problem in comparison to an NP-problem using the example of a sequence alignment versus protein folding with combinatorics.

## 8.3   Informatic Solutions for Computationally Intensive Bioinformatics Problems

Many interesting problems in biology and bioinformatics have a built-in combinatorics and thus a very large, difficult to understand solution space, which therefore has the difficulty NP (solution very difficult to find and computation time not foreseeable - if you show me the solution, I can usually confirm it relatively quickly). All in all, however, computational time problems are computer science problems, which can therefore also be tackled directly with tools from computer science and computer technology.

**Tip 1: Use Modern Computer**
This is often effective in practice. First, if you have a difficult or computationally intensive bioinformatics problem, you should not use a web server (otherwise you might wait until you black out!). However, most bioinformaticians have already taken this into account when designing their programs. Protein structure predictions, for example, are often not done online on the web server, but one receives (after a few hours or even days) the result by e-mail (for example, when using SWISS-MODEL for homology models or *ab-initio predictions* by the QUARK software from the Zhang lab). For own calculations I should first use a notebook or PC as up-to-date as possible. *Workstations* or small computer clusters have even more computing power at first. For larger calculations, local (university mainframes) or central computer clusters (e.g. Leibniz Computing Centers in Munich, etc.) are then available. Tier 1 or Tier 0 mainframes such as JUQUEEN in Jülich then provide the greatest performance (6 million billion floating point operations per second) with 5.9 petaflops per second (https://www.fz-juelich.de/ias/jsc/EN/Expertise/Supercom puters/JUQUEEN/JUQUEEN_node.html).

**Tip 2: Heuristics**

We have already seen in Chap. 6 that due to the large amounts of data in bioinformatics, one tries to use algorithms that are as fast as possible, even if this comes at the expense of accuracy (heuristics such as BLAST). For informatics tips and tricks for better and faster programming, the "numerical recipies" as described there are highly recommended. Another commendable and non-profit activity is "Project Jupyter" to advance freely available software, open standards and services for interactive work with dozens of programming languages such as Julia, Python and R. For this purpose, *Jupyter notebooks* and the *JupyterLab* were developed, which have a high reusability and good documentation.

**Tip 3: Parallelization**

Finally, an important technique for complex calculations is to use many processors in parallel. For this, the computational task must also be "parallelized", i.e. rewritten in such a way that the distribution to several processors (or computer nodes) actually saves time and does not lead to a mess and a lot of additional communication.

There are also particularly suitable programming languages for this purpose (e.g. Popjava, PopC or the network-friendly web-based environment from the Jupyter notebooks). Another programming language is Julia (https://julialang.org). It was released in 2012 after 3 years of development. It is a higher level Matlab-like programming language for numerical and scientific computing usable for Mac, LINUX and Windows alike with quite fast execution speed. The compiler with its own standard library was written in C, C + + and Scheme. Important are multimethods, LISP-like macros and metaprogramming, direct call to C and Python functions. Designed for parallel programming and distributed computing, co-routines allow easy multithreading by Julia.

These are important ways to equip and use a computer with many processors with appropriate operating software.

In general, it should be remembered that computers are stronger in a network. Even simple computers (PCs) can help solve difficult problems via networking on the Internet when their computing power is not otherwise needed (from SETI@home to Bitcoins to scientific projects, e.g. https://blog.exabyte.io/enabling-new-science-through-accessible-modeling-and-simulations-6710098a294).

## Other Possibilities Include

**Virtualization**

Alternatively, various LINUX or UNIX computers can be interconnected by suitable software to form a virtual, parallel computer (e.g. use of PVM, https://www.csm.ornl.gov/pvm/). In the meantime, there are also commercial providers of cloud computers, i.e. a virtual computer environment with many node computers is made available to interested customers by these providers via the Internet.

**Application Specific Chips (ASICs)**

Finally, it is also possible to use special computer chips on which exactly one computer program runs "hard-wired", so to speak, i.e. one computer chip for exactly one program. This is then an *Application Specific Chip* (ASIC). Field-programmable gate arrays (FPGAs) are much more expensive, but more flexible and allow to pre-test different ASICs in their properties after appropriate programming of the FGPR. ASICs have been and continue to be used for special programs. For example, the company Paracell had developed a chip for BLAST. The sequence comparison then runs much faster on this ASIC, and the Paracell computer was thus able to identify words very quickly and use them for BLAST (likewise the American secret service to monitor the Internet, s. Chap. 16). Even at present there are a number of such special computer chips for bioinformatics. However, these are used less frequently than the other solutions in this paragraph.

## 8.4 NP Problems Are Not Easy to Grasp

At least for mathematicians and computer scientists the difficult NP-problems exert a strong fascination. This is especially due to the fact that the existence of the correct (optimal) solution can usually be solved in a reasonable computing time (i.e. a so-called P-problem, with polynomial computing time), but nevertheless, without already knowing the solution, one does not know when the computer will stop searching for solutions if the correct solution is not yet known (non-deterministic polynomial). One can most easily understand this with the traveling salesman (TSP) problem already mentioned (Fig. 8.2). One can easily confirm an optimal solution at least very well. But the combinatorics of the cities, which makes the problem many times harder with each city more, leads to very long computation times for systematic trial and error. In addition, the distances between the cities can also be different, which makes the calculation more complex (Fig. 8.2, left: symmetric TSP with equal lengths; right: asymmetric TSP with different lengths between the cities).

Therefore, computer scientists, mathematicians and bioinformaticians keep trying to show that there is a way to trace NP-problems back to P-problems in general. So far,

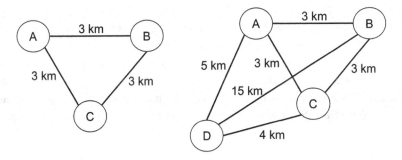

**Fig. 8.2** Simplified representation of the travelling salesman problem

however, this has been in vain. The list of failures or, in some cases, highly intelligent attempts to solve the problem is quite exciting to read. Even clearer and more exciting are the articles by Scott Aaranson (2003, 2005), which show quite amusingly what one can learn here about computers and complex problems. However, another aspect is perhaps even more fascinating: limitations of decisions, but especially formally exact computer-based decisions. This is masterfully illustrated in an article by Chaitin (2006), and the relations to Turing computability are also made well clear. The important point here is that humans, as thinking, feeling, and evaluating creatures, can obviously still make decisions that a computer, or more generally a Turing machine, can no longer make (see Chaps. 14 and 16). The Turing Award is the highest award for computer science. Laureates such as Martin Hellmann (pretty-good-privacy-encryption *of* e-mails) show that they are fully aware of this human responsibility (https://nuclearrisk.org; see Chap. 16).

**Conclusion**
- Alan Turing has generally reproduced all computable problems with the help of the Turing machine. All non-Turing computable problems cannot be solved by computers and remain tasks for humans. The question of when a bioinformatics problem will be completed is difficult to answer for problems with built-in combinatorics.
- Unfortunately, many particularly interesting problems in bioinformatics are NP (nondeterministic polynomial complexity) problems, for example, protein structure prediction as well as most network computations (e.g., the traveling salesman problem: How does he optimally plan his city route?). Computer clusters are needed for processing large omics datasets and in modeling genome-wide metabolic networks, but also for modeling complex signaling cascades, for *ab initio protein folding simulations,* and for complex image processing (e.g., 3-D tomograms, *deep learning*), as well as for large *in silico drug screens* and molecular dynamics simulations.
- *In* general, more powerful computers, the bundling of many computer nodes (parallelisation) and *application-specific chips* can also directly increase computer performance. In addition, the search for faster heuristics and new, clever algorithm strategies and procedures is a current task in bioinformatics, since the data are rapidly becoming more and more complex. Simpler problems (P-problems), on the other hand, require very manageable computing time, for example all sequence analyses, because a database search or query only grows linearly with the size of the database and the length of the query sequence, i.e. quadratically overall (quadratic polynomial problem P), as do predictions on RNA folding.

## 8.5    Exercises for Chap. 8

**Task 8.1**

How much does the computation time increase with different algorithms?

Compare the RNA folding algorithm RNAfold, a BLAST search, and protein folding. With BLAST, also try to clarify at the same time how the E-value moves favorably downward, to smaller values, with a smaller database.

Just try out the different calculation times with your own test examples.

**Task 8.2**

So how do you deal with the hard problems that biological systems present you with? Please list some different search strategies that you have learned about in the book or that you can think of (don't worry, the best ones will be discussed in a moment).

**Task 8.3**

What general search strategies for complex problems in bioinformatics do you know?

**Task 8.4**

Explain what is meant by NP-problems or P-problems in bioinformatics? How is a difficult computational problem defined informatically? Make this clear with an example.

**Useful Tools and Web Links**
https://baba.sourceforge.net
- Here basic algorithms of bioinformatics like local and global alignment are presented very nicely and exemplarily.
  https://discrete.gr/complexity/
- This page gives a nice introduction to computing complexity.
  Turing machine:
  https://www.alanturing.net/turing_archive/pages/reference%20articles/what%20is%20a%20turing%20machine.html
- There are many representations of this, but this one is right on the Turing network and descriptive.
  NP problems pitfalls:
  https://www.win.tue.nl/~gwoegi/P-versus-NP.htm
- This page shows a bit of how not to do it (or how easy it is to fail at this problem). For solid work, see Aaranson 2003, 2005, respectively.
  Introduction to parallel programming (for time-consuming calculations):
  *Parallel Programming with C+ + :*
  https://gridgroup.hefr.ch/popc/doku.php
  Message Passing Interface (MPI):
  Parallelization (Introduction to parallel programming)
  https://mpitutorial.com/tutorials/mpi-introduction/

## Literature

Aaranson S (2003) Is P versus NP formally independent? Bull EATCS 2003(81):109–136

Aaranson S (2005) NP-complete problems and physical reality. SIGACT News Complex Theory Column 36(1):30–52. arXiv:quant-ph/0502072v2 (* Both works are not only enjoyable to read, but take care of NP problems solidly and accurately)

Chaitin GC (2006) Limits of reason. Scientific Am 294(3):74–81. (* Very nice introduction to boundaries for human and computer decision making)

Cook S (1971) The complexity of theorem proving procedures. In: Proceedings of the third annual ACM symposium on theory of computing. ACMS, New York, pp 151–158

Hodges A (2014) Alan Turing: the enigma vintage. Random House, London

Levin L (1973) Universal search problems (Russian: Универсальные задачиперебора, Universal'nyeperebornyezadachi). Problems of Information Transmission (Russian: Проблемыпередачиинформации, ProblemyPeredachiInformatsii) 9(3):115–116 (pdf) (Russian), (Englische Ausgabe: Trakhtenbrot BA (1984) A survey of Russian approaches to perebor (brute-force searches) algorithms. Ann Hist Comput 6(4):384–400. https://doi.org/10.1109/MAHC.1984.10036)

# Complex Systems Behave Fundamentally in a Similar Way

**9**

**Abstract**

Biological systems are self-regulating and maintain their own system state (attractor). Negative feedback loops help to prevent overshooting, while positive activation loops (feedforward loops) activate the system when it is too weak (e.g. heartbeat). Bioinformatics is able to selectively tap central key elements (e.g. central signalling cascades; highly linked proteins in the centre, so-called "hubs"; sequence and system structure analyses, e.g. with interactomics and gene ontology), through whose concurrence the system behaviour essentially comes about ("emergence"). The starting point is the machine-readable description of the system structure (software Cytoscape, CellDesigner, etc.), which is then used to simulate the dynamics (e.g. SQUAD, Jimena, CellNetAnalyzer), whereby the comparison with experiments requires many ("iterative") model improvements. Systems biology is the most important future field of bioinformatics, especially in combination with molecular medicine, neurobiology and systems ecology, modern omics techniques and bioinformatic analysis (R/statistics; read mapping and assembly; metagenome).

## 9.1 Complex Systems and Their Behaviour

Now that we have become acquainted with the basic limitations of computer calculations, we can next consider how the computability of living systems looks in general. In principle, there is a clear contrast here: although biological systems are virtually digital in structure, and therefore consist of clear building blocks, the emerging system is difficult to manage because of chaotic system effects, although this "natural chaos" and the underlying principles can be very fascinating (Gleick 2008).

© Springer-Verlag GmbH Germany, part of Springer Nature 2023
T. Dandekar, M. Kunz, *Bioinformatics*,
https://doi.org/10.1007/978-3-662-65036-3_9

So there are the clear letters and units of information in the cell that can be determined by sequencing RNA and DNA molecules. There are technical limitations, but with today's modern sequencing methods it is possible to sequence almost any amount of nucleic acids and thus have any amount of this form of information available in a short time to answer a question. For example, "transcriptomics", i.e. the reading out of the RNA inventory of a cell, enables us not only to find out globally which information is stored in all mRNA molecules of a cell, but also to read out very precisely the inventory of switched-on genes ("expressed genes") that are active in this cell. In this way, a rapid inventory of the system status of an immune cell or a cancer cell is obtained. In the future, this will be used more and more intensively, for example to better design chemotherapy against cancer in patients, or to know whether the immune defence is in good condition. So: No problem, a lot of information about the living cell can be measured, at least with regard to DNA and RNA.

Nevertheless, there is a fundamental limitation for biological systems and even for all sufficiently complex systems. Their behaviour is said to be "chaotic", i.e. predictable only over short periods. This is perhaps easiest to see if you think of the best-known chaotic system: the weather. There, too, we can only predict what the weather will be like tomorrow in, say, Würzburg, Erlangen or Amberg. Accordingly, this can only be described with a certain probability, and over several days such a forecast is always relatively uncertain. On the other hand, we know that the climate here in Lower Franconia, Middle Franconia and the Upper Palatinate is a typical Central European one, we will neither expect a tropical storm nor deserts or glaciers here. This can be generalized: biological and more generally, so-called chaotic systems, can only be described exactly over relatively short periods of time. Their long-term behaviour, however, is kept within fixed limits. In the case of weather, this is called climate. More generally, such a confined system state is called an "attractor" because it draws nearby system states into this stable ground state. A good example from biology is our own health. Even there it is clear, sometimes I can be out of breath or sweating, have a fast pulse etc., after a few minutes everything is back to normal. On the other hand, if I catch germs, live unhealthy and that over longer periods, my system state can also change radically, especially I can get sick. That is then a different attractor. Because once you are sick, it takes some time and some effort to change from the sick system state back to a healthy one. Many people, especially older people, nevertheless remain chronically ill: the pathological condition is too strong, even with medicine the person remains ill.

With this we already have the most important terms for the system description together and can state: Biological systems can only be described exactly for a short time, but remain attached to stable system states, so-called attractors, over longer periods of time. However, if the system is disturbed or changed just enough, a new system state can then suddenly exist, which then reinforces itself again. A so-called *tipping point* is reached. For example, the forest has suddenly become a savannah or even a grass steppe or desert, to name a few ecological examples at this point. It is therefore important to understand systems in terms of their behaviour. Whenever they have feedbacks (positive, negative) and reinforcements, small changes can build up – and this is exactly the reason why systems are then called "chaotic". Only by measuring the state of the system with very high accuracy can I accurately describe how my system is evolving for short periods of time. But any error grows over time. And in "chaotic" systems it doubles within a short time, so that already after ten such time units the error is more than 1000 times larger after these ten doubling steps (2 to the power of 10 = 1024). For this reason, the behaviour of such systems over longer periods of time cannot be described exactly. On the other hand, it is precisely the negative feedbacks that keep the system within fixed limits (climate in the case of the weather at a particular location, health in humans). Only if strong positive feedbacks transform the respective system state, it is possible that it changes rapidly (*tipping point*) and one then suddenly has a new state (climate change or in humans a disease). The sudden system change when crossing tipping points was considered mathematically by Rene Thom (catastrophe theory, because systems then change catastrophically and rapidly).

It is interesting that we can in principle also understand a chronic illness well with this. Because here, too, strong feedbacks must be at work that prevent a return to the healthy state. If we recognize and treat these causes, a return to the healthy system state will also be possible. For example, by changing the way of life (stress reduction, more exercise and sport, reducing overweight) I can bring high blood pressure, recognised in time, back into balance and become healthy again. However, if no lifestyle change is made or possible and treatment is lacking, then subsequent regulation of blood pressure is often only possible through chronic medication. However, this is again clearly a symptomatic treatment, because I have to keep taking my medication, the disease-producing feedbacks will otherwise cause my blood pressure to skyrocket again and again. Unfortunately, at this stage, a (causal) therapy based on the actual causes, such as a permanent correction of the blood pressure regulation, is not yet possible. However, modern systems medicine can use large amounts of data, for example on gene expression, to show exactly what the main effects (intended: here blood pressure reduction) and side effects (harmful, e.g. liver damage) of a drug are and thus help to improve these drugs (Fig. 9.1).

The main effect of a drug is often the blocking of a receptor, i.e. the blocking of signal transduction via this receptor molecule. In Sect. 5.1 we gave the example of receptors in the cardiac muscle cell, which then lead to heart failure via phosphorylation of Erk kinase.

Ideally, that would then be all the effects of the drug. Would we then also see this in the gene expression experiment, i.e. only a down-regulation of the messenger RNA for the ß-adrenergic receptor, if, for example, we carefully creep in a beta-blocker against the increased blood pressure and the heart failure? Interestingly, we wouldn't see that exactly because the receptor is made ("expressed") via the mRNA, just as it is without the drug administration. If this were not the case, the drug could not bind to it at all. If the receptor does not transmit its signal, the heart has less work to do and the patient feels better. This is the intended and proven heart-protecting effect of beta-blockers. This can only be done by carefully increasing the dosage. Unfortunately, beta-blockers do lead to an improvement in symptoms, but not to a prolongation of life. This is due to the fact that the cause, the ageing heart, is not really combated without future methods such as stem cells for new cardiomyocytes. This is exactly why we are doing intensive research on stem cells in our

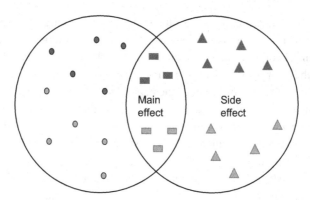

**Fig. 9.1** Illustration of the effect of a drug. A drug usually shows a main effect, i.e. an optimal therapy related to the cause of the disease (centre). However, there are also other genes (light and dark circles) that are altered by the intended molecular main effect, but do not show a change due to the disease itself (left). In addition, however, there are also side effects, so-called side effects, whereby, among other things, other receptors (light and dark triangles) are affected by the drug (Right). Nowadays, it is standard practice, for example in newly developed therapeutic approaches, to investigate the effect of a drug by means of gene expression experiments and then to analyse it bioinformatically in order to identify the changes in gene expression (e.g. mRNA upregulated [light] or downregulated [dark] after drug administration, compared with the untreated state [disease]). In this way, it is easy to overlook whether the actual main effect of the drug is achieved and which other genes are additionally affected (positively as well as negatively) by the therapy in order to develop a drug effectively. The aim is always to develop a drug as specifically as possible and to keep the side effects as low as possible for the patient

own department, especially since there are increasing possibilities to generate them from adult cells, especially the old cells of the patient and the patient (ethically safer method, but more difficult). But this is still a long way off. Therefore, let us now look at the further effects of the beta-blocker on gene expression, because these can already be given (just as, for example, the lowering of blood pressure by ACE inhibitors, which even now has a favourable effect on life expectancy). The beta blocker, amazingly, changes numerous other genes in expression, namely because the signaling cascade is now downregulated and this downregulates many genes as well as upregulating some others. Even stronger (and a little slower in its effect) is the heart-protective effect: genes for the further growth of the heart are somewhat down-regulated by the heart failure. However, some genes are transcribed more strongly again.

Finally, there is another factor when giving a drug (a pharmacon): very often these drugs hit the intended receptor, but more or less fit other receptors as well. The resulting gene expression changes are the side effects. Applied to our example, it is particularly the case that there are beta receptors not only in the heart, but also in many other organs, for example in the lungs. Although there are slightly different beta receptors there, namely beta2 receptors as opposed to beta1 receptors of the heart. Still, the risk of getting a bad side effect in the lungs this way is high enough that people try not to give beta blockers in asthma, for example. This is because they would seal the lungs because the beta2 receptors that keep the airways clear would then be blocked. So often the side effects of a drug come from other receptors besides the intended receptor being hit and blocked by the drug. However, when I measure gene expression, I only see the side effects if I also measure in a tissue where such side effects come into play. For example, especially in the lungs, but also in other tissues where beta2-receptors are present, these effects would cause the receptors to be less active, again changing numerous genes.

Of course, one can still more generally require that the main effect only fixes exactly the defect (causal therapy) and does not change anything else (no side effect). But this is not the case for most drugs because the body is too complex. A good example is diabetes treatment (*diabetes mellitus, diabetes*) by insulin. Actually, this is exactly the substance that the diabetic lacks. But since even insulin pumps cannot control insulin as precisely as the healthy body can with the help of the pancreas, the sick person has to deal with many small over- and underdoses of insulin all the time and in every cell of the body at the moment.

Bioinformatics can therefore be used to effectively evaluate the large amounts of data (DNA: so-called genomics, RNA: so-called transcriptomics, proteins: so-called proteomics, metabolism: so-called metabolomics) that describe in detail how biological systems react to drugs or environmental influences. There are fundamental limits to the short-term exact describability that apply to all systems controlled with feedback loops, such as living cells or even our weather. Therefore, it is important to know the range to which such systems are set and into which they always fall back, the attractors of the system. You have already learned about these in Sect. 5.1. There we introduced them simply as "stable system states". Stewart Kaufmann is an important researcher and founder of system sciences who has described natural and biological systems in general terms.

## 9.2   Opening Up Complex Systems Using Omics Techniques

Figure 9.2 illustrates how genomics, transcriptomics, proteomics and metabolomics all contribute together, for example, to accurately infer the effects and side effects of pharmaceuticals. In addition to our gene expression measurements from Sect. 9.1 (called transcriptomics, but any measurement of RNA, for example by large-scale RNA sequencing), we can measure exactly what happens to the proteins in the treated heart muscle cells (proteomics), how the metabolites, for example the sugar level, change under treatment (metabolomics). And of course we can also look at the patient's gene sequence (genomics, e.g. genetic predisposition to heart failure).

Genome sequencing using *ultrafast sequencing* technologies, such as the 454 or Solexa technology, is now a common method that enables the rapid and cost-effective sequencing and annotation of genomes (nucleotide sequence in DNA). The ever-improving sequencing technologies also allow for increasingly high-resolution sequencing, which means that newer and newer genes can be annotated. Numerous genomic data are accessible through

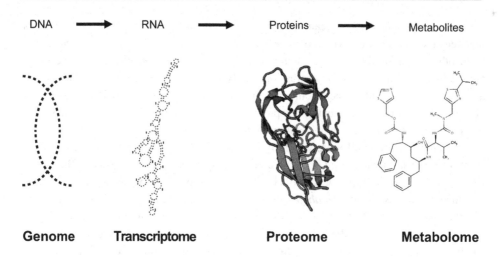

| Genome | Transcriptome | Proteome | Metabolome |

**Fig. 9.2** Omics techniques

genome browsers (e.g. Ensembl or UCSC). Specific genomic mutations, for example in human tumors or heart failure, are also deposited in various databases (e.g. OMIM) and can be used by users. DNA sequencing thus makes it possible to sequence unknown genomes, such as new resistant bacterial strains, or to determine the underlying mutations in diseases in medical diagnostics.

Transcriptome sequencing (gene expression sequencing) provides insights into gene expression, i.e. into the activation of gene transcripts. Common methods are *microarray experiments* or newer high-throughput methods such as RNA sequencing. These measure gene expression (mRNA level) and thus provide information on the corresponding changes in mRNA (up- or down-regulated), for example after infection or treatment. Meanwhile, there are increasingly efficient methods that can, for example, measure the expression of the host and the pathogen in parallel in one cell and thus provide insights into the changes in both organisms after an infection (dual RNAseq). Subsequent bioinformatic gene expression analysis can then examine the RNA secondary structure (e.g., RNAfold), the RNA sequence for regulatory RNA elements such as IRE (e.g., RNAAnalyzer) or in more detail with regard to possible interaction partners, for example RNA-protein (e.g., catR-APID, NPInter) or miRNA-mRNA interactions (e.g., miRanda, TargetScan). Numerous databases already contain gene expression datasets (e.g. GEO, cBioPortal, TCGA or GENEVESTIGATOR), information on RNA sequence, structure and binding motifs (e.g. Rfam) or information on specific RNA classes (e.g. miRNA [miRBase], lncRNA [LNCipedia]) and can be used for own analyses.

Protein sequencing can be done with mass spectroscopy or protein microarrays and provides information on the amino acid sequence in the protein. It is often of great interest how the proteome changes under certain conditions, for example after an infection or therapy. However, one is usually also interested in the changes or modifications in the amino acid sequence, for example in the functional side, and their effect on protein function. For this purpose, one can bioinformatically perform a domain annotation, i.e. which binding domains and functional sites are present, which thus provide information about binding factors, but also the regulation and function of proteins. Databases such as SMART, Prodom and Pfam provide information on proteins and domains and can also be used to search a protein sequence for existing domains. Other important tools are the BLAST algorithm, conserved domain or ELM servers, which allow the analysis and prediction of domains in unknown sequences.

Information on the metabolome (metabolism) can be obtained using mass spectroscopy or gas chromatography. Metabolome sequencing is of interest to see how, for example, metabolites change after a pathogenic infection or a drug, or how the metabolism of humans and the pathogen differs. This is important, for example, for a potential pharmaceutical to specifically affect the metabolism of a bacterium, but without producing a toxic effect in humans. Important databases on biochemical metabolism include Roche Biochemical Pathways, KEGG. The Metatool, YANA, YANAsquare or PLAS (Power Law Analysis and Simulation) software are useful for investigating metabolism in more detail, e.g. which metabolic fluxes are present or what effect changes in metabolic pathways have.

The large amounts of data that we can generate with modern techniques obviously help much better to describe a biological system, such as the heart muscle.

On the other hand, it is clear that the crucial thing is to understand the underlying principles, as just explained for main and side effects and further illustrated by other central system building blocks in this chapter. Therefore, one has two possibilities to describe a complicated biological system:

First of all, *knowledge-based* research is used to elucidate the basic principles of the biological system (for the myocardial cell in heart failure, see Figs. 5.1 and 5.2). Next, one uses new data, preferably a great deal of it (nothing else is meant by "*big data*"), to substantiate or modify the insights and hypotheses gained.

As you can see, relying only on the amount of data and large data sets is more a sign of bias or inexperience. If I don't have a clear hypothesis about the behavior of the system, I have a much harder time reading the right thing from the data, or better yet, verifying it.

*Even* worse: "*hypothesis free*" *research* is mostly bad, even if advocates claim that one would then be unbiased towards the results, because it is very easy to fall prey to chance.

Let's illustrate this again with the gene expression dataset in heart failure. Let us assume that we have measured 20,000 mRNAs and now want to understand, without a clear hypothesis, which ones are increased in heart failure. Now, even if no objective differences can be shown between drug and no drug, given 20,000 mRNAs, we would then purely by chance find 1000 mRNAs that show a difference in expression between the two groups with a p-value $<0.05$. Bioinformaticians and statisticians or experimenters, as experts in large data sets, know this and therefore correct the statistics for such large data sets. This is the correction for multiple testing, for example according to Bonferroni. In this correction for many comparisons, the *p-value* is divided by the number of tests (n). For example, for the 20,000 mRNAs, one would only accept differences with a p-value $<0.0000025$ (adjusted p-value). This is a very hard correction, but it applies to any distribution of measurements. Other "*multiple testing*" corrections are not quite so hard, because the distribution of the results usually satisfies a normal distribution.

Nevertheless, it can be generally stated that it is much easier to carry out such evaluations with clear hypotheses and less likely to fall prey to random deviations or come to false conclusions from the large data sets.

## 9.3    Typical Behaviour of Systems

How do systems behave in general? Surprisingly, there is a simple answer strategy. One distinguishes between the different kinds of systems that can exist at all. There are three types of systems: ordered, random and chaotic systems. These differ fundamentally (Table 9.1):

**Ordered systems** can be described by simple mathematical equations, for example the flight behaviour of a rocket or an airplane (function in time as independent variable, with the x, y and z coordinates for the position) or of a train (route plan). As we can see, this behaviour is predictable and can be described exactly for the entire period of the flight or train journey.

In addition, the system can also be easily controlled, for example by the aircraft pilot using the joystick or the acceleration/deceleration of the train by the train driver. The so-called state space of the system (where the train or the plane is at which point in time) can be described exactly, for every hour, for every minute.

A **random system** cannot be predicted at all for the next moment. The ideal example is a dice roll. No one can predict whether the next roll will be a one, a two, or a three, or even a six. And it stays that way. Also, the next roll is just as random as the previous one. This

**Table 9.1**  System behaviour (ordered, random, chaotic) with typical properties

| System | Order | Mayhem | Random |
|---|---|---|---|
| Example | Clocks, planets | Clouds, weather | Noise (sound), dice |
| Single event predictable | Very accurate | Only briefly (weather forecast) | Not at all → simple laws |
| Effect of small disturbances | Very small | Escalating over time, explosive | No effect, random disturbances are averaged out |
| Possible states | Few pure states | Many: Circling around attractor | Noise of all possibilities (1 to 6 on the dice) |
| Dimension | Finally | Low, e.g. circular orbital plane, healthy pulse | Infinite (any sequence is possible) |
| Control | Simply | Difficult, but effective | Barely (dice) |
| Attractor | Clear point, exact circular path (strange, fractal) | Scattered around the attractor | No attractor: Any state possible |

is called a random Markov chain (also *Markov* process, after Andrei Andreyevich Markov; other spellings Markov chain, Markoff chain, Markof chain). By knowing only a limited antecedent history (e.g. last litter), only as good prognoses about the future development will be possible as by knowing the whole antecedent history. But which ones? Here we see amazing things: while we can't predict the next litter at all (it's random, after all), we can predict the outcome space for all futures: It can only be a one through six.

So we see: with random systems, everything is not predictable at all in the short term. But the long-term prediction is surprisingly simple. The entire value frame is swept according to a random function (in our case: one sixth of the dice is a one, a two, a three ... a six; a more complex solution for a different system could be a Gaussian distribution) - this is the description of the entire future. Finally however, the result can never be controlled, a random system is and remains random.

**And what about the biological systems?** The fascinating thing is that the biological systems are a mixture between both extremes, between randomness and total order. These systems are called **chaotic systems.** We have already seen an example from everyday life: the weather. Here there is only a more or less certain weather forecast for the next day or even 2 weeks, but no certainty for longer periods. On the other hand, the result space is quite fixed, namely the climate of the place, which is e.g. temperate and sets the frame of possible weather forecasts.

This applies analogously to biological systems, for example heart rate and blood pressure (Fig. 9.3). Both can change rapidly, but as long as we are healthy, a fixed frame remains (e.g. pulse between 60 and 80, blood pressure between 80 and 120 mmHg). What is important for all such systems is that it takes considerable effort to move the system

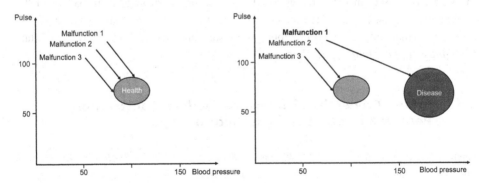

**Fig. 9.3** Representation of the biological system state healthy versus diseased state. Biological systems are, as long as we are healthy (left: system state health as green circle), in a stable, fixed frame, e.g. pulse between 60 and 80, blood pressure between 80 and 120 mmHg. The system moves within its tolerated limits (tolerance range; here pulse and blood pressure) and can compensate for external influences (e.g. exertion or anger, as disturbances 1–3). However, if the external disturbances are too strong, the system exceeds its tolerance range and the previous, stable system state (health) is abandoned. The patient becomes ill (right: system state illness, e.g. high blood pressure, as red circle due to too strong disturbance 1)

away from the healthy state. This is because after an effort (pulse) or anger (blood pressure) the system returns to the normal state. That is why the healthy state is called a system attractor ("system-attracting"). The attractor of a system can only be left in case of stronger disturbances (e.g. continued anger, stress at work), which can lead to a disease (e.g. high blood pressure). If this is a chronic condition, everything is now reversed. One must constantly exert force (such as taking an antihypertensive) to keep blood pressure within tolerable limits. An attractor for a sick state has then been reached, in the example it is the disease high blood pressure. Only a causal (cause-oriented) therapy can change this again. In particular, the system control must be adjusted to a different blood pressure value again, which is not yet possible in practice, since blood pressure regulation is extremely complex.

Why are such systems called chaotic? The reason is, they are only predictable for a short time. This is because biological systems, but also the weather and chaotic systems in general, are not controlled linearly. That is, a small change in control, just like a small error in description, doubles with each time step. For example, if I have only 1 per mil error in the description, just ten time steps later I have more than 100% error and can no longer predict the system state. The time scale on which this no longer describability happens varies among systems and is a characteristic time. However, the result is the same for all chaotic systems. Even for relatively short periods of time, one no longer knows what their concrete state is, since one never knows the starting state with infinite precision, and small errors always build up exponentially (the definition of a chaotic system). On the other hand, controlling such a system is very effective (the so-called butterfly effect, since even the smallest changes are always amplified exponentially). Finally, we now also know that the result space of the system sets clear bounds, as does the climate of a place. Even if I can't predict the system in the short term, I can predict what the system will stay within in the long term based on the attractor. For the same reason, stronger disturbances of the system are very dangerous. Then it can happen that the system not only gets out of balance, but permanently leaves its previous system state and changes into a new "sick" state (crossing a *"tipping point"*, see Chap. 16).

## 9.4    System Credentials: Emergence, Modular Construction, Positive and Negative Signal Return Loops

Even if we analyse large amounts of data with these *"omics"* technologies, there are a number of recurring concepts that help us to understand such biological systems – regardless of the level, i.e. whether we are looking at molecules at the lowest level, cells, tissues, organisms or even entire ecosystems. "Scale invariance" in this context means that at each size scale, the same phenomenon occurs in a similar way. Benoit Mandelbrot, for example, has looked at how self-similar at large and small scales such chaotic systems often are. Well-known examples include clouds (which repeat at every scale from the smallest cloud to huge weather fronts) and coastlines (which also look the same when viewed at every scale). Since all of these effects, which recur at different levels of order, rely on existing system control through regulation, we need only consider in detail what control capabilities the system has. Of particular importance are feedback loops (Fig. 9.4) that return an output signal back to the system. Subsequently, this signal given back (feed-back) can amplify the system response ("positive *feedback loops*" or loops), for example, faster cell division, stronger excitation, the system becomes more and more excited. However, other feedback signals can also dampen the system ("negative *feedback loops*" or loops), thereby preventing an excessive response or excitation, the system is stabilized. In addition, all biological systems are made up of many identical units ("modular"). At the lowest level, these are the building blocks for nucleic acids and proteins, i.e. nucleotides and amino acids respectively. But this is how it continues to larger and larger building units in the cell (filaments, organelles are formed from molecular networks). The cells in turn form tissues, these then form the organism and many individual organisms then networks of interacting organisms and whole ecosystems. The building blocks therefore alternate and thus always form new patterns and properties ("combinatorics").

Therefore, another general phenomenon is the study of network effects that arise anew as components come together, called emergence. A system is much more than the sum of its parts. At each new level, fundamentally new effects and phenomena occur that did not exist at the lower level and that breathe new properties into life through the interaction of the components. One example is the circulatory system, which is more than the many individual blood and heart muscle cells and supplies the body with nutrients and oxygen, resulting in system properties such as blood pressure and pulse. Another fine example is

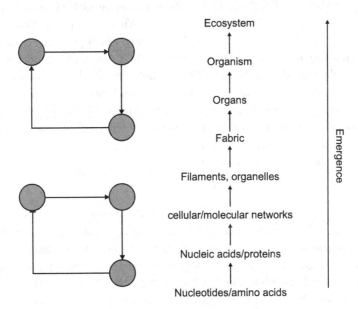

**Fig. 9.4** *Feedback loops:* positive, negative; modular structure

human consciousness, where instead of individual nerve cells or brain regions, a whole new quality of existence and perception emerges.

Finally, in this chapter we would like to introduce five important representatives of complex systems theory. We begin with Alan Turing, who thought in general terms about the computer and its limits. This is followed by Rene Thom, who founded the general theory about the behavior when crossing system boundaries. In addition, Benoit Mandelbrot is introduced, who founded his own field with his fractals, to elucidate natural system behavior and structures with simple principles, the fractals, as well as Leroy Hood, who is an example as a highly recognized representative of the younger systems biology and its application in medicine (there are many who could also be mentioned, not only in the USA, but also from Europe. So he provides here only a strong example). Reinhart Heinrich concludes the book, representing all non-US efforts in systems biology, which already produced exciting initial results in the Soviet Union and the former GDR, for example on phosphorylation cascades.

## 9.5    Pioneers of Systems Science

### Alan Turing

Alan Turing has the merit (Hodges 2014) of having already mathematically thought out what a computer would be capable of even before most computers were built. This makes him at least one of the greatest computer scientists who ever lived. In his memory there has been the "Turing Award" for the best computer scientist since 1966. Turing was English and lived from June 1912 (London) to June 1954 (stigmatized for his homosexuality since 1952, died of a cyanide overdose, suicide/accident). With the help of the concept of the Turing machine (Fig. 8.1) he was able to show clearly which problems computers and formal systems can decide and which cannot ("On Computable Numbers, with an Application to the Decision Problem"; 1936; Alonso Church's lambda calculus showed something similar earlier, but without this ingenious machine). In particular, it is impossible to decide algorithmically when a Turing machine holds. Of his many other contributions to mathematics, logic and computers, it is worth mentioning that he deciphered the code of the German cipher machine "Enigma" with the help of the first English large-scale computer "Colossus" during the Second World War.

### Rene Thom

Professor Rene Thom (02.09.1923–25.10.2002) was a mathematician. His "catastrophe theory" attempts to describe specifically the discontinuous, erratic behavior of dynamical systems (Poston and Stewart 1998). His theory studies the branching behavior of solutions (bifurcations) as parameters vary as a mathematical treatment of chaos theory. His "theory of singularities of differentiable mappings" means here the seven possibilities for mathematical functions to change suddenly and abruptly its seven "normal types".

**Fig. 9.5**   Fractals (Yami89, CC BY-SA 3.0). (https://upload. wikimedia.org/wikipedia/ commons/d/d2/M2_1024.png)

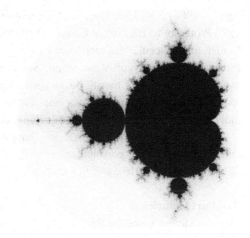

**Benoit Mandelbrot**

Benoit Mandelbrot (20.11.1924–14.10.2010) was born in Warsaw. As an intelligent Jew he had to flee from Poland to Paris in 1936, first learning mathematics from his uncle Szolem Mandelbrojt. He then lived with his family in Tulle, undetected by the Nazis. He was able to return to Paris to study at the École Polytechnique from 1944, and at the California Institute of Technology from 1947 to 1949, graduating with a master's degree in aeronautics (aircraft design). He received his PhD in mathematics from the University of Paris in 1952, and was a CNRS member (French national researcher) from 1949. He was an IBM Fellow from 1958 and did research there for 35 years, although there were also stays at Harvard and many other scientific honours during this time.

In 1975 he created the term fractals for scale-free structures (Fig. 9.5), which therefore look the same at any distance. Popular examples are coastlines, clouds, trees, the lungs, blood vessels and many more. He was able to show convincingly that these are mathematical objects that are often assumed in nature (but then always with a lower and upper limit) and not computer errors (technical term: "artefacts", i.e. caused by programming and not occurring in reality). He explored numerous properties of these fractals. Important books of him are "*Fractals: Form, Chance and Dimension*" (Mandelbrot 1975) as well as "*The Fractal Geometry of Nature*" (Mandelbrot 1982). With these books he founded an entire field of research that is still being intensively investigated. His creativity and ability to give himself time to let ideas mature are exemplary. He would never have called himself a bioinformatician. But he used computer programs to first study and describe fractals, and then showed what profound insights about nature, especially biology, are possible when one sees how fractals determine many processes in nature or are simply beautiful (e.g. snowflakes).

**Leroy Hood**

Leroy Edward Hood (born October 10, 1938 in Missoula, Montana) is an American systems biologist. B.Sc. degree from the California Institute of Technology (1960), medical degree (1964) from John Hopkins University (MD), then Ph. D. in biochemistry in 1968. 1975 Professor of Biology at Caltech, 1989 Director of the NSF Science and Technology Center for Molecular Biotechnology at the National Science Foundation (NSF), 1992 Professor University of Washington. Since 1999, Hood has directed the Institute for Systems Biology. His merits lie in particular in his work on the automation of DNA sequencing (see Automated Sequencing) and on the diversity of possible antibodies of an individual through recombination of genetic information (see V[D]J recombination).

This short selection of papers shows that Leroy Hood cares a lot about molecular medicine (Qin et al. 2016), personalized medicine (Sagner et al. 2016), systematic proteomics (Kusebauch et al. 2016), and *Big Data analysis* (Toga et al. 2015).

**Reinhart Heinrich**

Reinhart Heinrich was an East German biophysicist, born in Dresden on April 24, 1946, died in Berlin on October 23, 2006. He first lived with his parents in Kuibyshev in the Soviet Union, received his doctorate in 1971 on solid state physics, B doctorate in 1977 (then habilitation), lecturer in 1979 and worked as a professor of biophysics at Humboldt University from 1993. He is one of the pioneers of systems biology. His contributions to metabolic control theory, which date back to the 1970s, are exciting and instructive. Fundamental works appeared together with Tom Rapoport. His theoretically elegant and impressive works on regulation (Schuster and Heinrich: *The Regulation of Cellular Systems.* Springer Verlag New York, 1996) and signal cascades (Heinrich et al. 2002). In particular, he succeeded in defining more precisely general properties of phosphatases and kinases for signal processing. This concerns phosphatases for the regulation of signal amplitude, signal frequency, signal duration (the phosphatase must switch off fast enough for both) and kinases for the regulation of signal amplitude and signal height (the kinase must amplify strongly enough). His work has influenced a number of German systems biologists in their school days (e.g. Edda Klipp, Stefan Schuster, Thomas Höfer) and even more bioinformaticians.

Comparable work with different terminology has been done by Henrik Kacser and Jim Burns at the University of Edinburgh. This also applies, for example also to German research personalities such as Jens Reich (emeritus group leader at the Max Delbrück Center for Molecular Medicine Berlin; member of the Berlin-Brandenburg Academy of Sciences), Peer Bork (EMBL in Heidelberg, internationally perhaps the best-known German bioinformatician, many strong publications on genomics, metagenomics; *Nature Mentoring Award*), Thomas Lengauer (MPI for Informatics Saarbrücken, member of the presidium of the Leopoldina) and Martin Vingron (MPI for Molecular Genetics Berlin, member of the Leopoldina). In addition, there would be a long list of bioinformatician friends in Berlin, Bielefeld, Dortmund, Freiburg, Greifswald, Hamburg, Jena, Heidelberg, Mainz, Munich, Tübingen etc. (in alphabetical order of cities). (alphabetical order of cities) and elsewhere. For each of them, I can only stress that the high quality of all these contributions only becomes fully apparent when you really get to grips with the individual papers.

This already shows: our selection of persons can only be exemplary. Especially in bioinformatics, it would fill a book of its own if one wanted to acknowledge all the important contributions of even just the important people.

With this overview we can see that systems biology can be traced back to clear basic principles (first part), but is also characterised in a very interdisciplinary way by important principles from neighbouring fields (second part: presentation of the five research personalities in systems biology).

But how do we use these insights in practice? Modeling software is available for this purpose, but we only understand its results if we do not forget the principles and keep them in mind.

## 9.6   Which Systems Biology Software Can I Use?

As we have already learned, one typically proceeds in two steps. First, one assembles the necessary components for the system description and then proceeds to a systems biology modeling of the dynamics, i.e. the time course in a semiquantitative model. Semiquantitative here also means that we learn from the model the sequence of processes, i.e. what is stronger and what is weaker, but not the absolute strength of the signals or the exact "kinetics", i.e. the precise pace of the processes. This requires yet more data, especially experiments that accurately measure the speed. These data can then be used to incorporate them into the models as accurately as possible. This is then the final third step, the exact mathematical modelling. There are numerous ways to do this (see also the nice textbook "Systems Biology" by Klipp et al. (2016)). Here, only particularly well-known and easy-to-use tools can be mentioned, without claiming to be exhaustive. Above, we have already presented some tools that can be used for metabolic modeling, but which also work well for signal cascades:

> CellNetAnalyzer, COPASI, COBRA (Table 4.2) and Odefy (Krumsiek et al. 2010) should be mentioned here. SQUAD (di Cara et al. 2007) and Jimena (Karl and Dandekar 2015) have also been mentioned.

In particular, modeling with the convenient programming languages R and MATLAB is recommended. For the R language, as well as for MATLAB, there is an *R Systems Biology Suite,* and for the evaluation of gene expression data and systems biology based on it, there is the *Bioconductor Software package,* which also uses R.

**Conclusion**

- Complex biological systems are self-regulating and try to keep their own system state stable. The system state is therefore an attractor. Negative feedback *loops* help *to* prevent overshooting. Positive activation loops *(feedforward loops)* activate the system when it is too weak. For example, the heartbeat, pulse and body temperature of a healthy person remain stable within a narrow range and only oscillate around this range (limit cycle; so-called van der Pol oscillator), similar to the way a place has its fixed climate.

- Just as for the weather, exact predictions are only possible to a limited extent. Errors in system measurements increase exponentially. For this reason, complex systems can be described much better today using large amounts of data, for example with the help of omics techniques and statistics (scripting language R, important exercise, see tutorials). Alternatively, central key elements can be targeted (e.g. central signalling cascades, highly linked proteins in the centre, so-called "hubs", sequence and system structure analyses, e.g. with interactomics and gene ontology, important), through whose combination the system behaviour essentially comes about, i.e. in none of the components (modules) alone ("emergence"): the modules are correctly linked with each other, and the system properties only occur then.

- The systems sciences initially described important systems insights for physical systems (climate, chaos; Mandelbrot: fractals, Thom: catastrophe theory) and have since transferred them to biological systems (systems biology; e.g. Kaufmann, Hood, Reinhart) in order to place organisms, ecosystems, organ systems and brains (consciousness: extreme emergence, a fulguration), but also medicine and therapy on a new basis. Today's systems biology modeling software starts from the system structure described in machine-readable terms (Cytoscape software, CellDesigner and others), then recreates the dynamics in an easy-to-learn manner (e.g., SQUAD, Jimena, CellNetAnalyzer), with comparison to experiments requiring many ("iterative") model improvements. Systems biology is the most important future field of bioinformatics, especially in combination with molecular medicine, modern omics techniques (e.g. transcriptomics, metagenomics, *next generation sequencing)* and bioinformatic analysis (R/statistics, *read mapping* and *assembly; bar coding,* metagenome analysis), neurobiology (e.g. *C. elegans* conectome, *Blue Brain project:* Chap. 16) or ecology (systems ecology, e.g. modelling of climate change).

## 9.7   Exercises for Chap. 9

**Task 9.1**
Name and describe typical behaviors (ordered, random, and chaotic) of biological systems.

**Task 9.2**
Describe basic elements of biological systems.

**Task 9.3**
Name and describe various omics techniques.

**Task 9.4**
What is meant by emergence? Make this clear with an example.

**Task 9.5**
Draw a simple network consisting of a *feedback* or *feedforward loop*.

**Task 9.6**
Describe erythropoietin (EPO) production: (1) first qualitatively (set up control loop), (2) then identify possible quantitative relationships.

**Task 9.7**
Consider water: what different system states do you know?

**Task 9.8**
Cardiac hypertrophy simulation: which different system states could be distinguished for cardiac force stimulation if an increase in cardiac force can occur via a sympathetic and a hypertrophic stimulus?

| Useful Links | |
|---|---|
| Ensembl | https://www.ensembl.org/Homo_sapiens/Info/Index |
| UCSC | https://genome.ucsc.edu/ |
| OMIM | https://www.omim.org/ |
| RNAfold | https://rna.tbi.univie.ac.at/cgi-bin/RNAWebSuite/RNAfold.cgi |
| RNAAnalyzer | https://rnaanalyzer.bioapps.biozentrum.uni-wuerzburg.de/ |
| catRAPID | https://s.tartaglialab.com/page/catrapid_group |
| NPInter | https://www.bioinfo.org/NPInter/ |
| miRanda | https://www.microrna.org/microrna/home.do |
| TargetScan | https://www.targetscan.org/vert_71/ |

(continued)

**Useful Links**  (continued)

| | |
|---|---|
| GEO | https://www.ncbi.nlm.nih.gov/geo/ |
| GENEVESTIGATOR | https://genevestigator.com/gv/ |
| Rfam | https://rfam.xfam.org/ |
| miRBase | https://www.mirbase.org/ |
| LNCipedia | https://www.lncipedia.org/ |
| SMART | https://smart.embl-heidelberg.de/ |
| ProDom | https://prodom.prabi.fr/prodom/current/html/home.php |
| Pfam | https://pfam.xfam.org/ |
| BLAST | https://blast.ncbi.nlm.nih.gov/Blast.cgi |
| Conserved domains | https://www.ncbi.nlm.nih.gov/Structure/cdd/wrpsb.cgi |
| ELM | https://elm.eu.org/ |
| KEGG | https://www.genome.jp/kegg/ |
| YANA | https://www.bioinfo.biozentrum.uni-wuerzburg.de/computing/yanasquare/ |
| YANAsquare | https://www.bioinfo.biozentrum.uni-wuerzburg.de/computing/yanasquare/ |
| COPASI | https://copasi.org/ |
| COBRA | https://opencobra.github.io/ |
| SQUAD | https://www.vital-it.ch/software/SQUAD (Di Cara et al. 2007). |
| Jimena | https://www.bioinfo.biozentrum.uni-wuerzburg.de/computing/jimena_c/ (Karl and Dandekar 2013, 2015) |
| CellNetAnalyzer | https://www2.mpi-magdeburg.mpg.de/projects/cna/cna.html |
| PLAS | https://enzymology.fc.ul.pt/software/plas/ |
| Odefy | https://www.helmholtz-muenchen.de/icb/software/odefy/index.html |
| Roche pathways | https://www.roche.com/sustainability/what_we_do/for_communities_and_environment/philanthropy/science_education/pathways.htm |
| Metatool | https://pinguin.biologie.uni-jena.de/bioinformatik/networks/metatool/metatool5.0/metatool5.0.html |

## Literature

Di Cara A, Garg A, De Micheli G et al (2007) Dynamic simulation of regulatory networks using SQUAD. BMC Bioinform 8:462. (PubMed PMID: 18039375; PubMed Central PMCID: PMC2238325 * An excellent introduction to the SQUAD software, by the authors of the software themselves)

Gleick J (2008) Chaos: making a new science (Paper back). Penguin Putnam Inc., United States (Original from Viking 1987 * Highly recommended, nice introduction to the science of chaos. ISBN 10: 0143113453/ISBN 13: 9780143113454)

Heinrich R, Neel BG, Rapoport TA (2002) Mathematical models of protein kinase signal transduction. Mol Cell 9(5):957–970

Hodges A (2014) "Alan Turing: The Enigma" vintage. Random House, London

Karl S, Dandekar T (2013) Jimena: efficient computing and system state identification for genetic regulatory networks. BMC Bioinform 14:306. https://doi.org/10.1186/1471-2105-14-306. (PubMed PMID: 24118878; PubMed Central PMCID: PMC3853020 * Our representation of Jimena software, the 2015 work also explains how to find system control)

Karl S, Dandekar T (2015) Convergence behaviour and control in non-linear biological networks. Sci Rep 5:9746. https://doi.org/10.1038/srep09746. (PubMed PMID: 26068060; PubMed Central PMCID: PMC4464179)

Klipp E, Liebermeister W, Wierling C et al (2016) Systems biology: a textbook, 2. Aufl. Wiley, Weinheim (ISBN: 978-3-527-33636-4504 pages May 2016 * A good, in-depth textbook on systems biology)

Krumsiek J, Pölsterl S, Wittmann DM et al (2010) Odefy – from discrete to continuous models. BMC Bioinform 11:233. https://doi.org/10.1186/1471-2105-11-233

Kusebauch U, Campbell DS, Deutsch EW et al (2016) Human SRMAtlas: a resource of targeted assays to quantify the complete human proteome. Cell 166(3):766–778. https://doi.org/10.1016/j.cell.2016.06.041. (PubMed PMID: 27453469)

Mandelbrot B (1975) Fractals: form, chance and dimension. Freeman, San Francisco. (1977)

Mandelbrot B (1982) The fractal geometry of nature. Freeman, New York. (One New York Plaza Suite 4500 New York, NY 10004 ISBN 0-7167-1186-9)

Poston T, Stewart I (1998) Catastrophe theory and its applications. Dover, New York. (ISBN 0-486-69271-X * Ian Stewart writes catchy and well, so I recommend this book instead of studying Prof. Thom's original theory book [fine for more advanced students])

Qin S, Zhou Y, Gray L et al (2016) Identification of organ-enriched protein biomarkers of acute liver injury by targeted quantitative proteomics of blood in acetaminophen- and carbon-tetrachloride-treated mouse models and acetaminophen overdose patients. J Proteome Res 15(10):3724–3740. (PubMed PMID: 27575953)

Sagner M, McNeil A, Puska P (2016) The P4 health spectrum – a predictive, preventive, personalized and participatory continuum for promoting healthspan. Prog Cardiovasc Dis pii: S0033-0620(16)30078-0. https://doi.org/10.1016/j.pcad.2016.08.002 ([Epub ahead of print] Review. PubMed PMID: 27546358)

Toga AW, Foster I, Kesselman C et al (2015) Big biomedical data as the key resource for discovery science. J Am Med Inform Assoc 22(6):1126–1131. https://doi.org/10.1093/jamia/ocv077. (PubMed PMID: 26198305; PubMed Central PMCID: PMC5009918)

# Understand Evolution Better Applying the Computer

# 10

**Abstract**

The evolution of populations creates new species; the individual living being or protein is, after all, determined within narrow limits by the specific genome. New populations with ever new typical characteristics (through mutation and, in the case of sexual reproduction, through recombination) are always created, which allow an almost optimal adaptation to the prevailing environment, less environmentally related characteristics are less often passed on in the population (selection). However, many variants are also neutral or new structures only appear abruptly when enough mutations are present (neutral pathways in RNA structures; "punctuated equilibrium" according to Gould). Phylogeny helps to infer the evolution of different species on the basis of shared or non-shared characteristics via calculated predecessors. There are faster (neighbor joining) and more accurate methods (parsimony, most accurate maximum likelihood). Accompanying sequence and secondary structure analyses reveal conserved and variable regions as well as the evolution of functional domains. Most accurate phylogenetic trees require much practice and systematic comparison of all available information (e.g. alternative phylogenetic trees; marker proteins).

It is important for the understanding of evolution that it can only affect a whole population. In this respect, I have an even more complex task before me here than describing a complex individual system (Chap. 9): Cells or individual genes or proteins can change in an individual in the course of life. But since this mostly affects somatic cells ("somatic mutations"), they are not passed on to the next generation. But across generations, if you look at a whole population, there are changes over time due to the numerous mutations that happened in single genomes of germ cells and even passed on to new born individuals. Generally, this results in the population being better adapted to the environment that is

© Springer-Verlag GmbH Germany, part of Springer Nature 2023
T. Dandekar, M. Kunz, *Bioinformatics*,
https://doi.org/10.1007/978-3-662-65036-3_10

currently prevailing. Random changes (mutations), natural selection, and reproduction (replication) work together to achieve this. Sexual reproduction also allows new gene combinations in the offspring through recombination of the paternal and maternal genome. Depending on the environment, a mutation can thus be beneficial or detrimental or insignificant (neutral). Although this makes the evolution of genetic material a very complex process in populations, it is now possible to determine how these representative sequences change over time and how different they are in different populations by systematic sequence comparison of typical sequences for a population. One can then calculate from many such sequence comparisons (see Sects. 10.3 and 10.4) how different different populations are (one often compares species) and can then calculate back how the precursor populations looked (also extinct species).

This is all surprisingly difficult when looked at in detail, e.g. which sequences are representative of the population? Answer: The most frequent sequences in the population. So ideally you have to sequence many individuals, like in the 1000 genomes project for the human genome. However, when do I have a new species? This is not a problem for vertebrates and mammals, but it is not at all easy to determine with certainty for all other living organisms. Originally, individuals were classified morphologically (differing in appearance) and then simply called species. Later, a species was defined as a sexually fertile reproductive community between individuals in a population. This does not work for bacteria. Here, genetic exchange is complex with many transitions and reproduction is usually asexual, so that a three percent difference in the genome is often pragmatically defined as a new species. One also has to look carefully at the resulting phylogenetic trees in order not to make any mistakes, for example whether one can determine an original species ("root") or whether it is better not to do so because of the unclear data situation. Another example error in reconstruction is the tree-building error *(long branch attraction)*, since systematic errors can often occur, in particular distantly related species are wrongly considered closely related or closely related species are wrongly considered unrelated, which arises when sequences of different lengths are compared or when a single sequence is quite long and the taxa have a different number of mutations.

## 10.1    A Brief Overview of Evolution from the Origin of Life to the Present Day

Evolution always takes place in a population. The individual living being or protein is, after all, determined within a narrow framework by the specific genome. There are always new species (colloquially: "living beings always evolve over time"). In reality, there are always new populations with always new typical characteristics (by mutation and, in the case of sexual reproduction, by recombination) that allow near-optimal adaptation to the prevailing environment. Less environmentally related traits are less often passed on in the population (selection). However, many variants are also neutral. Even more exciting is that in RNA molecules, for example, new structures suddenly appear abruptly if enough mutations are present. These so-called neutral pathways in RNA structures can be thought of as follows: As long as the mutation does not change the base pairings, it changes the stability of the RNA only slightly. An identical base pairing, i.e. the replacement of an A-U pair by a C-G pair, is called a "*conserved substitution*". In addition, a structure can simply "endure" a series of minor mutations (i.e., remains stable enough) until eventually the stability is no longer enough, and then the RNA suddenly refolds. This then separates one RNA structure from the next via such a neutral pathway. So each pathway accumulates many neutral mutations. A similar thing happens when you look at protein structures in three dimensions (it's just more complex): Again, I can have a whole bunch of mutations associated with the same structure (*"neutral path"*), but eventually I have so many mutations that my structure suddenly flips. In evolution, this new structure only survives if it is beneficial to the prevailing environment. This has also led to a specific perspective on evolution: "*punctuated equilibrium*" according to Stephen J. Gould (1989) assumes that there are always "hot phases" of change in a population, because then suddenly mutations are no longer neutral and lead to new structures that are, among other things, advantageous. After that, everything remains similar for a long time (equilibrium). In reality, however, more and more neutral mutations accumulate in all structures until suddenly, due to decisive mutations, one or more RNA or protein structures "tip", i.e. change rapidly. This is followed by another period of quiescence in which mutations accumulate but no structural change occurs. This model explains at least relatively much about the predominant observed pattern of evolution. Over time, then, there has been no directed "higher development" in evolution, but rather the genetic material in the various populations continues to change through complex processes over time, while other populations die out altogether and new ones emerge.

Nevertheless, the overall effect on life as a whole in the 3.5 billion years since its origin has been considerable: about 450 million years ago, i.e. since the upper Silurian or lower Devonian age, higher (eukaryotic, see glossary) life spread to the land and shortly afterwards to the air. Over time, life has formed more and more species on average and the biomass has also grown more and more despite several mass extinction events.

The still numerically clearly dominant bacterial (prokaryotic) cells have consolidated. It can be shown that today's enzymes have been very well optimized in their catalytic activity. The same is true for metabolic pathways that have led to more and more, increasingly complex metabolites and have also become more and more efficient and robust. However, this then applies to the overall trend, across all bacteria (prokaryotes, i.e. eubacteria and archaebacteria). For a single species, specific environmental adaptations dominate, interspersed with neutral mutations, and the set of adapted mutations keeps changing as the environment changes. So, up close, variance and neutral change dominate. By the same token, evolution would never repeat itself the same way even when restarted, but would always find new species (or "solutions" if you will). Moreover, one must keep in mind that for every species alive now, there are 1000 others that are already extinct.

Each species lives about 1 million years, with a large range of variance: The pearl boat, an octopus, for example, has survived effortlessly through the last half billion years, including global mass extinctions in the Permian and Cretaceous, as has the cockroach.

With eukaryotes, there was also the possibility of investing in complexity. After all, these higher cells with nuclei are able to store about a hundred to a thousand times more genetic information than bacteria. This allows much more material for evolution. Splicing can also combine one and the same gene into numerous different proteins. And sexual reproduction also allows something new to be tried out in a diploid chromosome set in one allele of a gene (i.e. the variant from the father or the mother), since initially the already fairly optimal original variant of the gene in the other parent also pre-exists in the cell. This led to more and more complex organisms, and these also showed more and more complex behaviour. Until the appearance of humans, insects dominated on land among the higher organisms (eukaryotes) and among these the state-forming ants. With the appearance of humans, the total biomass of insects is still greater among higher organisms, but our civilization (including buildings and industry) has now become the dominant species on the planet for the ecological footprint and thus also at least for the necessary consumption of biomass (since 1950, is referred to by geologists as the "Anthropocene", new age). Before that, however, mammals gradually evolved higher since the Jurassic (200–140 million years before our era) and, with the extinction of the dinosaurs 65 million years ago, clearly outranked their present-day descendants, the birds, in occupying the ecological niches. But insects were still the dominant species. However, hymenoptera (bees, ants, wasps) only developed massively with the appearance of flowering plants, also in the Tertiary period (from about 65 million years ago).

The brief overview shows: It is not easy to interpret evolution correctly, and one also needs detailed data on the Earth's ages and the predominant species as well as the geological and climatic conditions. This book cannot do that. We will next look in more detail at how phylogeny (family tree science) can be used to infer the evolution of different species based on shared or unshared characteristics via calculated ancestors. The most accurate phylogenetic trees require a lot of practice and systematic comparison of all available information (e.g. alternative phylogenetic trees).

One should also know the species exactly in their macroscopic characteristics. It is also important to look at several molecular sequences, which are used for a phylogenetic tree, especially since proteins tolerate mutations at different rates. "Molecular clocks" go at different rates: Histone proteins hardly change at all because they are central and interact with many proteins. In contrast, less important proteins, or those that interact with few other proteins, can change much faster. Marker proteins can provide clarity here: frequently described molecules that occur in very many species, such as ribosomal RNA or, in the case of proteins, pyruvate kinase.

Phylogeny and other data from paleontology and molecular biology thus show, for example, how cytochromes (also important marker proteins) have evolved in comparison to hemoglobins. Such studies are supported by protein structure analysis. Interestingly, embryology can also often help: In order to form a new structure (for example, when a worm becomes a fly with wings), there must be matching mutations, which, after all, originate from the preform. This leads to the fact that in developmental biology, the earlier stages of evolution are often caught up (Häckel's law: every ontogeny, i.e. every individual development, recapitulates phylogenesis, i.e. the phylogeny). Genetics, for example with the help of the OMIM database, also helps to uncover gene relationships and mutation possibilities. More recently, better and better computer simulations have also become possible and allow insights (e.g. with regard to transposons) into a genome or, as an example from our own work, with regard to phage infection and cell wall metabolism (Winstel et al. 2013). Indeed, such processes accelerate evolution in the affected organisms (transposons jump and disrupt or modify genes) or slow it down (in our example, modified cell wall synthesis prevented infection with certain bacteriophages, which allowed the bacteria to evolve more separately from other staphylococci). In this way, both through phylogenetic trees and sophisticated new computational models, bioinformatics allows a new, detailed and more accurate analysis of evolution and its mechanisms (Connallon and Hall 2016). This also underpinned fascinating new insights into the evolution of life such as the endosymbiont hypothesis (e.g. bioinformatics analysis of organelle gene sequences transferred to the nucleus) and the RNA world (e.g. computational elucidation of ribosome structure, which revealed that peptide binding in the ribosome occurs through catalytic ribosomal RNA).

Since the advent of next generation sequencing, a very fast sequencing method, it has been possible to sequence environmental samples and characterise the mixture of organisms present in the sample without having to cultivate the organisms. The individual sequence fragments must be assigned to the individual genomes (metagenomics). The sum of the DNA in such an environmental sample is called a *metagenome*. With the usual culture methods, cultivation is only successful for 1–2% of organisms. Metagenomics thus significantly expands our knowledge of biological diversity. A synopsis of the new microbial diversity including detailed evolutionary analyses and new phylogenetic trees is given by Castelle and Banfield (2018). Five times more bacterial phyla ("strains", comparable to all vertebrates or all arthropods) are revealed than were recognized before these new methods. One can also prove very clearly with it that the higher cells (cells with real cell nucleus) represent indeed clearly a side branch of the Euryarchaeota, thus go back to the Archaebacteria and then have taken up additionally as energy factories with the mitochondria gamma-Proteobacteria or with the chloroplasts former blue-green algae, which drive then photosynthesis in the plants. With the higher cells (with cell nucleus, the eukaryotes), there are besides the animals and plants ("kingdoms") on the same level also the fungi. But this is only a small side branch of the archaebacteria in the phylogenetic tree. All bacteria (prokaryotes) make up the mass of the diversity of life, the archaebacteria seem only slightly less diverse than the eubacteria (the typical bacteria like gram-negative coliform, gram-positive like staphylococci and *Bacillus subtilis,* and completely new groups). All other life (animals, plants, fungi, higher protozoa) is just a small side branch. And to make matters worse, the impressive bacterial diversity is five times greater than was even thought possible just a few years ago.

## 10.2   Considering Evolution: Conserved and Variable Areas

The view of things just mentioned describes the fascinating control behaviour of biological systems and tries to understand this aspect as well as possible (systems biology). However, we can also ask ourselves how these wonderful adaptations of living beings came about in the first place. Bioinformatics also tries to better explore and understand this. The focus here is on sequence comparisons, evolution and phylogenetic trees. But this is perhaps a bit theoretical to start with. Evolution, in particular the common root of all life, on the other hand, can also be experienced first hand. All it takes is lunch. We can eat vegetables and meat, and indeed all vegetables, fruits, but also (at least as a non-vegetarian) all kinds of meat, fish and even crustaceans and shellfish. This is only possible because all of these creatures are related and even share the same genetic code. If this were not the case, the animal species that incorporate other amino acids into their proteins would be indigestible and probably even highly toxic to us. So the genetic code common to all living things can be tasted at lunch and is a strong indication of the "*last common ancestor*", the last common ancestor of all life.

Of course, evolution can be studied much more precisely by comparing the sequences of biomolecules. For example, all cells with a nucleus are related to each other (all animals, plants and fungi, more precisely: all eukaryotes). This can be seen, for example, by finding basic RNA molecules, such as U4 RNA, which is important for the production of messenger RNA from a long precursor (splicing process), in different organisms. Figure 10.1 shows the comparison of the U4 RNA between brewer's yeast and humans.

## 10.3   Measuring Evolution: Sequence and Secondary Structure

Figure 10.1 shows that even over great evolutionary distances RNA does not change its shape if it is important for function. Here, this is the very long time that has passed since humans and yeast cells still looked the same and the first cells with a cell nucleus emerged (about 1.5 billion years). Preserved is the shape of the RNA (secondary structure) and also single nucleotides (left, to be seen more clearly, each dot means preserved nucleotides). Boxes show that often even both have been preserved at this point. The figure also shows base pairings where there is also a base pair in the yeast cell, but a different base pairing than in humans (e.g., just above the GA-bulge, in the middle, right [box], the yeast cell has a GC pair, but humans have a UA pair). The whole molecule is much larger overall, but only the "front end", i.e. the 5'-part of the molecule, is shown. The high conservation stems from the fact that any mutation here could interfere with the splicing of the messenger RNA. For such important functions in the cell, base changes rarely occur because it is usually not advantageous to change anything else in the process.

Both the sequence and the base pairing possibilities that give RNA its structure are considered (both are highlighted in Fig. 10.1).

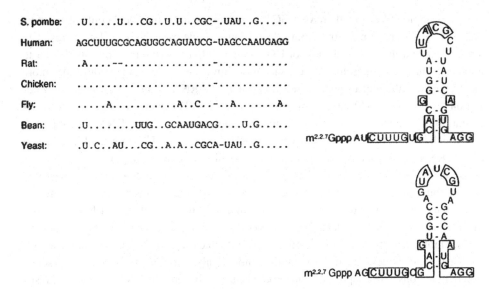

| | |
|---|---|
| S. pombe: | .U.....U...CG..U.U..CGC-.UAU..G..... |
| Human: | AGCUUUGCGCAGUGGCAGUAUCG-UAGCCAAUGAGG |
| Rat: | .A....--.................-........... |
| Chicken: | .....................-.............. |
| Fly: | .....A..........A..C..-..A........A. |
| Bean: | .U........UUG..GCAAUGACG....U.G..... |
| Yeast: | .U.C..AU...CG..A.A..CGCA-UAU..G..... |

**Fig. 10.1** Comparison of the U4 RNA secondary structure and primary sequence from brewer's yeast and humans. Conserved features of secondary structure (right, boxes) and primary sequence (left, capital letters) are highlighted. Only the 5'terminus is shown (full structure is about 128 nts)

Interestingly, it turns out that both important elements of structure, but also of sequence, have been conserved in this important RNA molecule over the huge evolutionary distance of two billion years during which the three organisms baker's yeast *(S. cerevisiae),* brewer's yeast *(S. pombe)* and humans have evolved from each other. Of course, this is not so much the case with less important molecules, and there are enough molecules that are only found in humans but not in either of the yeasts and so on.

Importantly, however, looking at RNA folding and RNA sequence allows us to see important conserved structures in organisms by comparison, and based on that, how evolution causes such structures to form and adapt.

It is easy to imagine a mutation swapping one or more letters. And it is indeed the case that the sequence already changes much faster than the structure in a relatively short time ("short evolutionary distance"). Yet short evolutionary distances mean millions of years. If a letter of such an important molecule is successfully exchanged, it takes thousands of generations until this happens by chance and is not immediately eliminated by disadvantages for the cell (negative mutation) (because the organism dies). Thus, over "short evolutionary distances" (typically millions of years, many thousands of generations), a few nucleotides can be exchanged, but the structure remains the same (as can easily be seen in the figure). Over even further distances (many millions of years – like humans and yeast cells, which separated their evolutionary lineages about two billion years ago), even the structure can change. This happens so slowly because when the structure changes, the partner molecules must also adapt. Such combined mutations take time to occur randomly in the generational sequence. The easiest to understand is a combined mutation that does not change the structure of the single molecule at all, a so-called compensatory mutation.

For example, if I originally have a GC base pairing in my molecule, I could also use an AU, UA, or CG base pair for it, but all other nucleotide exchanges no longer result in strong base pairings (a "weak" GU or UG pair can help transition from one stable pair to another). These compensatory base pairings within a molecule happen somewhat more easily, but everything else happens over time, so the U4 RNA changes in structure depending on the organism, and interacting partner molecules also change to a greater or lesser extent. In our example, this is in particular the catalytically active, mRNA-splicing RNA U6, which is initially kept inactive by the U4 RNA because the U4 RNA fits like a cap on the U6 RNA (this structure was called the Y model because of its shape).

By analyzing many U4 RNA structures in this example, we can see how evolution works. Thus, one can see how first (over short periods of time, in closely related organisms) single mutations change the sequence already in a short time and then over longer periods of time (in more distantly related organisms) the structure also changes, perhaps even new partner molecules are found or simply the gene doubles so that the second copy can perform a completely new function and mutates more easily. Evolution by mutation and selection of mutations with adaptive advantage can be traced in detail by RNA structure analysis. The comparison of the RNA structure in many organisms helps in this process.

## 10.4   Describing Evolution: Phylogenetic Trees

To do this, one only has to calculate phylogenetic trees for a widespread gene, i.e. on the one hand see which organisms are closely or distantly related according to their sequence and also try to work out the earlier branchings and precursor molecules. Although these are very rarely actually handed down (only if, for example, the already extinct mammoth can be thawed from the ice and re-sequenced), the information about the precursor molecules is hidden in the existing sequences. In this context, bioinformatics allows us to work out the precursors. There are several ways to do this. The easiest to calculate is the *neighbour joining* method. Here, one first sorts the molecules that one wants to connect in the phylogenetic tree according to their similarity and then always calculates the respective ancestors for direct neighbours.

A somewhat more elaborate procedure is *"parsimony"*, i.e. starting similarly, but calculating the mostly not directly observable ancestors of today's molecules in such a way that one can generate all observed today's sequences with as few mutations of these precursor sequences as possible. This reflects the actual conditions surprisingly well, because each individual mutation is very rare. A phylogenetic tree that introduces an unnecessarily large number of mutations is therefore *a priori* less likely than a phylogenetic tree that manages with as few mutations as possible.

It stands to reason that a pedigree that does not simply consider the most exact probabilities possible for the ancestors, but calculates them for each individual mutation, is the most accurate. This can be done by means of the so-called *maximum-likelihood* method, i.e. the calculation of the most probable path for all mutations. For this, one has to estimate from the observed sequences how probable which mutation is at which position, i.e. compile a very large table (more precisely: a matrix of transition probabilities) and then calculate the phylogenetic tree. This third method is particularly computationally intensive and time-consuming, but of course particularly accurate.

In practice, it is still important that the faster methods are also more easily off the mark when things get complicated. Depending on the calculation rule used, the result is more or less easily falsified. This happens especially when sequences of different lengths are compared or when a single sequence is quite long *("long branch attraction")*. The infobox summarizes a number of tools.

---

**Phylogenetics Tools**
**Phylogeny**
Family trees resemble real trees if there is a clear root (origin), for example by including a distant species *("outgroup")*.
**Basically, there are three ways to calculate family trees:**

- Always merge and calculate direct neighbours: *neighbor joining*. This can be done quickly and is implemented excellently and efficiently in the CLUSTALW software, for example.
- Parsimony tries to calculate the family tree with as few mutations as possible. This is already more computationally expensive.
- *Maximum likelihood* considers the most computationally expensive procedure. Each nucleotide exchange is considered according to its (often estimated) probability and then the most probable phylogenetic tree is calculated.

---

Thus, bioinformatics enables us to describe evolution more precisely and to understand important aspects of it by analysing many such phylogenetic trees, but also genomes and, in particular, by taking a detailed look at individual gene families. In particular, by analysing the amino acid sequences involved, but also the available structural data of important enzymes, it is possible to describe and analyse exactly how they function, which amino acid residues are important for the chemical reaction they catalyse and which functional subunits they consist of. These subunits are also known as protein domains. They are typically 100–150 amino acids long, fold stably (hence their size – if they were longer they would fold into multiple sections, if they were smaller they would not fold at all) and each has a specific function. For example, there are catalytic domains, regulatory domains, interaction domains, those that bind cofactors (often vitamins), and those that allow for a solid structure in the protein (e.g. fibrils or fibers). Looking at protein families can shed light on how a protein function changes or adapts across different organisms and how, for example, additional mutations can turn a catalytic domain into a regulatory domain.

## 10.5    Protein Evolution: Recognizing Domains

As we have just learned, protein families are subject to evolutionary processes such as natural selection. However, most mutations have no direct influence on fitness, i.e. they do not lead to any advantage or disadvantage (neutral evolution). In addition, random changes in the sequence (gene drift) or larger sequence regions or even entire genes (gene shift) can also occur, which can influence the function, e.g. catalytic domain or functional side. Thus, it is possible that a domain occurs in several proteins, but the remaining functional domains in the protein differ, which may contribute to new functions, for example. Domains thus provide an important clue to the origin and function of a protein. Proteins are thus grouped into protein families on the basis of their domains and similar functions and stored in databases or can be used bioinformatically to predict domains and functions of unknown sequences (see tutorials). Known functional domains of proteins can be found, for example, on UniProt (https://www.uniprot.org/), bioinformatically predicted with InterPro (https://www.ebi.ac.uk/interpro/), Pfam (https://pfam.xfam.org/), SMART (https://smart.embl-heidelberg.de/) or with Eukaryotic Linear Motif (ELM; https://elm.eu.org/).

In addition, proteins and metabolites are interconnected in signalling cascades *(pathways),* some of which are different, and which have also evolved over time. There are several theories about the evolutionary processes responsible for the emergence of signalling pathways and their cellular functions (Fig. 10.2; from Schmidt et al. 2003). Initially, there is *de novo* evolution (all enzymes and reactions evolved independently). Soon (after two to three billion years), the available material led to new possibilities: duplication of existing signalling pathways, retroevolution (selective pressure leads to the formation of

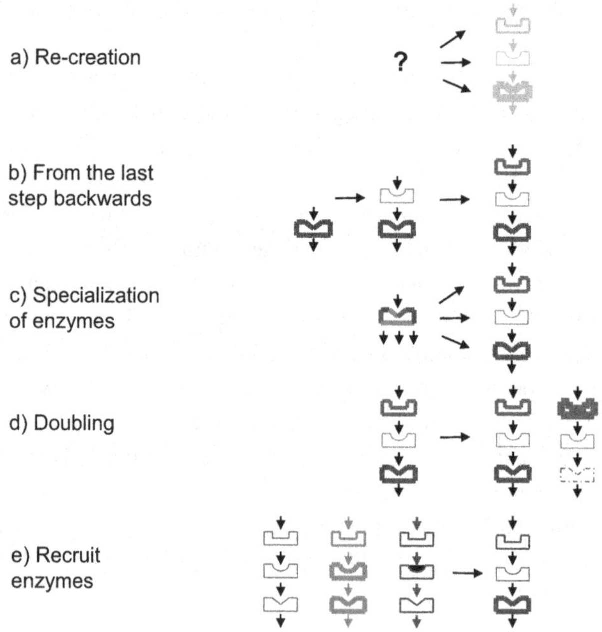

**Fig. 10.2** Metabolic pathway evolution. (Figure from Schmidt et al. 2003)

a) Re-creation

b) From the last step backwards

c) Specialization of enzymes

d) Doubling

e) Recruit enzymes

an optimal end product), by gathering (recruitment) of enzymes from different signalling pathways or by specialisation of a multifunctional enzyme.

One can also compare different engines of evolution in bioinformatic analyses, e.g. gene duplication for new enzyme activities that are simply tried out, with other methods of recruiting new enzymes to a metabolic pathway (Schmidt et al. 2003).

So we have learned some nice examples of processes that drive and shape evolution, first in domains, then in metabolic pathways. There is really a lot of research on engines of evolution, and the two examples only illustrate this.

Apparently, evolution via metabolites and adaptable protein structures (so-called *changer folds* that can bind and process many substrates in related enzyme structures) is rapid and efficient. On the other hand, this evolution has a necessarily limited starting material: simple molecules that serve as metabolites and 20 amino acids. Can this starting material constrain evolution? Interestingly, Stephen J. Gould, among others, has been able to show how building blocks determine and constrain evolution (Gould 1997). Certainly, an exciting new possibility is that the chemist, or humans in general, can use their intelligence and also try evolution on a completely new chemistry. Here a very well-known example is the SELEX process, i.e. the breeding of new RNA molecules by enrichment and subsequent propagation via the polymerase chain reaction. Jeff Szostak was awarded the Nobel Prize in 2009 for such experiments on telomerase (protection of the chromosome ends). In addition, however, he was able to show that RNA molecules can be grown for any basic function that a protein enzyme also performs (Adamala et al. 2015). However, such RNA experiments demonstrate the adaptability of RNA and are important evidence for an early phase of evolution in which both information storage and enzymatic catalysis were particularly carried by RNA. Here bioinformatics is enabling exciting new design experiments with ever new building blocks, for example unprecedented protein folding types (Garcia et al. 2016; Huang et al. 2016; Bhardwaj et al. 2016 are three recent papers from David Baker's internationally leading group), new amino acids (Wang et al. 2016), artificial new ribosomes (Neumann et al. 2010) or new nucleic acids (Chen et al. 2016). We can throw off the shackles of our building blocks!

However, there is still a lot to discover about evolution itself with the help of bioinformatics, and the subject of evolution itself is quite a broad field. For example, one can also look at the already mentioned "neutral evolution" (Maruyama and Kimura 1980; Prof. Kimura was one of the first and very great in this field). This has already been suggested by Fig. 10.1 of the U4 RNA, that many evolutionary changes simply happen because time passes, but without changing fitness, so that even without special selection organisms split up and become more and more dissimilar. This has been well studied in the case of influenza viruses, for example, since we always want to have the best possible influenza vaccine, but the influenza virus continues to change by chance due to individual mutations *("drift")*, sometimes even more strongly *("shift")*. Depending on the point of view, one can be more interested in this neutral evolution (since a neutral mutation naturally occurs much more frequently as a single event than a positive mutation) or in the positive and negative selection processes (which then accumulate more and more over millions of years as existing, "fixed" mutations with a fitness advantage for the organism).

Quite other important processes of evolution are, for example, genome modifications by selfish DNA, repetitive DNA and jumping gene elements (transposons). For illustration, there is a recent review of mobile genes in the human microbiome and how they are constructed (Brito et al. 2016). Other important factors in evolution include sexual selection (Connallon and Hall 2016), parasite-host interplay (Tellier et al. 2014), and the newly discovered important role of RNAs in genome evolution (e.g., pi-RNAs; Vourekas et al. 2016).

What is significant for the fascination of bioinformatics is that with the new data and their evaluation by bioinformatics, but also with new simulations and calculations about evolution, the formative diversity of these processes of evolution is revealed.

**Conclusion**
- Evolution is central to understanding the development of life. It always takes place in a population. The individual living being or protein is, after all, determined within a narrow framework by the specific genome. There are always new species (colloquially: "living beings always evolve"). In reality, there are always new populations with always new typical characteristics (by mutation and, in the case of sexual reproduction, by recombination) that allow a near-optimal adaptation to the prevailing environment. Less environmentally related characteristics are less often passed on in the population (selection).
- However, many variants are also neutral, or new structures only appear abruptly when enough mutations are present (neutral pathways in RNA structures; *"punctuated equilibrium"* according to Gould). Over time, there has been no directed "higher evolution". But there has been spread of life to the land and air, more species and biomass formed. Bacterial (prokaryotic) cells, still clearly dominant in numbers, have consolidated and become increasingly robust. In the case of eukaryotes, in addition to many new species (99.9% are extinct!), more and more complex organisms and complex behaviour emerged (dominant on land: insects, from the Tertiary onwards the state-forming insects; from Holocene onwards: humans and civilisation).
- Phylogeny (family tree science) helps to infer the evolution of different species based on shared or non-shared traits via calculated ancestors. There are faster *(neighbour joining)* and more accurate methods (parsimony, most accurate *maximum likelihood*). Accompanying sequence and secondary structure analyses reveal conserved and variable regions as well as the evolution of functional domains. Basic techniques for this are easy to learn (see tutorials). Most accurate phylogenetic trees require much practice and systematic comparison of all available information (e.g. alternative phylogenetic trees, also macroscopic features, molecular sequences, marker proteins). Phylogeny and other data from paleontology and molecular biology as well as from protein structure analyses, embryology, genetics and simulations also allow the analysis of evolution. This provides fascinating new insights into the evolution of life, such as the endosymbiont hypothesis and the RNA world, but also into the mechanisms of evolution.

## 10.6    Exercises for Chap. 10

### Evolution Tasks

**Task 10.1**
Describe what is meant by evolution. Also elaborate on different mechanisms of speciation.

**Task 10.2**
**Eigen's ball games simulate evolution:**
A chessboard, at least two colors of game pieces, and two octahedron cubes are needed. Additional information and game instructions can be found in Eigen and Winkler (1975).
**Game variant normal evolution:**
If a die roll hits a color but no square is empty, that piece is removed from the game. If, on the other hand, a die roll hits the piece and a square is empty, then that colour is also placed in the previously empty square.
Just observe what happens when you've rolled a total of about 64 times (hitting each square once on average) or when you've rolled a total of about 700 times (an afternoon, worth it). Feel free to try multiple colors as well.

**Task 10.3**
**For advanced players:**
Interpret the observations obtained in terms of neutral evolution, directed evolution, *"survival of the fittest"*.

**Task 10.4**
**Hypercycle evolution:**
Same playing field, but four colors. Two always form a tandem of information store and replicating enzyme (a so-called *"hypercycle"* is such a tandem of enzyme and information store). Now play the game according to the rule so that whenever a DNA (blue or red chips) is hit and a square is free, it makes the corresponding enzyme (yellow or green chips). Whenever an enzyme is hit and a field is free, the corresponding DNA is polymerized (so if yellow, the blue DNA; such a chip into the field or if green enzyme then the red DNA). Whenever all fields are occupied, a color is randomly thrown out by an octahedron roll.
Question: What happens now in the game, which tandem wins and how fast? Also test whether it is now easier or more difficult (compared to game 10.2) for a tandem to grow up. To understand this, let a rare tandem compete against a dominant tandem that has already occupied many fields.

## Phylogeny Tasks

### Task 10.5
Describe various methods of phylogenetic tree analysis.

### Task 10.6
You want to build an HIV pedigree on HIV-1 polymerase, how do you go about it?

### Task 10.7
How would you construct a phylogenetic tree for RNA polymerase II?

### Task 10.8
Calculate a phylogenetic tree using CLUSTAL and MUSCLE. What are the similarities/differences?

### Task 10.9
What is multiple *alignment* and what can I use it for?

### Task 10.10
Familiarize yourself with the SMART/Pfam database and look at a domain family/seed alignment using an example of your own choosing.

| Useful Tools and Web Links | |
|---|---|
| UniProt | https://www.uniprot.org/ |
| InterPro | https://www.ebi.ac.uk/interpro/ |
| Pfam | https://pfam.xfam.org/ |
| SMART | https://smart.embl-heidelberg.de/ |
| ELM | https://elm.eu.org/ |
| ITS2 | https://its2.bioapps.biozentrum.uni-wuerzburg.de/ |

## Literature

Adamala K, Engelhart AE, Szostak JW (2015) Generation of functional RNAs from inactive oligonucleotide complexes by non-enzymatic primer extension. J Am Chem Soc 137(1):483–489

Bhardwaj G, Mulligan VK, Bahl CD et al (2016) Accurate de novo design of hyperstable constrained peptides. Nature 538(7625):329–335. https://doi.org/10.1038/nature19791. (PubMed PMID: 27626386)

Brito IL, Yilmaz S, Huang K et al (2016) Mobile genes in the human microbiome are structured from global to individual scales. Nature 535(7612):435–439. (PubMed PMID: 27409808; PubMed Central PMCID: PMC4983458)

Castelle CJ, Banfield JF (2018) Major new microbial groups expand diversity and alter our understanding of the tree of life. Cell 172. https://doi.org/10.1016/j.cell.2018.02.016

Chen TL, Chang JW, Hsieh JJ et al (2016) A sensitive peptide nucleic acid probe assay for detection of BRAF V600 mutations in melanoma. Cancer Genomics Proteom 13(5):381–386. (PubMed PMID: 27566656)

Connallon T, Hall MD (2016) Genetic correlations and sex-specific adaptation in changing environments. Evolution 70(10):2186–2198. https://doi.org/10.1111/evo.13025. (PubMed PMID: 27477129)

Eigen M, Winkler R (1975) Laws of the game: how the principles of nature govern chance. ISBN: 9780691025667 Published: Apr 11, 1993 Princeton University Press. (* The book has been translated into six languages and is worth reading both for its clear presentation of how evolution works and for the "ball games", selection experiments with octahedron cubes and a chess board, which anyone interested can recreate).

Garcia KE, Babanova S, Scheffler W, Hans M, Baker D, Atanassov P, Banta S (2016) Designed protein aggregates entrapping carbon nanotubes for bioelectrochemical oxygen reduction. Biotechnol Bioeng 113(11):2321–2327. https://doi.org/10.1002/bit.25996. (PMID: 27093643)

Gould SJ (1989) Wonderful life: the Burgess Shale and the nature of history. Norton, New York. (* This book is a fluent read and shows that evolution has no direction and that the specialization into which the original abundance of basic forms leads is also based on a series of random selection processes for environmental adaptation, so would probably not be repeated a second time in exactly the same way)

Gould SJ (1997) The exaptive excellence of spandrels as a term and prototype. Proc Natl Acad Sci USA 94(20):10750–10755. (PubMed PMID: 11038582; PubMed Central PMCID: PMC23474) (*Building blocks set fundamental limits for evolution. This is the main statement of Gould. It is also true, for example, for central metabolites (see Schmidt et al. 2003). On the other hand, our technology helps us to jump over these limits, to bring completely new building blocks into play.)

Huang PS, Boyken SE, Baker D (2016) The coming of age of de novo protein design. Nature 537(7620):320–327. https://doi.org/10.1038/nature19946. (PubMed PMID: 27629638)

Maruyama T, Kimura M (1980) Genetic variability and effective population size when local extinction and recolonization of subpopulations are frequent. Proc Natl Acad Sci USA 77(11):6710–6714. (PubMed PMID: 16592920; PubMed Central PMCID: PMC350358 (*Kimura is a great evolutionary theorist who has particularly explored all aspects of neutral evolution)

Neumann H, Wang K, Davis L et al (2010) Encoding multiple unnatural amino acids via evolution of a quadruplet-decoding ribosome. Nature 464(7287):441–444. https://doi.org/10.1038/nature08817. (Epub 2010 Feb 14 PubMed PMID: 20154731)

Schmidt S, Sunyaev S, Bork P et al (2003) Metabolites: a helping hand for pathway evolution? Trends Biochem Sci 28(6):336–341. (Review. PubMed PMID: 12826406 * Here we studied how key metabolites drive metabolism in evolution. It also showed that the only 20% variable enzyme structures catalyze 80% of all reactions, but the 80% not so adaptable share the remaining only 20% reactions.)

Tellier A, Moreno-Gámez S, Stephan W (2014) Speed of adaptation and genomic footprints of host-parasite coevolution under arms race and trench warfare dynamics. Evolution 68(8):2211–2224. https://doi.org/10.1111/evo.12427

Vourekas A, Alexiou P, Vrettos N et al (2016) Sequence-dependent but not sequence-specific piRNA adhesion traps mRNAs to the germ plasm. Nature 531(7594):390–394. https://doi.org/10.1038/nature17150. (Epub 2016 Mar 7. PubMed PMID: 26950602; PubMed Central PMCID: PMC4795963)

Wang Y, Yang YJ, Chen YN et al (2016) Computer-aided design, structural dynamics analysis, and in vitro susceptibility test of antibacterial peptides incorporating unnatural amino acids against microbial infections. Comput Methods Programs Biomed 134:215–223. https://doi.org/10.1016/j.cmpb.2016.06.005. (Epub 2016 Jul 6 PubMed PMID: 27480745)

Winstel V, Liang C, Sanchez-Carballo P et al (2013) Wall teichoic acid structure governs horizontal gene transfer between major bacterial pathogens. Nat Commun 4:2345. https://doi.org/10.1038/ncomms3345. (* Here we show how the cell wall can be altered by teichonic acids to allow other phages to infect the staphylococci. This then leads to a strong genetic barrier and accelerates the development of different staphylococci and bacterial strains.)

# Design Principles of a Cell

# 11

**Abstract**

The design principles of a cell can be bioinformatically decoded in detail by sequence analyses and more elaborate methods. Regulation, the localization of proteins, their transport and secretion are also precisely encoded in the cell and are crucial for the ordered structure of the cell. Modern imaging techniques and imaging software help to validate these predictions. It is also important to classify all cellular processes by analyzing the gene ontology. Combined with information on the protein-protein interactome, the resulting cellular network can be traced using software such as CellDesigner or Cytoscape, e.g. motor proteins and the actin-myosin cytoskeleton are crucial for cell movement. Metabolic "design" is quickly queried via databases such as KEGG or more accurately calculated via metabolic modeling (e.g., with YANA or Metatool). Complex signalling networks are important for fast responses (stress response, chemotaxis in bacteria) and especially for multicellularity. They are modelled in detail with dynamic modelling (cell differentiation, tumorigenesis, embryology, inflammatory processes, nervous system).

Evolution (last chapter) and systems biology (penultimate chapter) work together when the cell successfully asserts itself in the environment and organisms persist and reproduce. By combining both factors, cells are amazingly optimally engineered. Cells are and have always been exemplary "natural engineers", which once again makes it clear that the question of *"intelligent design"* posed by creationists is unfortunately always completely wrong. Just when living beings are so wonderfully organized by completely natural processes, this should arouse our admiration and sharpen our view for higher and highest levels of such processes and how wonderful our world is, in which such processes are naturally possible spontaneously and self-organized. This is no accident, but for this particular world, selection processes are again at work, but on a truly high level – a truly successful creation that subtly defies easy interpretation.

© Springer-Verlag GmbH Germany, part of Springer Nature 2023
T. Dandekar, M. Kunz, *Bioinformatics*,
https://doi.org/10.1007/978-3-662-65036-3_11

## 11.1    Bioinformatics Provides an Overview of the Design of a Cell

The design principles of a cell are derived from molecular biology: The flow of genetic information follows from the genome via RNA to individual proteins. This can be decoded bioinformatically in detail by sequence analyses and more elaborate methods. Bioinformatics thus enables a deeper, previously unattained insight into the molecular biology of the cell and the underlying design principles through the analysis of large sequence data and other types of data.

Let us now turn to the fundamental question: How can bioinformatics be used to understand what happens where with which molecule in a cell? The different molecules can be analysed, for example, on the basis of their sequence. In particular, the localisation of proteins within higher cells (eukaryotes: humans, animals, plants, fungi) can be determined by recognising signal sequences (software: SignalP) and cellular localisation signals (LocP, nucloc) as well as signals for attachment in the cell membrane (e.g. TMHMM), for example with hidden Markov models (Chap. 3). In the diagram (Fig. 11.1), *pyruvate carboxylase* is marked in black (localization in the mitochondrion) and nup36 in white (nuclear pore protein 36 in the nucleus). However, numerous predictions are also possible on the basis of linear sequence motifs (e.g. whether the protein is attached to cell membranes with a GPI anchor or is marked for lysosomes or other locations with the aid of glycosylations). Phosphorylation positions can also be readily identified, although too many are easily predicted. The ELM server (Eukaryotic Linear Motif Server) bundles together a whole set of such predictions. This bundling (one server for server) is called a metaserver in this context. Even without the sequence, one can predict properties of proteins and their functional building units (the independently folding protein domains) based on the biochemical features (tool: domain databases such as Pfam and SMART), especially about the mediated function (e.g. catalytic – specific enzyme family; regulatory domain, structural domain and cofactor-binding domains). However, this can only be done if one is expert enough to nail down the domain finely enough about the function (e.g. SH2 or SH3 domain if protein interactions between proteins are mediated by these domains). Once the domain has been determined, domain databases come up with a lot of additional information, e.g. about the sequence, three-dimensional structure, important motifs, interactions and detailed description of the function.

## 11.2    Bioinformatics Provides Detailed Insights into the Molecular Biology of the Cell

Complementary to Fig. 11.1, I can also start from the main cellular processes (Alberts 2013; Simon et al. 2013; Watson 2013) and use the main bioinformatics algorithms for analysis for each of these processes (Table 11.1).

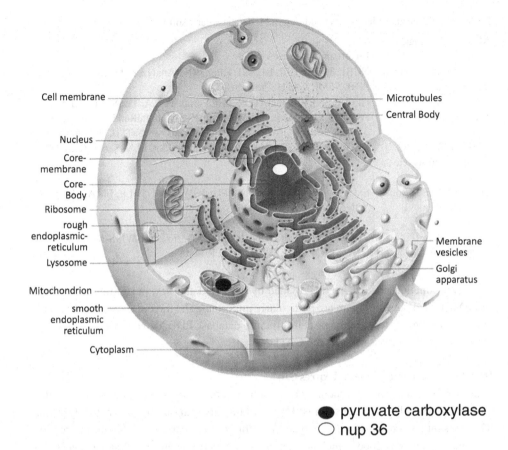

● pyruvate carboxylase
○ nup 36

**Fig. 11.1** Bioinformatic view into the interior of a cell. (Image from https://commons.wikimedia. org/wiki/File:Aufbau_einer_Tierischen_Zelle.jpg)

### How Do Cells Read Their Genome?

Well, cells read their genome with the help of special enzymes called polymerases. They use them to translate the genome into RNA molecules and then into new proteins via protein translation. This is where the NCBI website first comes in handy in order to find DNA sequences that have been sequenced before or are currently being sequenced. There are now millions of such sequences and billions of base pair sequence information. The European EMBL database also has comparable information. The EBI website still has ready-made program modules for the bioinformatician. For all sequence comparisons, the BLAST sequence algorithm is a very good starting point. It quickly compares sequences from genomes, but also from RNA molecules and proteins with the large relevant databases for these molecules. Subsequently, a functional overview of all RNA molecules is obtained using the Rfam database. The RNAAnalyzer software identifies subtleties of RNA such as regulatory elements, secondary structure, energy and sequence motifs. RNAfold is used to understand structures in RNA. An overview of protein functions based

**Table 11.1** Molecular biology foci and important databases and software

| Molecular biology focus | Important databases/software |
|---|---|
| How do cells read their genome? | NCBI, EMBL, EBI, BLAST, Rfam, RNAAnalyzer, RNAfold, SMART, PDB, SCOP, CATH, ProDom, Pfam |
| How do cells control gene expression? | GEO, GENEVESTIGATOR, cBioPortal, TCGA, TESS, ALGGEN PROMO, Genomatix, MEME Suite, iRegulon, miRanda, TargetScan, STRING, KEGG, Roche Biochemical Pathways, STITCH, DrumPID |
| How cells localize, transport and secrete proteins | KEGG, PyMOL, RasMol, Ramachandran Plot, ELM Server, TMHMM |
| How cells build a solid skeleton and move actively | ExPASy, PROSITE, ProDom, PlateletWeb, MUSCLE, EMA, Metatool, YANAsquare |
| How do cells communicate? | STRING, iHOP, PRODORIC, SQUAD, Jimena, SWISS-MODEL, I-TASSER, LOMETS, QUARK, Rosetta |

on a given function or sequence is provided by domain databases; in addition to SMART and EMBL, ProDom and Pfam are particularly important. The three-dimensional structure of many proteins is stored in the protein structure database PDB, details of the architecture in the structure databases SCOP and CATH.

## How Do Cells Control Gene Expression?

Interestingly, at any given moment, only a fraction of the genome information is translated into RNA molecules. The question is: How do I quickly find out bioinformatically which RNA is synthesized in which cell type? For this purpose, the GEO (Gene Expression Omnibus) database is good, which holds numerous data from gene expression experiments for different organisms, tissues and diseases in detail. A similar database is GENEVESTIGATOR. The cBioPortal and The Cancer Genome Atlas (TCGA) databases focus on cancer. In particular, because usually all transcripts of a cell are measured, these experiments can also be used to infer from previous data how one's desired gene is regulated. For this purpose, GEO, GENEVESTIGATOR, cBioPortal and TCGA also hold statistical analysis. Next, there is promoter analysis software. This allows me to determine which regulatory sequences regulate the turning on and off of a gene. There are simple programs for this, such as TESS or ALGGEN PROMO, which simply reveal numerous binding sites for transcription factors, and usually far too many possibilities. In addition, there are better, but often commercial programs such as Genomatix, which, among other things, compare which of the many binding sites within a gene family are conserved and thus presumably actually regulate transcription, so-called modules (e.g. consisting of three specific transcription factors), for example to specifically transcribe liver genes, *such as Liver-specific-transcription-factor-1 modules. Ab initio approaches* such as MEME Suite and iRegulon offer another possibility to find unknown TF motifs and regulatory TF factors.

For regulation in the cell, it is also important that proteins control each other. For this, the protein interaction database STRING (EMBL) is very good and broad (and there are numerous other protein interaction databases). Similarly, RNA molecules control each other. For this, TargetScan and miRanda indicate possible binding sites between mRNAs and miRNAs (small RNAs that regulate mRNAs). Finally, cell compartments through membranes control that certain reactions take place in specific cell regions. This can be easily reconstructed using biochemical databases, e.g. KEGG or Roche Biochemical Pathways. It is also of interest to look at the interaction between a drug and its target. The STITCH database or the DrumPID database developed by us are helpful for this. Regulation, in particular through control of gene expression, is thus an important design principle that is bioinformatically analysed through analyses of RNA and statistical analyses of gene expression, on which network analyses are then based.

**How Cells Localize, Transport and Secrete Proteins**
This is already described in Fig. 11.1 (identification of the corresponding signal sequences). Individual proteins can be viewed in detail using visualization software such as PyMOL or RasMol. The localization of proteins, their transport and secretion are also precisely encoded in the cell and is crucial for the ordered structure of the cell. This can be elucidated in particular by sequence analyses (localization signals, secretion signals, transport signals).

Modern *imaging* techniques now even achieve optical resolution down to 1 nm (Stefan Hell, Nobel Prize winner for *superresolution microscopy*): Using clever tricks, namely combining flashing (with the DSTORM technique down to 10 nm) and quenching fluorescence signals (also goes down to 10 nm), where integrating over time is critical, one can actually resolve structures much smaller than half the wavelength of visible light (400–800 nm), which was the classical lowest resolution limit of optical microscopy. In these techniques, software is indispensable for high resolution. Further bioinformatics software is required for localization. For example, one can use common microscopy image processing software such as ImageJ and write scripts for it (called macros) that allow one to filter out individual features of the image from large amounts of data, for example for the detection of synaptic vesicles. An introduction to such techniques is provided by Kaltdorf et al. (2017), including a tutorial on how to learn the software.

It is also important to classify all cellular processes by analyzing the gene ontology. Combined with information about the protein-protein interactome, the resulting cellular network can be traced using software such as CellDesigner or Cytoscape. For example, motor proteins and the actin-myosin cytoskeleton are crucial for cell movement.

The Gene Ontology Consortium has hierarchically classified all processes in the cell according to three criteria (https://www.geneontology.org): molecular function (e.g. enzyme and which enzyme), cellular compartment (such as in the cytoplasm or in an organelle) and cell biological process (e.g. a signalling cascade such as apoptosis). Thus, with an analysis of the gene ontology of the proteins involved, an overview of the design of a protein network can be quickly obtained by bioinformatics. For example, one can easily evaluate the proportions of the proteins involved in the processes determined with the help of the Gene Ontology (as also described in Chap. 5, Task 5.9 BiNGO analysis and Task 5.10).

**How Cells Build a Solid Skeleton and Move Actively**

Interestingly, cells also need a solid skeleton to be able to move actively. This is composed of various structural proteins, in particular actin, tubulin and myosin. Bioinformatically, structural proteins can be identified by the fact that only a few amino acids are always repeated, resulting in a stable structure in that protein. Such *repeats* can be found with the help of *repeat recognition software* (available, for example, on the ExPASy website). Protein-specific signatures can also be detected using PROSITE software, which, for example, displays actin signatures when the actin sequence is entered. Similarly, one can recognize structural proteins again using protein domains and matching databases, for example, the ProDom domain database. Equally important for active movement is enough energy. This energy is available in the form of ATP, which is provided from metabolism. To calculate the resulting metabolic fluxes, we naturally use metabolic modeling software (see penultimate chapter). An orderly and efficient metabolism in the cell is essential for survival. Its "design" is quickly queried via databases such as KEGG or calculated more precisely via metabolic modelling (e.g. with YANA or Metatool).

**How Do Cells Communicate?**

The considerations and algorithms from Chap. 5 are particularly useful here. However, the messages that cells exchange are also subject to the principles of Chap. 7. And, of course, each passing of messages from one protein to the next (for example, in a signalling cascade with kinases to switch on and phosphatases to switch off) can also be analysed with the aid of interactome software and databases (e.g. STRING database at EMBL, "*information hyper-linked over proteins*" website, iHOP), for example, checked for completeness and function of the components and simulated with a mathematical model (e.g. SQUAD, Jimena).

Looking a little closer at the communication of individual organisms, bacterial cells have quite direct communication, with mRNAs mostly being translated directly into proteins. Besides the standard promoter, there is a second binding site (PRODORIC is a very good database for bacterial promoters for this), the sigma factor, which determines whether "everything is fine" (70-S sigma factor) or whether different types of stress or lack of food are present and then other raw factors are used.

**Eukaryotic cells** in higher organisms, on the other hand, have much more complex communication. First, hormones circulate in the bloodstream. These in turn excite receptors in specific organs. Now a second messenger is often sent off, e.g. cAMP, which then sends signals into the nucleus. There, a complex combination of transcription factors (three and more) first determines the cell type, then the metabolic situation and then the general transcription (a good software for the analysis of such promoters is the Genomatix software).

**Differentiation** as well as all switching processes in the **nervous system** are based on the fact that cellular communication first determines the cell types and tissues involved. Subsequently, different brain regions, but also all different differentiation pathways, result starting from stem cells (and again this can be modeled fully dynamically or semiquantitatively, see Chap. 10).

First, the genetic material in the cell nucleus contains all the information necessary for the survival of the cell. Depending on the conditions, this information is used to produce proteins that are always optimally adapted to the environment via RNA molecules that migrate into the cell plasma (mRNA) with the help of the protein factories (ribosomes) of the cell. Complex signalling networks are important for rapid reactions (stress response, chemotaxis in bacteria) and in particular for multicellularity (cell differentiation, tumour development, embryology, inflammatory processes, nervous system). These are modelled in detail with dynamic modelling (see Chaps. 5 and 9), but can also be clearly described again with the aid of protein networks and, taking the processes into account, with the aid of gene ontology.

What does this look like, for example, for a cell from our own body (i.e. a cell with a nucleus, a so-called higher or eukaryotic cell in contrast to the bacterium, which does not yet have a nucleus)? The deeper insight provided by bioinformatics (Fig. 11.1; diagram of eukaryotic cell) enables us to understand (Table 11.1) what is contained in the genome ("annotation" of the genome), to understand when gene information is read out ("promoter analysis"), but also how the fine regulation of gene expression works (for example, via the analysis of large RNA data sets). Subsequently, proteins are synthesised. The function of each individual protein can be analysed in detail using domain databases, and often also in terms of its three-dimensional structure. Another important aspect is to ensure that each protein reaches its correct location (*protein sorting*). For this purpose, too, there are now powerful algorithms that can accurately predict the location within the cell nucleus, but also in individual organelles or in the membrane, as well as individual protein modifications. Cells can also deform and move. In general, cell biological functions are the result of cellular networks. These can be studied in much greater detail than before, both bioinformatically (structure, components) and systems biologically (dynamics, new emergent effects, over time). Particularly active research is also being conducted into how signals are processed in the cell itself, how different cells communicate with each other, how the cell transforms into other cell types (differentiation) and where this process goes wrong (cancer, but also diseases of old age).

Since bioinformatics precisely detects and examines the signals in the cell using various algorithms, it is also particularly powerful in uncovering details in cellular communication, in differentiation and in disturbed cell communication, but also in the computer-assisted search ("*in silico* screening") for new drugs.

We should note, however, that these groundbreaking achievements in modern molecular biology are always team efforts. It is true that all these results are also based on the analysis of large amounts of molecular data and are inconceivable without this analysis, but progress requires equally state-of-the-art machines for generating data in experiments and is, however, in my opinion not sufficiently appreciated today, intelligently planned experiments as well as intelligent interpretation of the results seen.

Were there any surprises in the design of the cell? Of course countless, but many concern the details of the individual *pathways* involved (e.g. Wnt signalling in differentiation and cancer) or important genes (e.g. the P53 gene in apoptosis). It was great and also surprising that for all the systems biology properties discussed in Sect. 5.1, we can now tell through concrete data analysis which molecules are involved in the individual *feedback loops*, in signalling cascades and in the individual building units (modules).

But larger contours are also becoming visible. For example, the importance of RNA as an important level of cell regulation had previously been underestimated, as has only been fully realised in recent years with the discovery of lncRNAs (long non-coding RNAs) and miRNAs (microRNAs) in higher cells and sRNAs *(small RNAs)* in bacteria. For example, an important lncRNA inactivates the second X chromosome in females (xist RNA) and is therefore involved in this fundamental difference between males and females. In contrast, miRNA-21 stops phosphatases such as PTEN and stimulates tumor growth, thus being an important tumor marker. For understanding this new level of cellular regulation, integrative bioinformatic analysis of the transcriptome (and its interplay with other omics domains) is a crucial prerequisite (e.g. two of our papers Fuchs et al. 2020 and Stojanović et al. 2020 showing a link of RNA and proteome to miRNA regulation in cardiac and pulmonary fibrosis).

A second example for a deeper understanding of the design principles of our cells is tissue replacement by artificial tissue or stem cells. Here, bioinformatics is essential to uncover signaling pathways and generate suitable tissue or reprogram stem cells.

Another current application of the cell's design principles is **protein design:** bioinformatics and experiments that systematically change protein structures to investigate how a protein acquires new properties. This now works well enough with the large number of protein structures (e.g. 3D coordinates from the PDB database) that this is being used more and more actively. First of all, the protein structure has to be predicted. This can be done particularly well using a template (protein with a known structure; "homology modelling"), for example using the SWISS-MODEL software (Waterhouse et al. 2018). All known structural domains in a protein can be found with AnDOM (3D domain annotation). If there is insufficient (approximately 62% same/similar amino acids) similarity to a known protein structure, one can determine the best matching structure by threading the sequence on all known structures ("threading"; e.g., server I-TASSER; Zheng et al. 2019a) or LOMETS (Zheng et al. 2019b), or by protein folding simulations (*"ab-initio"*; e.g., QUARK server; Zheng et al. 2019a).

This is followed by the design step: for about three decades, ligands and pharmaceuticals have been optimized to better fit the protein structure, e.g. the receptor. Drugs against HIV infection have often been achieved by design. More recently, one actively incorporates protein structures into simulations and predictions, using high-throughput experimental methods (Lam et al. 2018; Dominguez et al. 2017), and also understands catalysis in enzymes or receptor function better and better (Mahalapbutr et al. 2020; Sgrignani et al. 2020). However, protein structure can also be used to selectively alter protein structure itself, for example to improve enzyme activities (Leman et al. 2020; Rosetta software) and to systematically change protein building units, even to combine them into logic circuits (Chen et al. 2020), where it is now easy to add or swap secondary structure in particular.

**Conclusion**

- The design principles of a cell are derived from molecular biology: The flow of genetic information follows from the genome via RNA to individual proteins. This can be deciphered bioinformatically in detail by sequence analyses and more elaborate methods. Regulation, particularly through control of gene expression, is an important design principle that is bioinformatically analyzed through analyses of RNA and statistical analyses of gene expression, upon which network analyses are then built. Protein and drug design use protein building principles.
- The localization of proteins, their transport and secretion are also precisely encoded in the cell and are crucial for the ordered structure of the cell. This can be elucidated in particular by sequence analyses (localization, secretion, transport signals). Modern imaging techniques and advanced imaging software help to validate these predictions. It is also important to classify all cellular processes by analyzing the gene ontology. Combined with information on the protein-protein interactome, the resulting cellular network can be traced using software such as CellDesigner or Cytoscape. For example, motor proteins and the actin-myosin cytoskeleton are crucial for cell movement.
- An orderly metabolism is important. Its "design" is quickly queried via databases such as KEGG or calculated more precisely via metabolic modelling (e.g. with YANA or Metatool). Complex signalling networks are important for fast reactions (stress response, chemotaxis in bacteria) and especially for multicellularity (cell differentiation, tumorigenesis, embryology, inflammatory processes, nervous system). These are modelled in detail with dynamic modelling (see Chaps. 5 and 9), but can also be clearly described with the aid of protein networks and, taking the processes into account, with the aid of gene ontology.

## 11.3    Exercises for Chap. 11

**Task 11.1**
Describe how a linear RNA code becomes a three-dimensional protein structure.

**Task 11.2**
Describe how to bioinformatically perform protein structure analysis.

**Task 11.3**
Name methods for predicting a protein structure from a sequence.

**Task 11.4**
What is a Ramachandran Plot and what can I use it for?

**Task 11.5**

Name databases/software where they can find information on proteins.

**Task 11.6**

Name software they can use to visualize and analyze protein structures and interactions.

**Task 11.7**

What can be found in the SCOP and CATH databases? What are the similarities and differences?

**Task 11.8**

To get a conclusion about the possible function, it is helpful to examine a protein sequence for specific protein domains and sequence motifs. How can the function of a protein sequence that has not yet been assigned a function from experiments be investigated using different software or database queries? In doing so, highlight differences between the programs.

**Task 11.9**

Name and describe databases/software that can be used to screen a protein sequence for conserved regions/domains.

**Task 11.10**

Describe how to identify conserved motifs using multiple *alignment*.

**Task 11.11**

You have ten different sequences and want to examine them for conserved sequence regions. Name databases/software that you can use to perform such a multiple *alignment*.

**Task 11.12**

Example:

Now download the protein sequence for the "TAR protein" and perform a search with the PROSITE database (https://prosite.expasy.org/) and the AnDom software (https://andom.bioapps.biozentrum.uni-wuerzburg.de/index_new.html) in the next step.

Which of the following statements are correct (multiple answers possible)?

A. Both programs find a *double stranded RNA-binding domain* (dsRBD).
B. Neither program finds a match for a protein domain.
C. AnDom performs a structural analysis based on the SCOP classification.
D. Based on the dsRBD domain I found, I can assume that my protein binds to double-stranded RNA molecules.

**Task 11.13**

Example:

   Visualize and model the inhibition of HIV-1 protease with indinavir (PDB ID: 1HSG) in PyMOL (please download PyMOL [https://www.pymol.org/] and the sequences from PDB [https://www.rcsb.org/pdb/home/home.do] beforehand). Answer the following questions:

- How to bring together the complex of protease and ligand?
- How are the charges distributed at the point of contact?
- Where is the catalytic center? What holds the center together?

**Task 11.14**

How can we bioinformatically predict the localization of a protein?

**Task 11.15**

Describe how to bioinformatically screen a protein for its localization (e.g. membrane protein, transcription factor) in eukaryotes/prokaryotes (name software/database and briefly describe).

**Task 11.16**

Describe what is meant by neural (machine) learning. Are there any differences between this and human learning?

**Task 11.17**

Name databases/software where you can get information on signal peptides (e.g. sequence or localization).

**Task 11.18**

Develop a simple program that examines a sequence for its localization. Also enumerate what parts this program would consist of.

**Task 11.19**

Cellular communication can also be viewed informatically. Briefly explain how to calculate the information content of a message and show an example.

**Task 11.20**

Biological communication has problems with the transmission of information, what? How are these problems solved biologically by the cell (explain with an example)?

**Task 11.21**

Describe examples of cellular communication (e.g. what can I look at/describe bioinformatically?).

**Task 11.22**

The TMHMM server is used to detect transmembrane motifs in proteins that are partially located in a membrane. Find the link to it and analyze a matching sequence.

**Task 11.23**

Locate NucPred/LocSigDB for nuclear localization and analyze a matching sequence.

**Task 11.24**

Find SignalP for secretion signals in proteins and analyze a matching sequence.

**Task 11.25**

PROSITE finds motifs in proteins. Find the PROSITE database and analyze a matching sequence.

**Task 11.26**

The ELM server outputs all this in bundles. Find this one as well and analyze a matching sequence.

**Task 11.27**

The Gene Ontology annotation (GO annotation) brings bioinformatics "order" to a cell. Learn about the GO annotation and the three classifications used: molecular function, biological process and cellular compartment.

**Task 11.28**

Describe how to draw a network in the cell using Cytoscape.

**Task 11.29**

How to study the drawn network with the BiNGO plugin?

**Task 11.30**

Using the PlateletWeb database as an example, how can you study a particular cell type? Perform a simple query, looking through the design (where do I put or insert the anticoagulant?) and export the network (then you could reconnect the previous network analysis and mathematical modelling tasks).

**Useful Tools and Web Links**

| | |
|---|---|
| ALGGEN PROMO | https://alggen.lsi.upc.es/cgi-bin/promo_v3/promo/promoinit.cgi?dirDB=TF_8.3 |
| AnDOM | https://andom.bioapps.biozentrum.uni-wuerzburg.de/index_new.html |
| BLAST | https://blast.ncbi.nlm.nih.gov/Blast.cgi |
| CATH | https://www.cathdb.info/ |
| cBioPortal | https://www.cbioportal.org/ |
| DrumPID | https://drumpid.bioapps.biozentrum.uni-wuerzburg.de/compounds/index.php |
| ELM | https://elm.eu.org |
| EMBL-EBI | https://www.ebi.ac.uk/services |
| ExPASy | https://www.expasy.org |
| GENEVESTIGATOR | https://genevestigator.com/gv/ |
| Genomatix | https://www.genomatix.de/ |
| GEO | https://www.ncbi.nlm.gov/geo/ |
| iHOP | https://www.ihop-net.org/UniPub/iHOP/ |
| iRegulon | https://iregulon.aertslab.org/ |
| I-TASSER | https://zhanglab.ccmb.med.umich.edu/I-TASSER/ |
| Jimena | https://www.bioinfo.biozentrum.uni-wuerzburg.de/computing/jimena_c/ |
| KEGG | https://www.genome.jp/kegg/ |
| LOMETS | https://zhanglab.ccmb.med.umich.edu/LOMETS/ |
| MEME Suite | https://meme-suite.org/ |
| Metatool | https://pinguin.biologie.uni-jena.de/bioinformatik/networks/metatool/metatool5.0/metatool5.0.html |
| miRanda | https://www.microrna.org/microrna/home.do |
| MUSCLE | https://www.ebi.ac.uk/Tools/msa/muscle/ |
| NCBI | https://www.ncbi.nlm.nih.gov/pubmed/ |
| Nucloc | https://www.nucloc.org/ |
| PDB | https://www.rcsb.org/pdb/home/home.do |
| Pfam | https://pfam.xfam.org/ |
| PlateletWeb | https://plateletweb.bioapps.biozentrum.uni-wuerzburg.de/plateletweb.php |
| ProDom | https://prodom.prabi.fr/prodom/current/html/home.php |
| PRODORIC | https://prodoric.tu-bs.de/ |
| PROSITE | https://prosite.expasy.org |
| PyMOL | https://www.pymol.org/ |
| QUARK | https://zhanglab.ccmb.med.umich.edu/QUARK/ |
| Ramachandran Plot | https://mordred.bioc.cam.ac.uk/~rapper/rampage.php |
| RasMol | https://www.openrasmol.org/ |
| Rfam | https://rfam.xfam.org/ |
| RNAAnalyzer | https://rnaanalyzer.bioapps.biozentrum.uni-wuerzburg.de |
| RNAfold | https://rna.tbi.univie.ac.at/cgi-bin/RNAWebSuite/RNAfold.cgi |

(continued)

**Useful Tools and Web Links** (continued)

| | |
|---|---|
| Roche Biochemical Pathways | https://www.roche.com/sustainability/what_we_do/for_communities_and_environment/philanthropy/science_education/pathways.htm |
| ROSETTA | https://www.rosettacommons.org/software/ |
| SCOP | https://scop.mrc-lmb.cam.ac.uk/scop/ updated latest version at: https://scop.berkeley.edu; SCOPe (enhanced) |
| SignalP | https://www.cbs.dtu.dk/services/SignalP/ |
| SMART | https://smart.embl-heidelberg.de/ |
| SQUAD | https://www.vital-it.ch/software/SQUAD |
| STRING | https://string-db.org/ |
| STITCH | https://stitch.embl.de/ |
| SWISS-MODEL | https://swissmodel.expasy.org |
| TargetScan | https://www.targetscan.org/vert_71/ |
| TCGA | https://www.cancer.gov/about-nci/organization/ccg/research/structural-genomics/tcga |
| TESS | https://www.cbil.upenn.edu/tess/ |
| TMHMM | https://www.cbs.dtu.dk/services/TMHMM/ |
| YANAsquare | https://www.bioinfo.biozentrum.uni-wuerzburg.de/computing/yanasquare/ |

# Literature

Alberts B (2013) Essential cell biology, 4. Aufl. Garland Sciences (Taylor & Francis Group), United States (ISBN-13: 978-0815344544, ISBN-10: 0815344546)

Chen Z, Kibler RD, Hunt A et al (2020) De novo design of protein logic gates. Science 368(6486):78–84. https://doi.org/10.1126/science.aay2790

Dominguez M, Alvarez S, de Lera AR (2017) Natural and structure-based RXR ligand scaffolds and their functions. Curr Top Med Chem 17(6):631–662. https://doi.org/10.2174/1568026616666160617072521

Fuchs M, Kreutzer FP, Kapsner LA et al (2020) Integrative bioinformatic analyses of global transcriptome data decipher novel molecular insights into cardiac anti-fibrotic therapies. Int J Mol Sci 21(13):4727. https://doi.org/10.3390/ijms21134727

Kaltdorf KV, Schulze K, Helmprobst F et al (2017) FIJI Macro 3D ART VeSElecT: 3D automated reconstruction tool for vesicle structures of electron tomograms. Comp Biol 13(1):e1005317. https://doi.org/10.1371/journal.pcbi.1005317

Lam PC, Abagyan R, Totrov M (2018) Ligand-biased ensemble receptor docking (LigBEnD): a hybrid ligand/receptor structure-based approach. J Comput Aided Mol Des 32(1):187–198. https://doi.org/10.1007/s10822-017-0058-x

Leman JK, Weitzner BD, Lewis SM et al (2020) Macromolecular modeling and design in Rosetta: recent methods and frameworks. Nat Methods. https://doi.org/10.1038/s41592-020-0848-2

Mahalapbutr P, Lee VS, Rungrotmongkol T (2020) Binding hot spot and activation mechanism of maltitol and lactitol towards the human sweet taste receptor. J Agric Food Chem. https://doi.org/10.1021/acs.jafc.0c02580

Sgrignani J, Fassi EMA, Lammi C et al (2020) Exploring proprotein convertase subtilisin/kexin 9 (PCSK9) autoproteolysis process by molecular simulations: hints for drug design. Chem Med Chem. https://doi.org/10.1002/cmdc.202000431

Simon EJ (New England College), Dickey JL (Clemson University), Reece JB (Berkeley, California) (2013) Campbell Essential Biology, 5th Edition Benjamin Cummings, Pearson Education, San Francisco, S 544 (ISBN13: 9780321772596)

Stojanović SD, Fuchs M, Fiedler J et al (2020) Comprehensive bioinformatics identifies key microRNA players in ATG7-deficient lung fibroblasts. Int J Mol Sci 21(11):4126. https://doi.org/10.3390/ijms21114126

Waterhouse A, Bertoni M, Bienert S et al (2018) SWISS-MODEL: homology modelling of protein structures and complexes. Nucleic Acids Res 46(W1):W296–W303. https://doi.org/10.1093/nar/gky427

Watson JD (2013) Molecular biology of the gene. Cold Spring Harbor Laboratory Press, Cold Spring Harbor

Zheng W, Li Y, Zhang C et al (2019a) Deep-learning contact-map guided protein structure prediction in CASP13. Proteins. 87(12):1149–1164. https://doi.org/10.1002/prot.25792

Zheng W, Zhang C, Wuyun Q et al (2019b) LOMETS2: improved meta-threading server for fold-recognition and structure-based function annotation for distant-homology proteins. Nucleic Acids Res 47(W1):W429–W436. https://doi.org/10.1093/nar/gkz384

# Part III

# What Is Catching and Fascinating About Bioinformatics?

Bioinformatics helps biologists with the increasingly needed integration and analysis of large amounts of data, but of course the deeper connections that numerous individual bioinformatic analyses reveal about life are much more fascinating. So in this part, we want to do more *"computational biology"* and get to the bottom of biological information processing, and here we describe some fascinating insights emerging in the currently tumultuous fields of bioinformatics omics, synthetic biology, artificial intelligence and neurobiology, and ecosystem modeling. Other current areas of bioinformatics, such as image processing and *drug design,* also provide exciting new insights, but seem somewhat less central as other pacesetters in modern bioinformatics.

Biology is the key science of the twenty-first century and bioinformatics is its computational spearhead. Sequence analysis of DNA, RNA and proteins as well as the analysis of metabolic and regulatory networks lay the foundations of today's bioinformatics (Part I).

Heuristics and good databases, encoded molecular information and clever strategies to solve NP problems with combinatorics approximately and quickly, along with increasingly powerful computers, are the informatics arm that has made modern bioinformatics so strong. The insight into the molecular biological design of the cell and the system effects, the knowledge of biological signals and their decoding with neural networks, sequence analysis or hidden Markov models enabled a systems biology view in bioinformatics that can model or at least analyze almost any process in the cell. Today, the observation of evolution, for example of protein sequences, allows the functionally important conserved regions in an enzyme to be determined and labelled in a matter of seconds and also the domain composition to be understood, at least functionally (Part II).

The upswing in bioinformatics can be seen very tangibly in current **genomics**. For example, we can now describe much better how individual our genome is. Every person has his or her SNPs, *indels* and *copy number variations* (a total of several percent individual differences). Fascinating insights into the individuality of each person are only now possible through modern genome informatics (entry).

One can also say that bioinformatics does not decipher *the* language of life, but *all of them*. And these languages are characterized by the fact that they are all defined by the context, the coherence of meaning in the living cell (Chap. 12). In fact, life is always inventing new levels of language. Humans and our civilization do similar things: modern communication media, natural languages, programming languages, and even the Internet. We now know the molecular language of life so well that we can use it deliberately (**synthetic biology**), for example, to make new biological computer chips from nanocellulose, DNA, and light-guided proteins (Chap. 13). Mutations lead to misunderstandings in biological signalling resulting for example in cancer (Chap. 14).

Modern bioinformatics benefits greatly from advances in **artificial intelligence research.** We can describe more and more precisely how and where humans differ from computers and how this can again be used for new bioinformatic insights (Chap. 14). Even the presumably most complex object of our universe to which we have direct access, our brain, its **neurobiology** as well as even higher brain processes as a prerequisite for our consciousness can be described, modelled and analysed much better by bioinformatics and simulations and models than before (Chap. 15). More generally, the **systems biology insights** of bioinformatics, the unimagined large amounts of data now available to us, also allow global insights, for example, into bioinformatics **models of ecosystems** (e.g., climate and population dynamics). Increasing digitization will soon lead to the **"Internet of Things."** Where is each thing and how do all the components interact? However, global digitalisation can also lead to control, synchronisation and steering via the internet. Countermeasures are transparency, protection of personal freedom and personality on the Internet. However, bioinformatics also helps to positively translate the **"Internet of Things"** into modern molecular biology, supports new biotechnology, **molecular medicine** and accelerated *drug design*, and sheds light on **pandemics** such as COVID-19 (Chap. 16).

## 3.1    No Black and White: Fascinating Shades of Individuality

As an introduction, Fig. 1 shows an individual genome in an artistic representation. The author and computer graphic artist is Dr. Beat Wolf (University of Fribourg, Switzerland). His artistic images were presented at the VIZBI 2014 congress, among others. In his main profession, he is a bioinformatician and works on genome analysis pipelines. His work is based on the NA12878 *exome sequence* from the 1000 Genomes Project.

The page https://software.broadinstitute.org/gatk/guide/article?id = 7869#1.3 shows a tutorial on how to bioinformatically process this genome sequence.

Everyone, all of us like that, carries around a significant amount of small, medium, and more severe genetic "errors" (in various shades in the figure). However, whether these come to fruition depends on (i) the unpredictable combination of parental chromosomes, often the diploid chromosome sets rebalance, and (ii) the environment in which we live.

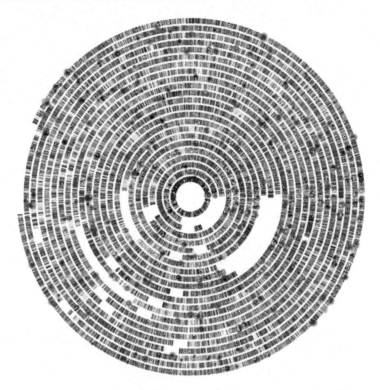

**Fig. 1** No black and white: A look at the fascinating shades of individuality. In this artistic representation, all mutations of a human genome (identifier: NA12878) are displayed. They are organized on several circles, representing the different chromosomes, according to their position on the chromosome. The size and shading were chosen according to the severity of the effects on the function of the genome. For example, one can see the many gray variants that do not fall on any gene and are therefore difficult to classify. This contrasts with the black and dark variants that cause a severe defect in the affected genes. This shows how a considerable number of gene defects can be found even in human reference genomes. Therefore, in each body cell there is almost always a double set of chromosomes, so that only rarely do gene defects dominate the other copy. (Coloured version of Fig. 1: coloured cover picture on page 2)

For example, in one gene combination a singled out "gene defect" can lead to schizophrenia and mental illness, in another to a creative original, in a third, archaic one to a shaman who talks to the dead and advises the tribe.

Genetic diversity therefore means artistic freedom, and it is something to behold. We should by no means sort into "good" and "bad" people. No genome is black or white, everyone carries his or her shades. We would do well to place each person in his or her appropriate environment, where all his or her abilities and colors are brought to bear as positively as possible.

# Life Continuously Acquires New Information in Dialogue with the Environment

**12**

**Abstract**

All molecules of a cell are closely related to each other. Only this context-related information has real meaning. It conveys the cell's behavior, which is important and correct for survival. Print errors are constantly selected away in the population. Database searches and sequence comparisons unlock this biological meaning (in practice, usually the function of the compared molecule). This is strongly tied to sequence elements and a defined structure; random sequences make no biological sense. Even the domains in an enzyme relate to each other, e.g. in the case of glutathione reductase: For the catalytic domain, there are the matching two cofactor domains (for FAD, NAD), the optimal regulatory domain and also the dimerisation domain, otherwise the enzyme would not function. Similarly, one checks the consistency of sequence analyses. Everything must fit together; if contradictions arise, one of the partial analysis results was not yet correctly classified. Also on the level of protein networks everything relates to each other, it can be deciphered by network analyses: Central proteins ('hubs'), signaling cascades and interfering signals, and modifying input ('cross-talk'). A fascinating and illustrative example are the *KEGG maps of cancer pathways*.

First of all, it is fascinating that the central molecule of life, DNA, does nothing but store information. Obviously, storing information is an important aspect for living beings. Information is encoded in genes via the DNA molecule. Then the information is transcribed into RNA, transported out of the cell nucleus, and these are then the building instructions for proteins with which the cell performs its tasks. Originally, this main direction of information processing in living cells was called the central dogma of molecular biology. In the meantime, information processing in the opposite direction is also known, in particular from RNA to DNA, for example via the enzyme reverse transcriptase (for

© Springer-Verlag GmbH Germany, part of Springer Nature 2023
T. Dandekar, M. Kunz, *Bioinformatics*,
https://doi.org/10.1007/978-3-662-65036-3_12

example in HIV infection; see Sect. 1.1). Above all, it is clear that life is information, which is the only reason why bioinformatics works at all! The success of bioinformatics and computer-assisted modelling of life processes is not due to the recent abundance of data and the discovery of the computer, but exactly the opposite. Because we are information in our innermost being, we have been able to discover all this and then produce so much data recently with self-built machines.

## 12.1   Molecular Words Only Ever Make Sense in the Context of the Cell

But in the following we want to come a little closer to the nature of stored information in living beings. Of course, there is a quick answer that could be interpreted esoterically in any direction. It is living information, as distinguished, for instance, from the dead, "cold" information in a computer. However, since Friedrich Wöhler (who in 1828 became world famous for his synthesis of urea from ammonium cyanate) showed that organic molecules could be easily synthesized in a retort, one has been skeptical about simply ascribing different properties to building blocks in living things than outside (such as by a mysterious *"vis vitalis"*, a force of life; McKie 1944).

But incidentally, the special properties of information in living organisms can also be experienced very nicely in concrete terms, and we want to trace this right now. In the first chapter, you have already become acquainted with the rapid sequence comparison using the "BLAST" algorithm (Altschul et al. 1990). This can be used, for example, to track down protein sequences in a database. Our Figs. 12.1 and 12.2 illustrate this using the protein sequence of glutathione reductase (Q03504_HUMAN glutathione reductase from UniProt database), which, as the result shows, also occurs in humans (Fig. 12.1). In addition to the annotation, the BLAST search also provides other helpful information, such as protein classification or conserved domains (Fig. 12.2). Thus, it can be seen that glutathione reductase belongs to the class of *FAD-dependent oxidoreductases,* which influence iron-sulfur proteins. According to the BLAST result, it has five conserved domains, e.g. glutathione reductase domain (gluta_reduc_1, Accession: TIGR01421, Position 1–229) or FAD/NAD binding domain (Pyr_redox; Accession: pfam00070, Position 108–171) (follow the link in BLAST for a better representation). Thus, it can be seen that it is a rust protector against oxidation and thus one of the biochemical helpers that allow us to reach old age.

Are there also differences in the information that BLAST would recognize? If one were to blast any German text at this point, e.g. "Bioinformatik macht jede Menge Freude" (Bioinformatics is a lot of fun), compared to a biological sequence and with a normal German text, the six letter codons, namely J, U, Z, B, O and X, would not be assigned to any codon and would simply be omitted from the BLAST server during the comparison (mnemonic "JUZBOX", i.e. *juke-box*). In this case, BLAST would find no similarity and say that this is not a protein in a cell (Fig. 12.3, here, for example, the O). So this is a first

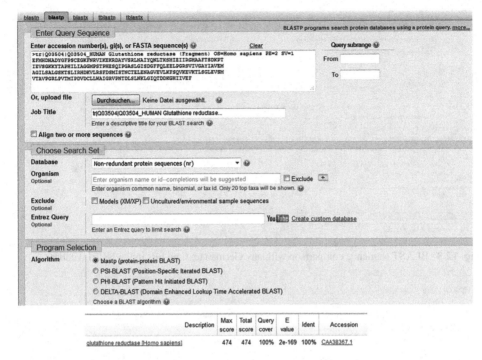

**Fig. 12.1** BLAST sequence comparison with a natural protein sequence. The figure shows the glutathione reductase found in humans

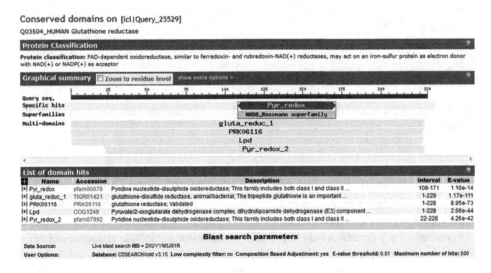

**Fig. 12.2** Important conserved domains of the glutathione reductase sequence are shown, e.g. FAD/NADH binding domain

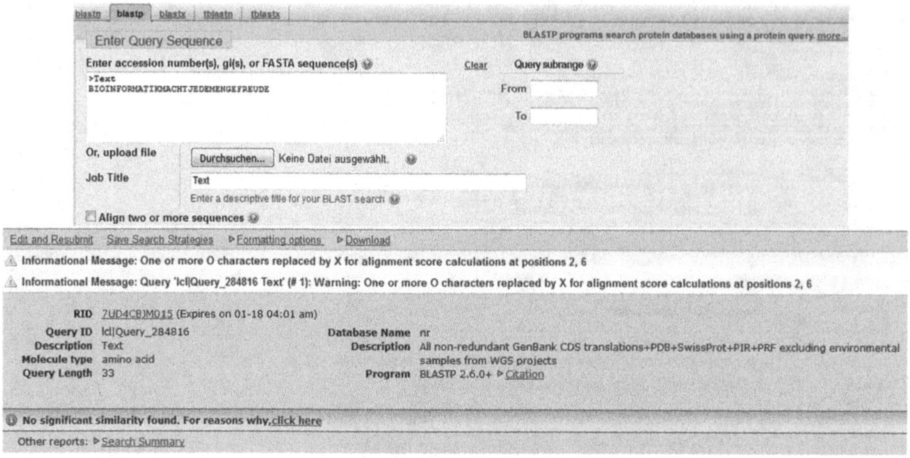

**Fig. 12.3** BLAST sequence comparison with any German text: no hits in the NCBI database

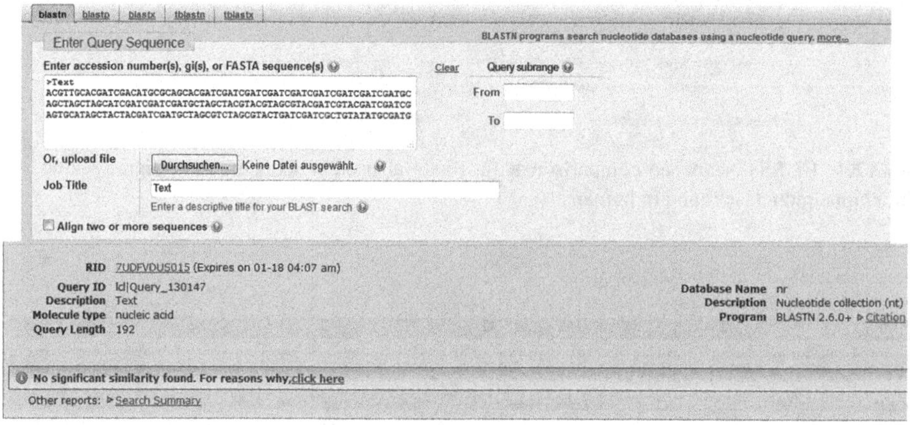

**Fig. 12.4** BLAST sequence comparison with a random nucleotide sequence: no hits in the NCBI database

difference of information in living cells and say in German texts, BLAST makes big differences here. Similarly, one can try this with nucleotides. A random nucleotide sequence is rarely found in the NCBI databases (Fig. 12.4), but a sequence corresponding to a biological signal is (see Fig. 1.1, HIV-1).

Apparently, only certain words in protein sequences or nucleotide sequences of cells are understood, namely those with a biological composition (i.e. sufficient similarity to biological sequences and biological signals). But this immediately reveals a second property of biological information units. Each word, each molecule is in a context, it is only evaluated as a biological signal and integrated into the cell metabolism if it fits the context of the other molecules. In addition to the sequence, the interaction context is equally important. This is illustrated in Fig. 12.5. Our useful example sequence, glutathione reductase (black box), interacts with many other proteins (e.g. shown here STRING database result) and in this way is only properly "understood" by the cell if no mutation disturbs these interactions or enzyme function.

In conclusion, biological information is always context-related, whereby the linguistic space that delimits the terminology and words used or understood is the respective cell itself (i.e. the totality of all stored information in this cell, represented by the many types of molecules: Nucleic acids, proteins, lipids, carbohydrates, but all with very specific sequences).

Of course, numerous experiments have already been carried out that have shown that a certain word (a gene or protein variant) is "understood" in one cell, i.e. that it triggers a certain effect that it does not trigger in other cells.

For example, the third Erk phosphorylation (Chap. 5) triggers cardiac enlargement only in the heart.

**The Language of Life**

In addition to the clear restriction of the vocabulary and the contextuality of all information, a third characteristic distinguishes biological information that is stored genetically. Information is only newly or additionally stored if this increases the overall probability of survival or the adaptability to the environment.

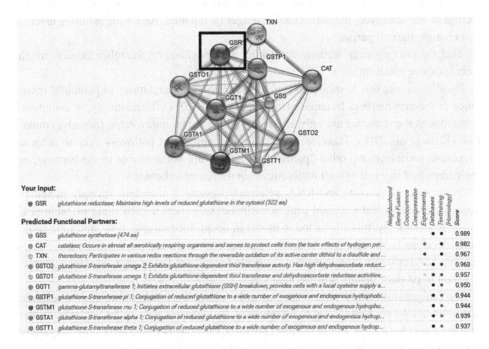

**Your Input:**

| | | | | | | | |
|---|---|---|---|---|---|---|---|
| ⊜ GSR | glutathione reductase; Maintains high levels of reduced glutathione in the cytosol (522 aa) | | | | | | |

**Predicted Functional Partners:**

| | | Neighborhood | Gene Fusion | Cooccurrence | Coexpression | Experiments | Databases | Textmining | [Homology] | Score |
|---|---|---|---|---|---|---|---|---|---|---|
| ⊜ GSS | glutathione synthetase (474 aa) | | | | | | | ● | ● | 0.989 |
| ⊜ CAT | catalase; Occurs in almost all aerobically respiring organisms and serves to protect cells from the toxic effects of hydrogen per... | | | | | | | ● | ● | 0.982 |
| ⊜ TXN | thioredoxin; Participates in various redox reactions through the reversible oxidation of its active center dithiol to a disulfide and ... | | | | | | | ● | ● | 0.967 |
| ⊜ GSTO2 | glutathione S-transferase omega 2; Exhibits glutathione-dependent thiol transferase activity. Has high dehydroascorbate reduct... | | | | | | | ● | ● | 0.963 |
| ⊜ GSTO1 | glutathione S-transferase omega 1; Exhibits glutathione-dependent thiol transferase and dehydroascorbate reductase activities... | | | | | | | ● | ● | 0.957 |
| ⊜ GGT1 | gamma-glutamyltransferase 1; Initiates extracellular glutathione (GSH) breakdown, provides cells with a local cysteine supply a... | | | | | | | ● | ● | 0.950 |
| ⊜ GSTP1 | glutathione S-transferase pi 1; Conjugation of reduced glutathione to a wide number of exogenous and endogenous hydrophobi... | | | | | | | ● | ● | 0.944 |
| ⊜ GSTM1 | glutathione S-transferase mu 1; Conjugation of reduced glutathione to a wide number of exogenous and endogenous hydropho... | | | | | | | ● | ● | 0.944 |
| ⊜ GSTA1 | glutathione S-transferase alpha 1; Conjugation of reduced glutathione to a wide number of exogenous and endogenous hydrop... | | | | | | | ● | ● | 0.939 |
| ⊜ GSTT1 | glutathione S-transferase theta 1; Conjugation of reduced glutathione to a wide number of exogenous and endogenous hydrop... | | | | | | | ● | ● | 0.937 |

**Fig. 12.5** Context of proteins: protein interaction networks of glutathione reductase from the STRING database

In this sense, cells are constantly storing and developing (from each generation to the next, after the selection of harmful mutations) new information that is essential for survival in dialogue with the environment.

Each word of this "language of life" only makes sense in the narrow context of the limited vocabulary of the respective cell. But what is newly learned or spoken is always determined by the success of adaptation to the environment. On higher levels, this leads to ever better coordinated processes of homeostasis, i.e. ever stronger abilities to keep cell conditions optimal in relation to the environment. By the way, it is possible to check for each real protein or nucleotide sequence whether it is really under a selection pressure or not.

Again, it is with the selection pressure on the individual words so similar to our considerations on evolution. Each word is more or less strongly neutral, i.e. variants are tolerated or strongly persecuted – how strongly this is the case is decided by the context of the cellular processes.

## 12.2    Printing Errors Are Constantly Selected Away in the Cell

The information in cells must be maintained and preserved. If biochemical reactions, oxidation, destruction by free radicals, etc. destroy individual proteins, these can be rebuilt on the basis of the building instructions in the cell nucleus. But if the building instructions change or are destroyed, the defect can no longer be repaired. And if the resulting error is bad enough, the cell perishes.

That the printing error leads to something new, positive, on the other hand, is much rarer (positive mutation).

For illustration, this is shown in Fig. 12.6 for observed mutations to penicillin resistance in the enzyme beta-lactamase (from Khan et al. 2011). Interestingly, in evolution, mutations in the sequence are only allowed if this is not an unfavorable (negative) mutation (Khan et al. 2011). Therefore, only a limited number of pathways exist to achieve maximum resistance. All other "paths of evolution" are not available to the bacteria, as their chance of survival against antibiotics then decreases in between.

In the case of proteins and RNA, selection pressure is therefore strongly linked to sequence elements and a defined structure; random sequences usually make no biological sense. Fascinatingly, this allows me to model in detail, for example, how antibiotic resistance develops in bacteria (combination of protein structure and phylogenetic tree analysis; Fig. 12.6) or how a protein code is optimally formulated for an organism (for example, if I want to produce insulin with optimal yield).

We can generalize this. Figure 12.7 uses the example of the genetic code to explain where mutations are most likely to be tolerated. The genetic code codes triplet by triplet one amino acid at a time. But in the third position, the code is characterized by frequently encoding the same amino acid for each nucleotide (Fig. 12.7).

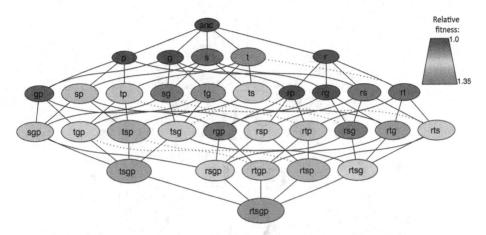

**Fig. 12.6** Antibiotic resistance: Only a path of always positive mutations with fitness increase is viable for evolution. The figure shows the 32 possible genotypic combinations for five mutations in the beta-lactamase gene. Colored gradation shows the fitness relative to the ancestral strain (anc). (anc = ancestral strain, r = rbs operon, t = topA promoter, s = spoT promoter, g = glmUS promoter, p = pykF). (Figure from Khan et al. 2011)

This type of genetic information, due to its defined vocabulary, its contextuality and its selection for survival advantage, thus differs significantly from other types of information, for example from physics (physical state bits, for example in transistors, quantum bits, Bekenstein entropy, physical entropy), computer science (Shannon, computer bits) or chemistry (Gibbs entropy, chemical energy).

As a bioinformatician, you can look at a protein sequence and calculate how much selection pressure is on the sequence. Since, according to the genetic code (Fig. 12.7), selection on the third nucleotide letter is weakest (usually a mutation there does not change the encoded amino acid), one can calculate for a given protein sequence, after phylogenetic comparison of the coding gene sequences, how much selection lies on the individual sequence regions, but also compare this with the phylogenetically conserved locations of the protein sequence.

Figure 12.2 illustrates this by an analysis of the conserved domains and sequence positions in glutathione reductase (result from sequence analysis with BLAST shown here, but domain analysis also possible with PROSITE and AnDom). The contextuality of biological information is thus repeated at all levels. The domains in an enzyme relate to each other, for example here in glutathione reductase: To the catalytic domain there are the matching two cofactor domains (for FAD, NAD), the optimal regulatory domain and also the dimerization domain, otherwise the enzyme would not function (FAD-/NADH-binding domain at position 22–228; *Pyr_redox_2,* Accession: pfam07992).

In the same way, the consistency of sequence analyses is checked for proteins: Everything must fit together. If contradictions arise, one of the partial analysis results has not yet been correctly classified or determined. For example, a transcription factor

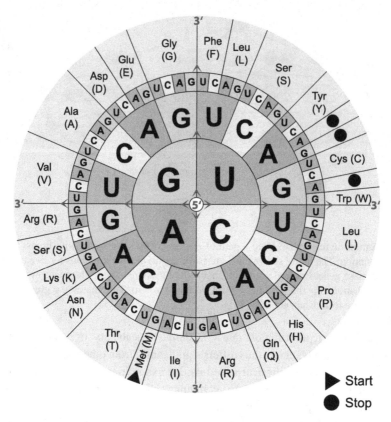

**Fig. 12.7** The genetic code – where is it fault-tolerant? (Figure from https://upload.wikimedia.org/wikipedia/commons/7/70/Aminoacids_table.svg)

(determined by BLAST sequence comparison with high sequence similarity to a biochemically verified protein) should then also have a nuclear localization signal (determined, for example, by the ELM server, the "eukaryotic linear motif server"), and the domain composition (determined by the SMART database; Letunic et al. 2015) should confirm the transcription factor found by a DNA-binding domain. After all, everything has to match because we always assumed the same sequence. Conversely, the different bioinformatics algorithms check and correct each other. In a living cell, the domains in the protein have to fit together correctly.

Learning to better understand this genetic "language of life" was, at least for me, a major reason to learn bioinformatics – and the computer is only one, albeit very powerful, tool for this.

Another way to approach this aspect of the language of life is through the proteins themselves. Their richness can be viewed directly with the Pfam database (all protein families; pfam.xfam.org) or UniProt (database of all known proteins and protein sequences; https://www.uniprot.org). This makes it much easier to understand the huge number of different protein families and associated proteins, which can be found in ever new combinations of domains. An overview of protein domain families can be found in the SMART database (https://smart.embl-heidelberg.de). Alternatively, one uses the "conserved domains", which enables independent verification (Lu et al. 2020).

If one wanted to look at the underlying genes, the "clusters of orthologous groups" (COGs) first gave an overview starting from bacteria (https://www.ncbi.nlm.nih.gov/COG/). Then eukaryotic gene groups were also considered (COGs; eukaryotic orthologous groups; https://mycocosm.jgi.doe.gov/help/kogbrowser.jsf). In this context, a group cluster of genes means that the same gene is found in very many organisms and thus the same protein is always required and encoded in very different organisms: An "ortholog" because the domain composition is the same. Eventually, these orthologous groups were systematically extended, called eggNOGs (Huerta-Cepas et al. 2017). Excitingly, it can also be well shown that the original richness of forms was much smaller, because the primitive cell underlying all present-day life (the "LUCA", last universal common ancestor; Weiss et al. 2018), already using the same genetic code, had only about 1000–1500 proteins, which are still found today as highly conserved protein families in virtually all organisms (similar, but not completely congruent, with the COGs). The protein language is universal and only grew from a relatively manageable inventory to its current richness over billions of years of evolution.

In order for everything to relate correctly to each other at the next higher level, the level of protein networks, there is considerable biological redundancy and robustness. This is necessary to ensure that every signal is correctly understood and does not get lost in the noise (see Chap. 7):

Signals are further amplified in signal cascades. All this can be deciphered by network analysis. This is a very efficient way of finding central proteins *(hubs)* that have a large number of neighbours (e.g. network analysis with Cytoscape). The structure of the network also detects interfering signals as well as modifying and reciprocal input *(cross-talk)*.

A fascinating and illustrative example can be found at the KEGG (Kyoto Encyclopedia of Genes and Genomes) pathway database. These are the *"maps of cancer pathways"*, which illustrate important stages of cancer (supporting and inhibiting *pathways*) for the user, whereby one can look at the different pathway inventory for different organisms. Building on these foundations, modeling *pathways* in cancer development and finding better drugs against them is certainly a fascinating topic in bioinformatics (see Chap. 13). Again, the contextuality of all molecules helps to systematically identify the promoting and inhibiting *pathways*, for example, by gene expression analyses of healthy cells and cancer cells (where thus almost all important observed changes in gene expression interact to further spark cancer).

Redundancy is also reflected by the fact that several synthetic pathways are possible in metabolic networks for important and many other metabolites. This simultaneously protects against numerous genetic mutations that could otherwise disrupt the network, but also allows us to cope much better with fluctuations in the metabolites present in the environment.

Moreover, this can also be used to define life as we know it very well. Our earthly life understands the genetic language of DNA.

The nice thing about this definition is that it can of course be extended to any other kind of life. But this (e.g. synthetic life; Bedau 2003) would then again use other storage molecules (e.g. PNA, so-called *peptide nucleic acids;* Nielsen et al. 1991; Nelson et al. 2000) because of *"once-for-all selection"* (Eigen and Schuster 1979; Eigen and Winkler 1975) during natural emergence. One then defines for these other types of living organisms again central storage molecules, survival-coded, cellular processes and analyzes the contextuality of the remaining molecules in the cell and what this then implies again evolution, metabolism and replication.

**Conclusion**
- Life is always developing new information in dialogue with the environment. It is important to see that all molecules of a cell are closely related to each other and help together so that all processes run in an orderly manner and metabolism and signalling cascades united help that the cell has optimal chances of survival. Only this contextual information has real meaning. It conveys the behavior of the cell that is important and correct for survival. Misprints are constantly selected away in the population.
- Molecular words only ever make sense in the context of the cell. Database searches and sequence comparisons reveal the biological meaning (in practice, usually the function of the compared molecule). This is strongly linked to sequence elements and a defined structure. Random sequences usually do not make biological sense. Fascinatingly, this allows me to model in detail, for example, how antibiotic resistance develops in bacteria (combination of protein structure and phylogenetic tree analysis) or how a protein code is optimally formulated for an organism (for example, if I want to produce insulin with optimal yield).
- The contextuality of biological information is repeated at all levels. The domains in an enzyme relate to each other, e.g. in glutathione reductase: To the catalytic domain there are the matching two cofactor domains (for FAD, NAD), the optimal regulatory domain and also the dimerization domain, otherwise the enzyme would not function. Similarly, I check the consistency of sequence analyses. Everything must fit together; if contradictions arise, one of the partial analysisv results was not yet correctly classified. Since everything relates correctly to each other at the protein network level, there is considerable biological redundancy and robustness. This can be deciphered by network analyses to very efficiently identify central proteins *(hubs)*, signaling cascades and interfering signals, and modifying input *(cross-talk)*. A fascinating and illustrative example are the *KEGG maps of cancer pathways.*

## 12.3 Exercises for Chap. 12

**Task 12.1**
Test senseless and meaningful oligonucleotide sequences: By now you already know BLAST. Now test more precisely which sequences are recognized and which are not: Compare a sequence from Genbank with a random sequence generated by random key strokes from the keyboard. Which different types of answers do you get from the server? What does the *E-value* mean in the result?

**Task 12.2**
Search the protein database for words from our language:

    (a) Search with a word, such as "DNA" or "JAMES WATSON".
    (b) Which letters never occur in the first letter amino acid code?
    (c) What are so-called wobble codons?

**Task 12.3**
What does the *"Universal Code"* look like (hint: https://www.ncbi.nlm.nih.gov/Taxonomy/Utils/wprintgc.cgi)?

**Task 12.4**
Applied example: Pick out the design of insulin expression for optimal expression in yeast cells. Put together the differences that directly catch the eye!

| Useful Tools and Web Links | |
| --- | --- |
| GATK | https://software.broadinstitute.org/gatk/guide/article?id=7869#1.3 |
| BLAST | https://blast.ncbi.nlm.nih.gov/Blast.cgi |
| UniProt | https://www.uniprot.org/ |

## Literature

Altschul SF, Gish W, Miller W et al (1990) Basic local alignment search tool. J Mol Biol 215(3):403–410

Bedau MA (2003) Artificial life: organization, adaptation and complexity from the bottom up. TRENDS Cognitive Sci 7:505–512

Eigen M, Schuster P (1979) The hypercycle: a principle of natural self-organization. Springer, Berlin

Eigen M, Winkler R (1975) Laws of the game: how the principles of nature govern chance. ISBN: 9780691025667 Published: Apr 11, 1993 Princeton University Press.

Huerta-Cepas J, Forslund K, Coelho LP, Szklarczyk D, Jensen LJ, von Mering C, Bork P (2017) Fast genome-wide functional annotation through orthology assignment by eggNOG-Mapper. Mol Biol Evol 34(8):2115–2122

Khan AI, Dinh DM, Schneider D et al (2011) Negative epistasis between beneficial mutations in an evolving bacterial population. Science 332(6034):1193–1196. https://doi.org/10.1126/science.1203801. (PubMed PMID: 21636772)

Letunic I, Doerks T, Bork P (2015 Jan) SMART: recent updates, new developments and status in 2015. Nucleic Acids Res 43(Database issue):D257–D260

Lu S, Wang J, Chitsaz F, Derbyshire MK, Geer RC, Gonzales NR, Gwadz M, Hurwitz DI, Marchler GH, Song JS, Thanki N, Yamashita RA, Yang M, Zhang D, Zheng C, Lanczycki CJ, Marchler-Bauer A (2020) CDD/SPARCLE: the conserved domain database in 2020. Nucleic Acids Res 48(D1):D265–D268

McKie D (1944) Wöhler's synthetic urea and the rejection of vitalism: a chemical Legend. Nature 152:608–610

Nelson KE, Levy M, Miller SL (2000) Peptide nucleic acids rather than RNA may have been the first genetic molecule. Proc Natl Acad Sci USA 97(8):3868–3871

Nielsen PE, Egholm M, Berg RH et al (1991) Sequence-selective recognition of DNA by strand displacement with a thymine-substituted polyamide. Science 254(5037):1497–1500. https://doi.org/10.1126/science.1962210. (PMID: 1962210)

Weiss MC, Preiner M, Xavier JC, Zimorski V, Martin WF (2018) The last universal common ancestor between ancient Earth chemistry and the onset of genetics. PLoS Genet 14(8):e1007518. https://doi.org/10.1371/journal.pgen.1007518

# Life Invents Ever New Levels of Language

# 13

**Abstract**

Life keeps inventing new levels of language: DNA, neurons, human language, computer code, Internet. Finally, the Internet makes bioinformatics software and biological knowledge (PubMed, open access publications), among other things, accessible worldwide. For this purpose, a Domain Name Server (DNS) is used to rewrite the Internet Protocol (IP) address into easily readable addresses. Synthetic biology uses all collected knowledge on biological processes for technical applications, e.g. classical biotechnology (microorganisms produce citric acid, erythropoietin or insulin), more modern are whole circuits (MIT parts list or Biobricks, IGEM competition). Such processes are described in the GoSynthetic database and the MIT BioBricks. Drug design using *in silico screens* and molecular dynamics simulations also noticeably shortens drug development. Natural and analog computing, for example, use slime molds for complex calculations. The nanocellulose chip is potentially superior to today's computer chips. It uses DNA for storage and light-controlled polymerases and exonucleases for reading in and out the stored information. Modulating proteins and processes act electronically across the nanocellulose membrane. New combinations of molecular biology, nanotechnology and modern electronics have huge future technology potential.

Apart from the genetic "language of life", another aspect is very fascinating: Life is always inventing new levels of language (see Sect. 12.1).

© Springer-Verlag GmbH Germany, part of Springer Nature 2023
T. Dandekar, M. Kunz, *Bioinformatics*,
https://doi.org/10.1007/978-3-662-65036-3_13

## 13.1   The Different Languages and Codes in a Cell

First, at the molecular level, we have the DNA code studied in Sect. 3.1, but of course then the next language is RNA transcription. Interestingly, the signal to transcribe or not this very RNA is encoded quite complexly in the DNA structure of the promoter upstream of the transcription start of the RNA and further "upstream" (5'-end to be precise) binding transcription factors. The next step, the translation of messenger RNA into proteins, called translation, also uses, after all, its own language, the universal genetic code, in which the 64 possible nucleotide triplets encode the 20 ("proteinogenic") amino acids (including dialects that assemble proteins in organelles such as mitochondria, for example, or in bacteria such as mycoplasmas; freely available from NCBI at the following link: https://www.ncbi.nlm.nih.gov/protein. This can be used to determine any protein sequence in any biological dialect).

For example, we can start from a normal human body cell (Fig. 13.1). Then there is first the genetic information in the cell nucleus, in the DNA. At first glance, this information is translated directly into protein sequences via the genetic code (Fig. 12.7).

Of course, we already know from the previous chapters that it is not quite that simple, there are really two steps:

DNA → RNA as well as RNA (in the cell nucleus) → mRNA at the ribosome. Only then is the mRNA translated into amino acids according to Fig. 12.7 and the **genetic code.**

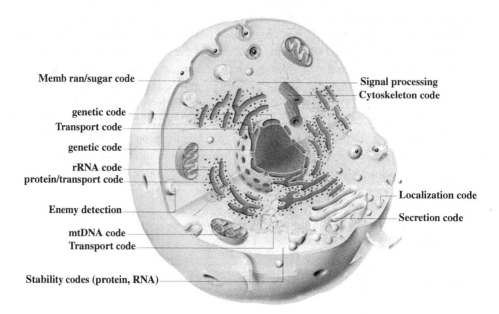

**Fig. 13.1** Different languages and "codes"in our body cells, e.g. regulatory RNA, translation, 3-D protein structure, stability and instability sequences in protein. (Figure modified from https://commons.wikimedia.org/wiki/File:Aufbau_einer_Tierischen_Zelle.jpg)

However, in order for these two other processes to take place in the cell in a regulated and regulated manner, they again have their own language or code. Figure 13.1 summarizes these different codes a little.

**Transcription Code**
The first code, transcription, determines when and how intensively a gene is read, in particular on the basis of promoter sequences.

Transcription factor binding sites are encoded by short nucleotide sequences, several of which act together to regulate readout in nucleated cells.

Some programs that can examine a promoter in more detail, such as TESS and Genomatix, have already been introduced in Chap. 11. However, there are also other databases, such as TRANSFAC (https://www.gene-regulation.com/pub/databases.html), MotifMap (https://motifmap.igb.uci.edu/) and JASPAR (https://jaspar.genereg.net/). Some of these are publicly available for reading and searching transcription factor binding sites. However, some have now become commercial and are no longer free to use.

But the closer one looks, the more unclear the transcription code is, in particular which transcription factors that are still unidentified must also be taken into account, but also more distant sequences that lead to increased ( *"enhancer"* ) or decreased transcription ( *"silencer"* ).

**RNA Codes**
However, the next step, the processing and *splicing* of precursor RNA, also follows its own codes. Here, the splicing sequences that distinguish between intron and exon have already been relatively well characterized. But it turns out that each organism has its own dialect for deciding what to splice and how. A good program that is adaptive and species-specific in predicting such sequences is the Augustus program (https://bioinf.uni-greifswald.de/augustus/). It can be specially trained on new species and uses hidden Markov models for prediction (Stanke et al. 2008).

However, one can look at numerous other codes in the RNA, in particular sequences that decide whether the mRNA leaves the nucleus or not (in the case of mRNA in general only one modified nucleotide, the 7-methylguanosine cap) and numerous other sequences that regulate the translation, localisation as well as stability of the RNA (see first part; a standard program to read these codes is the RNAAnalyzer: https://rnaanalyzer.bioapps.biozentrum.uni-wuerzburg.de).

**Protein Codes**
Once the protein has been translated according to the genetic code, the question arises as to whether it is modified post-translationally, i.e. whether sugar residues (e.g. aspartic acid residues), lipids or acetate groups (e.g. lysine residues) are added to individual amino acids.

Next, based on its sequence, the protein folds into a *molten* globule ( *"molten globule"* ) usually within milliseconds after its synthesis via the formation of a secondary structure, and then (seconds) it arranges itself into its final three-dimensional structure. This complex 3-D code has not yet been "cracked" either. Neither are the biophysical codes known in detail, nor do we have powerful enough computers to predict the structure accurately. In contrast, methods that first find out whether the protein sequence is similar enough to a known structure and then predict the 3-D structure after "copying" it are surprisingly powerful due to the sheer size of the data (tens of thousands of known protein structures with their x-y-z coordinates in the protein database).

Then, when the protein is completed to that point, its stability is determined by different amino acid codons at the 3'-terminus. For example, there are specific instability sequences at the C-terminus of the protein that determine its stability.

In general, it can be said that bioinformatics for deciphering these codes practically always starts with the sequence, but then uses other features, especially the structure, but for RNA, for example, the energy. For proteins, protein structure prediction is still computationally time intensive and very difficult for new folding types. Also, the decoding of transcription, DNA control sequences, and even new types of RNA (e.g., for lncRNAs (long non-coding RNAs) and miRNAs (microRNAs), one must correctly predict their *targets*) are only partially understood. On the other hand, increasingly complete large datasets of total transcription from a wide variety of cell types are available, gradually supplemented by proteome datasets and metabolite data.

## 13.2   New Molecular, Cellular and Intercellular Levels and Types of Language Are Emerging All the Time

The exciting thing, however, is that these types of languages are only the beginning. For example, at the molecular level there is also a sugar code (glycosylations and these sugar residue-binding proteins, so-called lectins), which regulates, among other things, which cells come together to form tissue associations and, for example, are simply ignored by the affected cancer cells when metastases form. There are also other codons for cell-cell communication (lipids, desmosomes and so on), until we finally arrive at one of the most complex systems of all, the immune system, which in each of us performs the task of reliably distinguishing between self and foreign. There is already a great deal of data on the immune system, for example on the white blood cells, where we can distinguish between lymphocytes (antibody-producing B cells and directly defending T cells; the latter are subdivided into helper cells, native killer cells and CD8 T cells and then into ever new subtypes), and on other defence cells, in particular monocytes, dendritic cells and macrophages. But that's the beginning. The immunologist and immunologist distinguish very fine subtypes depending on the surface receptors that white blood cells have and their specific subfunction. In addition, there are platelets that also support the immune response. We study these cell types intensively and find that for each of these defense cells, again, you can make a separate systems biology model. The language diversity and complex coding of the various immune responses are only surpassed in complexity by our nervous system. Both systems have only been deciphered in their various codes and language levels in rough outline. So there are still many open questions and exciting secrets that still want to be deciphered.

In evolutionary terms, the different levels of the languages of life can be simplified as shown in the box: Starting from preforms of life (about 3.3 billion years ago), as is still the case today with the virus, there is first a hypercycle between nucleotide genome and poly-merase. Then soon (about 2.8 billion years ago) the first prokaryotic organisms form, the typical bacterium then already has the three classical "language levels" of DNA → RNA → protein. Incidentally, we can only use the protein sequences to bioinformatically identify the *last common* ancestor of all life (LCA), which lived about 2.5 billion years ago. The earlier alternative forms are no longer preserved as sequences.

In the eukaryotic unicellular organism (about 2 billion years ago; at least brewer's yeast, baker's yeast and our ancestors separated in different directions at that time), everything is already considerably more complex (see Chap. 10). DNA is translated according to different codes. But also the RNA code is now already quite complexly divided into different lan-guage levels (precursor, splicing, export, translation, localization, stability). Proteins, too, are now already quite complexly coded and regulated (translation, modification, stability).

In multicellular organisms, we then already have a very high number of language levels, first of all, all those from the previous step within the cell. Then there are various external cell codes for communication with other cells (sugar, lipid codes). From this, tissues are formed, which then again establish new language levels and codes (tissue codes, locomotor system, immune, nervous, circulatory, digestive system) and finally together form an organism.

**Evolutionary View of the Language Levels of Living Systems**
**Virus**
- Nucleotide genome ↔ polymerase.

**Bacteria**
- DNA → RNA → Protein

**Eukaryotic protozoa**
- DNA → RNA (precursor, splicing, export, translation, localization, stability)
→ Protein (translation, modification, stability)

**Multicellular**
- Cell (sugar, lipid codes) →
Tissue (tissue codes, musculoskeletal, immune, nervous, circulatory, digestive system)
→ Organism

**Social community**
- Cell (sugar, lipid codes) →
Tissue (tissue codes, musculoskeletal, immune, nervous, circulatory, digestive system)
→ Organism (innate triggering mechanisms, body language, gestures, spoken lan-guages, etc.)

As already mentioned in Chap. 5 on systems biology, another astonishing basic property of life is to build up ever higher levels of regulation and thus of biological languages (see box). If you look closely, this principle is already indicated in Sect. 3.1. The ever higher levels allow us to adapt to the environment in an ever better and more far-sighted way. First jumps to a language level between cells are thus made possible by the sugar code, this then helps neurons, among others, to come together to form different brain tissues and thus to form a brain. The neuronal languages have been described quite well in many details, starting with chemical synapses (e.g. in our brain), electrical synapses (e.g. in insects) and the fast electrical conduction via ion channels along the long sprouts of the nerve cells via axons. However, as can easily be seen in well-studied brains of, say, humans, mice, ants and bees, it becomes very complex and unmanageable in the details. In each brain region, there are different mixtures of nerve cells, different glial cells appear, the ion channels vary (slow and fast, activating and inhibiting) and since also, really mysterious (see below) new processes become possible, such as our different types of memory.

At the next higher level, languages are used for communication between organisms - in other words, what we call language in everyday life. This requires a sufficiently complex brain. However, a few million neurons are sufficient in state-forming insects to use innate languages across the entire state and also to communicate new observations with this innate vocabulary, e.g. the already proverbial bee dance for honey sources.

We can thus finally attempt to decode these different language levels using the optimal bioinformatics methods in each case. Figure 13.2 compares important biological language levels and possible bioinformatics analysis options.

Human language is once again a significant step further developed, since it is newly learned by a sufficiently large brain, has a very broad vocabulary and also allows the

**Flow of genetic information**

| Level | Coding | Analysis option |
|---|---|---|
| DNA | 2-bit code | ATCG → Genome |
| ↓ | ↓ | ↓        Analysis |
| genetic code | Triplet code | 64 -> 20 amino acids → transcriptome analysis |
| ↓ | ↓ | ↓ |
| Protein (3D code) | Amino acid sequence | 3D structure, Interactions →Proteome analysis |

**Numerous other codes**

| | | |
|---|---|---|
| RNA code | Sequence, Structure, energy | IRE: CAGUGN, Stem-loop, -2.1 to -6.7 kcal/mol→RNA-/ functional analysis. |
| Sugar, lipid, cell-cell code | Membranes are create at membranes | Structure. Interaction partner → Functional analysis |

**Fig. 13.2** Examples of the levels of coding and possible bioinformatic analysis possibilities

formation of ever new concepts. The levels of abstraction that humans can manage are also far higher than in animals. It is true that in individual aspects other highly developed brains, especially in mammals and birds, can keep up here - but in the integrative overall performance definitely not.

## 13.3    Innovation: Synthetic Biology

Building on this background, one can of course also see our human society, including technology, as the field par excellence in which ever new levels of language are also created using technical aids. Strictly speaking, this is the reason why bioinformatics is possible at all. We are now consciously learning the molecular languages of life. But because this would otherwise be too complex for us, we use our own tool, the computer, to decode them and thus achieve an unprecedented direct link between these different languages. Interestingly, however, this can be applied even more strongly to biology.

In this chapter, we will first consider the new levels that technical communication brings, with both the computer and the Internet of course being particularly impressive examples of information processing, both of which are essential to bioinformatics.

In the meantime, however, bioinformatics has brought to light some astonishing cross-links between molecular information processing and computers. The technical use of biological processes is generally referred to as *synthetic biology*, a field of biology that is currently growing rapidly.

The focus is on achieving something new, on solving technical problems much better and innovatively by combining (molecular) biology and technology (usually computer technology, nanotechnology, modern chemistry or molecular sciences). It used to be innovative to use new organisms for biotechnology (since the 1980s, patenting of molecular cloning of genes in plasmid rings by Prof. Cohen, Stanford University (Stanford Universität)). This is indeed still being pursued and advanced. However, it has long been recognized that this is very useful for the production of substances (e.g. insulin, citric acid, antibiotics, etc.), but to see "new kinds of life" at the center is nonsensical. We do not have sufficient knowledge to achieve anything useful here, nor would we be able to ensure that there are not undesirable side effects on ecosystems or control of these new life forms. Even in the design of synthetic organisms, therefore, special attention is paid nowadays to these safety aspects, which are also easy to comply with in practice. For example, erythropoietin (a very useful hormone that stimulates blood formation and, for example, helps the sick person with severe kidney disease to be able to produce enough red blood cells) is produced in proprietary fermenters that provide the optimum temperature and medium for bacterial production, and the bacteria are not viable outside this environment. The focus of today's synthetic biology is thus the improved solution to a technical problem by merging different areas of technology with biology.

A first area we will look at is *"natural computing"*, i.e. computing with molecules or even with whole living organisms. Although this is in principle very unconventional, it promises to be superior to conventional computers for certain problems, at least in the future, since many molecules or organisms work on the computational problem in parallel and thus bring insights to light more quickly than a computer.

Next, we will use the nanocellulose chip developed in our own laboratory (Bencurova et al., 2022) to show how the next generation of computer chips could also function much better than at present by using biomolecules, in particular more environmentally friendly, more durable, faster and with better memory properties.

However, this is at the same time an illustrative example of synthetic biology, so that we will subsequently give an overview of other approaches from synthetic biology. It is important to keep in mind that the entire theory of design in synthetic biology is based on bioinformatics. This is because it is the only way to know which biomolecules should be assembled in what way, to know their properties and also to be able to use simulations, database searches and calculations to estimate which properties come into play in the technical problem so that a technically satisfactory and correct solution can also be achieved.

## 13.4  New Levels of Communication Through Technology

We have special features due to human civilization. In particular, humans develop devices for tool making, for example. We have culture and lore, mirror neurons and imitative instinct. Humans describe an emergent loop in that new inventions and forms of expression ("languages") lead to ever new inventions and forms of expression. Money and the general possibility of exchanging any commodity for another also accelerate innovation, creativity, and development (money as a "desire machine"; Ridley 2010). In this sense, steps in which a new transmission of information is achieved are always important for the advancement and continuation of our civilization (see box). The development of everyday human language into high-level, technical and scientific language, for example, is exciting.

However, new levels of data storage followed, first through writing (handwriting), then through letterpress, newspaper and typewriter. Finally, however, through electrical current, such as Morse code, telephone and telefax. Then, after the Second World War, transistors, *integrated chips* and finally the modern computer were added via electron tubes.

**New Levels of Human and Technical Communication**

- **Starting points** in animals: chemical communication, later sign language, vocalizations.
- Human **language.**
- First realizations of **writing** (bone carvings, cuneiform, papyrus, paper).
- Permanent storage realization of the writing: book, archive, microfiche, glass (LASER scratched).
- On the other hand, unstable: CD, DVD, stick (already unreliable after about 10 years).
- Technical realizations of **type** (letterpress, screen printing, rotary printing, 3-D printer ...).**Electronic communication**
- Power generator, Morse code, telephone, fax...
- Power grid, electronic tubes, first transistor ...
- First computer: Zuse, Enigma, Colossus (cracks Enigma)**Generations of semi-conductor technology**

First *integrated chip,* CMOS technology, X-ray lithography ...**New versions of the Internet**

- Internet: Conceptualization of the Internet as super-robust communications despite nuclear war by DARPA and Paul Brennan (1973).
- Minitel in France (early 1980s).
- Use of the Internet as WWW at CERN (1987).
- *World Wide Web Worm* (1990).
- Mosaic Browser (1994).
- Mozilla, Google, *cloud computing,* Android phones, *wearables, Internet of Things ...*

## 13.5   The Internet – A New Level of Communication

In the early 1990s, bioinformatics was still a matter for specialists. Programs were sent to acquaintances by mail, one installed the program or even received protein coordinates specifically by e-mail. Now, in the Internet age (since about 1995, local differences), anyone can use a program that is offered as software via the Internet. Also the data of the databases are available worldwide for every user. All computers are networked in such a way that via the Internet protocol information about data packets can safely reach the reader at the computer. For this purpose, a *Domain Name Server* (DNS) transcribes the Internet Protocol (IP) address into easily readable addresses. In particular, the Internet opens up the possibility of making bioinformatics software and biological knowledge and data, as well as scientific literature (PubMed), accessible to readers worldwide, especially via the *Open Access*   publications that can be found there.

Therefore, here are some key points about the Internet. After all, it is the basis of the worldwide availability of today's bioinformatics.

**The Internet Protocol**

This protocol is used to pass data packets from computer to computer, and to do so as quickly as possible. In order to maintain a communication option even in the event of atomic shocks, the Internet was originally theorized by DARPA in the early 1960s. It was only the intensive use at CERN around 1985 that led to the initial spark for worldwide distribution.

**Domain Name Server**

Each computer has a four-part number (e.g. 132.187.25.1), via which all computers that are open to the Internet are then connected to the Internet protocol. Here, 132 means Germany, 187 means the University of Würzburg (Universität Würzburg), 25 means a subnet, and the last number means a specific computer in the subnet. However, since people typically browse the Internet by name, the *Domain Name Server* (DNS) translates the names (e.g. www.ncbi.nlm.nih.gov) into the real IP addresses (i.e. the four-part numbers).

**Node Computer**

The information that is passed from computer to computer on the Internet is bundled at central points. These central computers are then called Internet nodes. Unfortunately, the node computers are not the property of all nations, but the most important ones are located in America and generally in individual countries. The allocation of domains and Internet names is also in the hands of an American company, ICAM. True, it would be nicer if the Internet belonged to everyone. But if it is already controlled by one country, America is always better than alternatives, such as authoritarian states, which then control information even more selectively than the USA.

The Internet has many emergent new properties, such as being super-resilient. Even when many computers are down, whether due to disasters, government oppression, or even war or even nuclear weapons, message packets continue to be transmitted efficiently through the Internet. Through intense communication, people are moving closer together to form the global village. In addition, the Internet has the property of not forgetting anything. Search engines, moreover, make it possible to keep track of this, but they themselves develop a life of their own, in particular which pages are offered to which user (i.e. which image of the world is reflected to the user).

https://www.icann.org: The Internet Corporation for Assigned Names and Numbers (ICANN) is a *non-profit organization*:

*Today, 1 October 2016, the contract between the Internet Corporation for Assigned Names and Numbers (ICANN) and the United States Department of Commerce National*

*Telecommunications and Information Administration (NTIA), to perform the Internet Assigned Numbers Authority (IANA) functions, has officially expired. This historic moment marks the transition of the coordination and management of the Internet's unique identifiers to the private-sector, a process that has been committed to and underway since 1998.*

## 13.6   A Parallel Language Level: Natural and Analogue Computation

Another exciting new area of bioinformatics uses natural processes to perform computations. An initial study by Prof. Adleman (1994) in the top journal "Science" showed that DNA molecules can be glued together like little sticks to perform addition, but also to solve simple traveling salesman problems (e.g., figuring out optimal routes for six cities so that the visit uses as little path length as possible) (Fig. 13.3, left, from Adleman 1994). More generally, natural computation uses the built-in parallelism of biomolecules, that is,

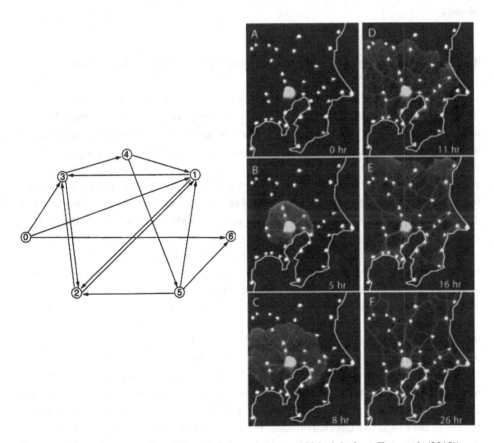

**Fig. 13.3**  Natural computation. (Figure left from Adleman 1994, right from Tero et al. (2010))

that many molecules simultaneously perform a large computation together, like a large computer cluster, but in molecular dimensions. Later studies showed that it is also possible to compute with RNA, proteins and cells. For example, one research group was able to logically switch RNA aptamers (i.e., RNA molecules that can recognize other molecules) (Nomura and Yokobayashi 2015). This then allowed the RNA switches to measure the concentrations of Valium and diazepam, compare them, and then perform a logical decision. Another very nice result was recently achieved with slime molds (Fig. 13.3, right, from Science News, January 22, 2010). Here, food was distributed to different locations, mimicking the exact position of stops on the Tokyo subway, the number of passengers by the amount of food. Then, by growing and collecting food, the slime mold manages quite accurately to recreate the optimal network of stops in Tokyo, as then shown by comparison with computer calculations and the real subway map (Tero et al. 2010).

Thus, "natural computing"(often also referred to as *analog computing*", when, as in the case of the slime mold, only an analog task is solved, but it is also calculated analogly and not digitally) has already achieved great success. However, the electronic computer is still used for difficult tasks. However, this may soon change due to advances in molecular biology.

## 13.7    Future Level of Communication: The Nanocellulose Chip

It is interesting to note that bioinformatics can be used to elaborate, but also to advance, the idea that a new level of language is within our grasp. However, it is essential that we also reach this level in concrete terms, so that the stability and security that the Internet already gives us for the transmission of information also applies to other aspects of our production, to computer technology and to everyday life. At the same time, the new technology is also sustainable, without electronic waste and will not be destroyed by radioactive radiation or nuclear weapons. Let us take a closer look at this example as a particularly gripping subject area for the new branch of bioinformatics (theory) and molecular biology (practice, experiments).

How can you know that and how is that possible? Well, it is a technique that uses DNA as a storage medium, polymerases to synthesize and read in sequences, and exonucleases to degrade and read out the sequences. Nanocellulose provides a matrix for the enzymes and DNA. Crucially, however, light-controlled protein domains allow the polymerases, exonucleases and other molecules to all be driven by light of different wavelengths. All of these building blocks already occur naturally in bacteria. The light-sensitive control domains are so-called blue light-sensitive protein domains (e.g., the BLUF domains in *Escherichia coli;* Tschowri et al. 2009), which in bacteria ensure that when blue light is incident, a stress response protects the bacterium against harsh environmental conditions and UV light. Similarly, polymerases and exonucleases from coli bacteria, for example, have long been used in a targeted manner. The fact that DNA enables a very high information storage density (1 g DNA stores up to 10 to the power of 18 bits, i.e. an exabyte) was already shown in 2012 by Church's research group in the USA (Church et al. 2012). A working group at the European Bioinformatics Institute (Goldman et al. 2013) then showed that it is possible to read longer texts, images and also sound recordings into DNA very well and also encode them securely with an appropriate error correction code. By considering the error codes and sequencing the DNA twice, all this information can also be read out very reliably. What's more, in 2015 the research group led by Prof. Stark at ETH Zurich showed that vitrification of the DNA makes it possible to store this information unchanged for up to a million years. Add in the fact that optical switches like the BLUF domain also switch very quickly (within femtoseconds), and the outlines of a technology that is much faster (a million times) and more durable (100,000 times longer) than our computer, and also has a much higher storage density (exabytes versus terabytes), become visible here. In the process, optical switches would replace transistors, DNA would replace memory disks, and the silicon matrix would be replaced by nanocellulose. Nature, especially bacteria, show us that this technology has been working smoothly in its components for billions of years, we just need to put them together efficiently.

This is what a finished nanocellulose chip could look like (Fig. 13.4a), which stores information for thousands of years, switches very quickly (femtoseconds, i.e. millions of times faster than today's computer chips, since light is used here instead of electronics for the switching processes) and also has an extremely high storage density (exabytes per gram of DNA; Church et al. 2012). It has already been shown that DNA is very good at storing information digitally, unit by unit of information (Church et al. 2012). This works for texts, phonograms and images alike, and for thousands of years (Goldman et al. 2013). The only important thing here is to sequence twice. In this way, one prevents, for example, that an image can only be seen blurred because errors have crept into individual image points ("pixels"). If the DNA is additionally conserved, for example by vitrification, this information can even last for millions of years (Grass et al. 2015). We are currently working (Fig. 13.4b; Dandekar et al. 2013) on an approach that rapidly reads in and reads out these storage capabilities of DNA (Goldman et al. 2013) using light-controlled polymerases and exonucleases (Dandekar 2013). Specifically, blue light-responsive (BLUF) domains turn on the enzymes; without blue light, they are turned off. Various nucleotides are also selected and incorporated in a light-controlled manner (rotating a histidine, which is important for substrate specificity) (Fig. 13.4c). Rapid readout by the exonucleases is most easily accomplished by using fluorescent nucleotides. The distribution of different DNA molecules to many small pots (storage sites) can be achieved using digital picoliter PCR (Hoffmann et al. 2012). Nanocellulose serves as the chip matrix or basis for this new computer chip (Jozala et al. 2016; Fig. 13.4a). Such an approach has many other advantages. In particular, all materials can be produced sustainably, and no electronic waste is generated. Moreover, unlike "microbes gone wild" ("*grey goo" syndrome:* living nanotechnology assimilates everything into nanorobots; Drexler 1986), such a technology can never take on a life of its own. The chip is dead and not a living thing, and all working molecules are directly directed and controlled by light. Pores (again light-controlled, but with specific sequences) in the nanocellulose membrane also yield electronic properties,

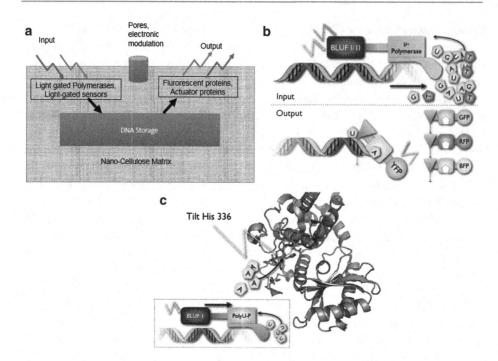

**Fig. 13.4** The nanocellulose chip. (**a**) Light replaces electronics (symbol: flashes, control by light signals) and does not control transistors but light-controlled polymerases (left, for reading in information) and exonucleases (right, for reading out). DNA is used for storage (blue rectangles in the middle). The results are read out via fluorescence or protein expression ("printer"). The nanocellulose matrix is responsible for durability and electronic modulation (pore; blue, top centre). (**b**) Light instead of electronics is used in various other processes in semiconductor technology, but special and new in the nanocellulose chip are proteins controlled with light. These can theoretically switch much faster than transistors (femtoseconds), which can be used to specifically synthesize RNA by fusing a light-control domain to an RNA polymerase that does not need a template (so-called mu polymerases; green) (BLUF domain, blue). **c** shows that the substrate pocket can be changed via light (tilting of a histidine residue) in such a way that an adenine is now incorporated instead of a uracil. Such changes can be mediated very quickly via light (petahertz chip). In addition, the high storage density (exabytes per gram of DNA) and long shelf life (thousands or millions of years) is a great advantage, as is the absence of scrap or other environmental damage, and the extreme robustness of the nanocellulose chip (even against various types of radiation or the electromagnetic pulse, as well as all computer viruses). We can also take advantage of the fact that digital picoliter PCR can already store a lot of different information in a very small space

so the new chip can be well interconnected with other conventional computer chips. One can go even further, and use self-assembled DNA and DNA conducting pathways (previously made light-controlled by the enzymes) and also make the nanocellulose itself semiconducting by iodine doping, and then even use single-electron transistors made of nanocellulose (Bencurova et al., 2022) - at least all of this can be bioinformatically modeled very convincingly and in principle reproduced in laboratory experiments. Finally, it can be shown that such a nanocellulose chip can also be easily manufactured in principle with the aid of a 3-D printer.

## 13.8   Using the Language of Life Technically with the Help of Synthetic Biology

These are all promising properties for our new nanocellulose chips. But it is not at all important to produce this new generation of bio-enhanced computer chips exactly the same way, but the great direction should be explored as intensively as possible in research to solve current problems of computer technology and electronics in a very sustainable, environmentally friendly, flexible, faster and better way than before. A current, simpler example than the nanocellulose chip shown above with new optogenetically switched enzymes and DNA as the storage medium is to achieve sustainable electronics by continuing to use commercially available electronic components and memory chips, but printing them thinly on nanocellulose paper in a much more environmentally friendly way than before (but still using electronic waste as opposed to above) (Jung et al. 2015). Another interesting marriage of electronics and proteins is electrically modifiable proteins (Ganesan et al. 2016; Hekstra et al. 2016). Even more generally, this is called synthetic biology, which we have already learned about. Biological molecules are recombined, allowing them to achieve new, technically desirable properties. As mentioned above, it can be problematic to create new organisms with new properties, since such organisms are capable of reproduction. On the one hand, beer as well as bread and cheese have been produced with biotechnologically bred organisms for centuries, i.e. with the help of organisms systematically genetically modified through breeding. But since this has been going on for centuries, it is perceived by the population as "natural" (a bit irrational). On the other hand, however, the potential danger posed by radically new synthetic organisms (such as new viruses, fusion of very different cell types, etc.) is already significantly higher than that posed by centuries-old biotechnology. However, since typical synthetic biology processes focus on design rather than on a whole organism, there are of course ways and means of keeping the risks within limits. The safest way is simply to use parts of an organism. These can then achieve new properties, but are not themselves capable of reproduction. One can also deliberately incorporate further control steps (as explained above with the BLUF domains).

Some useful links on the topic of synthetic biology are presented again below (see box). This shows above all that there are many efforts in this field that are innovative and often already relatively successful.

If, on the other hand, one turns to the design of individual proteins, the necessary bioinformatic techniques and the experiments coupled to them are called protein design.

Figure 13.2 shows how the work on individual building blocks and proteins is designed, which bioinformatic steps are important here so that desired properties can be achieved. After analyzing the function and domains in the protein, one tries to determine its structure and then finally to further develop the protein specifically in the direction of desired properties. This fascinating interdisciplinary field, in which bioinformatics is not an auxiliary science but helps to plan and design new technology like an engineer, is called protein design for individual proteins and synthetic biology for new, unprecedented processes. One has to be particularly careful with whole organisms. On the other hand, biotechnologically used organisms have been part of our culture for centuries, from beer and baker's yeast to citric acid from aspergilli and insulin from bacteria. There is also interesting work extending the language of life itself, such as introducing two additional entirely new nucleotides into the genome and passing them on stably (Zhang et al. 2017), or even fitting in new amino acids via recoding of the genetic code (via matching tRNAs or an adapted ribosome). This allows the incorporation of entirely new substances into DNA or proteins and thus even synthetic chemistry in microorganisms for biotechnology and drug production.

So much for future technologies and efforts in the field of synthetic biology.

Why should we develop synthetic biology so intensely? Well, for one thing, to achieve technological progress. Nanotechnology, molecular biology and electronics are our future technologies, and if we can make electronics much faster with optical methods, we should strive for this. The steps towards this are already showing great progress, such as picoliter computer PCR, which would make it possible to place a million or so different DNA molecules on a slide and thus greatly speed up vaccine production, for example. On the other hand, such efforts have the general advantage of simultaneously merging information storage, cellular programming, and also synthesis and microfabrication. Precisely this also results in a very robust and very environmentally friendly way of producing, as bacteria and blue-green algae have been demonstrating to us for billions of years. The introduction of light-controlled protein switches, however, makes it possible to switch each molecule on and off in a very targeted manner and thus also to achieve a previously unattained precision of synthesis and information processing. In particular, the construction principle prevents the technology from taking on a life of its own, something that was not considered in the 1980s and 1990s when nanotechnology was propagated with living bacteria. On the other hand, our current technology is not very robustly designed, always has to contend with raw material problems (today's electronics, for example, have a shortage of rare earths), produces dangerous waste (electronic waste) and is very susceptible to disruptions, interruptions in world trade and, even more so, to catastrophes or armed conflicts. Reason enough, therefore, to intensively pursue this molecular technology with the help of bioinformatics, which has combined three particularly strong exponents of nanotechnology here for illustration (own proposal: DNA, nanocellulose and light-controlled protein domains; Dandekar 2013), but has also already achieved very considerable success with other biomolecules.

## Synthetic Biology: Important Links and Applications for Bioinformatics

**GoSynthetic Database**
https://gosyn.bioapps.biozentrum.uni-wuerzburg.de/index.php

Previous work of my own has led to this database, which examines various biological circuits and the resulting engineering properties.

**MIT BioBricks**
https://biobricks.org/

At MIT, important molecular circuits and their descriptions are bundled and made available to the general public as a *"repository"*, i.e. a collection of construction manuals.

**Rep-Repro/Darwin 3-D Printer**
**Rep-Repro/Darwin 3-D Printer**: https://bigrep.com/de/

There is also a race for the 3D printers to get and program better and better instructions for such three-dimensional printer templates. It is also possible to print various cells and tissues. Finally, people are also trying to make a 3-D printer themselves with a 3-D printer. I'm somewhat skeptical of the latter enterprise, but it's amusing to read (because of the danger of overly active self-reproducing machines, but also because of fundamental limitations of this approach, e.g. plastic remains plastic, other components are missing).

**"Eternal" (Thousands or Millions of Years) Permanent Storage via DNA**
**DNA storage:** https://www.3sat.de/wissen/nano/dna-als-datenspeicher-100.html

The possibilities of DNA as an extremely good and very compact digital storage for information are explained in the film. The concept was first demonstrated using *Next Generation Sequencing* by Church et al. (2012). Ultra-long storage was published by Grass et al. (2015), and the encoding of images, sounds, and texts was analyzed by Goldman et al. (2013). Thus, image files can be effortlessly determined from long DNA sequences using *double* sequencing and *next generation sequencing*.

## Emerging Technology Competition

**Emerging Technology Competition of the Royal Society**
https://www.rsc.org/competitions/emerging-technologies/

This is a Europe-wide competition for new technologies, in which our approach to DNA storage using light-guided polymerases and exonucleases was also presented in the final in 2015 (Dandekar and Lopez 2015; see Dandekar et al. 2013). The Emerging Technology Competition also pits a number of other fascinating new approaches against each other each year.

This is a competition on synthetic biology to use new molecular circuits to trigger new technically helpful developments. Harvard and MIT's global competition is deliberately aimed at students who want to advance molecular design and synthetic biology with new ideas.

**Active DNA Storage by Light-Controlled Proteins**
https://register.dpma.de/DPMAregister/pat/PatSchrifteneinsicht?docId=DE102013004584A1

Helpfully, Dandekar (2015, 2016) and Shityakov et al. (2019) describe all the details and our current experiments on the nanocellulose chip for replication.

**Synthetic Biology Competition**
https://igem.org/Main_Page

### Analysis of the Function and Domains in the Protein

It is important not only to consider the function of the domains of the natural protein, but also to look for suitable domains for the new functions that the protein is to have. For example, a BLUF domain is used for control by light, and a kinase domain (for phosphorylation), a cofactor-binding domain or a DNA-binding domain (so that transcription can be specifically activated) is used for other new enzyme properties.

The insertion of individual sequence sections can also be pretested in this way.

https://www.uniprot.org/

https://smart.embl-heidelberg.de/

**Protein Structure Analysis**

It is not uncommon that the basic type of protein structure is already known. Then you can get the three-dimensional structure coordinates from the protein database and then use a protein visualization program to insert the mutations that you would like to plan.

We have already got to know RasMol as a visualization program and the PDB database for the three-dimensional protein structure coordinates.

For proteins that have similarity to a protein sequence, we can make a homology model (see Sect. 1.2), for example with SWISS-MODEL or MODELLER. Again, remember to be conscientious about comparing original sequence and modification.

**Prediction of the New Protein Properties**

Next, one considers how the protein then changes overall. Various methods for mutation prediction exist, e.g. large protein alignments of related proteins enable the comparison of correlated mutations to identify even distant regions that influence each other in cooperative structural changes (e.g. working group of Prof. Ranganathan). This method is described in detail in Poelwijk et al. (2016). A case study investigates how allostery and protein structures can be specifically changed using this method (Raman et al. 2016).

A nice, simple start-up software package to plan molecular biology experiments is the Geneious software or the Husar software at DKFZ. Both software packages allow a lot of molecular biology for new proteins and the individual steps for this: the planning and recognition of DNA interfaces for cloning experiments, the prediction of secondary structure, amino acid composition, optimal protein expression, translation of nucleotide sequences, protein modification, etc. (we have already used the Expert Analysis System as a general introduction). (As a general introduction, we already got to know the Expert Protein Analysis System (ExPASy) from Switzerland in Part I).

https://www.geneious.com/

https://genius.embnet.dkfz-heidelberg.de/menu/biounit/tools_db.shtml

More ambitious protein design experiments are building on this. The current world leader is David Baker's lab at the Institute for Protein Design (Institut für Protein-Design) at the University of Washington (Universität Washington; https://www.ipd.uw.edu). A nice introduction is the game "Fold it" (first link below), the second link below gives a complete overview of modern protein design.

https://fold.it/portal/info/science

https://www.bakerlab.org

One step higher, one would like to link different protein components with each other to form new networks and circuits. This is made possible by our GoSynthetic Database, which then also directly compares natural networks with technical processes.

https://gosyn.bioapps.biozentrum.uni-wuerzburg.de

Of course, one can also look the other way around, that a molecule works well as a drug to influence a receptor, or a protein is itself used as a drug (e.g. the tissue plasmin activator, TPA, after effect extension by removing an inactivation loop). This type of *protein engineering,* which aims at a *drug* effect, is therefore called *drug design.* The first step is to look in detail at the three-dimensional protein structure in order to mark promising changes in the protein or to see how well a drug (or, more generally, a molecular ligand) fits into the protein structure. We can practice this with programs like RasMol or PyMOL by using this software to turn the three-dimensional structural coordinates into a colorful, spatial image of the protein structure and look at how well ligand molecules fit in here (Fig. 1.2). On the other hand, if you want to test a large number of molecules, you can do this automatically using a substance database with the computer. In this way, thousands or even millions of compounds are pretested in the computer (technical term *in silico* screening) in order to then test the best molecules in the screen (those with the lowest binding energy) pharmacologically in the experiment. Other criteria (e.g. whether the substance is easy to synthesise) are usually taken into account when selecting substances. Such an *in silico* screen is bioinformatically a work of several weeks or longer. In addition, molecular dynamics can be considered: The computer starts from the protein and ligand structure and now systematically samples how the interaction between the two changes over time. This too now allows accurate predictions of the effect and how it is best achieved on the protein by the ligand molecule. However, we can recreate such complex calculations in an introductory way by querying a database where many such results are systematically stored. The DrumPID database (https://drumpid.bioapps.biozentrum.uni-wuerzburg.de/compounds/index.php), for example, thus gives numerous substance suggestions for a protein for which one is looking for pharmaceuticals and also takes into account filter criteria such as tolerability (e.g. *Lipinski's rule of five*) and whether already approved by the FDA or experimental (Kunz et al. 2016; a tutorial here also explains its use in more detail).

**Conclusion**

- Life is always inventing new levels of language. Starting with the cell with molecular codes, higher levels are cellular and intercellular codes, then neurobiological via nerve cells, and at the level of individuals, scent signals, behaviour, gestures and language. In the case of humans, this is followed by the rapidly developing levels of technical communication – up to and including the Internet. In particular, the Internet opens up the possibility of making bioinformatics software and biological knowledge (PubMed, *open-access* publications) accessible worldwide. All computers are networked in such a way that information via data packets securely reaches the readers at the computer via the Internet protocol. To this end, a *Domain Name Server* (DNS) transcribes the Internet Protocol (IP) address into easily readable addresses.

- Synthetic biology profits from this world-wide knowledge and attempts to describe and understand biological processes so well that they can be used for technical applications, such as biotechnology (microorganisms produce citric acid, erythropoietin or insulin). However, numerous circuits and parts from cells that are interesting in their effects are now also being used (MIT parts list, BioBricks, iGEM competition). Bioinformatics is crucial to describe and directionally modify these parts and processes, for example through database tools (*work benches*) such as the GoSynthetic database and MIT BioBricks. Synthetic biology can be quite fruitful this way. In contrast, our knowledge of "artificial life" is too limited, and if one really wants to produce artificial organisms (e.g., modified viruses), it is important here to have sufficient, strong safeguards against release as well as built-in controls on the organisms (genetic engineering laws and regulations). The structure of individual proteins is optimised by protein design. This, too, can now show a number of successes (e.g. removal of a loop region in the *tissue plasminogen activator* leads to an extension of the effect). *Drug design* using *in silico screening* and molecular dynamics simulations also significantly shortens the development of drugs because only the best compounds then need to be tested experimentally in a time-consuming manner.

- Natural and analog computing uses biological or even physical processes to perform complex calculations by having many molecules working in parallel. This allows, for example, the Tokyo subway map to be efficiently reproduced using slime moulds. Nevertheless, no convincing application of such techniques has yet succeeded in being superior to a normal computer made of silicon chips. The nanocellulose chip, on the other hand, is potentially superior to today's computer chips. It uses DNA to store information and, via a BLUF or LOV domain, light-controlled polymerases and exonucleases to read in and out the stored information. Further modulating proteins and membrane pores are used for electronic signals via the nanocellulose membrane. This promises higher storage density (exabytes), longer storage (millennia or more) and faster switching (by light, up to petahertz) than conventional silicon chips. But more generally, the combination of molecular biology, nanotechnology and modern electronics offers huge future technological potential.

## 13.9  Exercises for Chap. 13

In addition to this part, please work on the exercises in Chaps. 10 and 11.

**Task 13.1**
Give examples of domain databases and find them on the web.

**Task 13.2**
Where do I find protein structures? Which database do I use? Get an overview of the wealth of forms. Also find out about SCOP and CATH.

**Task 13.3**
Protein design: locate artificial *folds* in the PDB database.

**Task 13.4**
Can you show the *Tissue Plasminogen Activator* and the engineering of the *loop* structure with RasMol to understand the design?

**Task 13.5**
How does the inhibition of the HIV protease actually work? Please refer to the following figure: https://www.hiv.lanl.gov/content/sequence/STRUCTURE/PROTEASE.HTML
    1A30: Biochemistry. 1998 Feb 24;37(8):2105–2110.
    HIV-1 protease complexed with tripeptide inhibitor from HIV-1 trans-frame region.
    Now describe exactly what you see, that is, how inhibition works.

**Task 13.6**
Protein helix permutations: Might there be general design principles for proteins? Search for relevant papers by David Baker in Nature or Science in PubMed.

**Task 13.7**

1.  Use the GoSynthetic database. How would one find the GoSynthetic database on the net?
2.  *Oncolytic virus*

Oncolytic viruses are a fine example of successful synthetic biology. The idea is that the cancer-dissolving virus multiplies preferentially in cancer cells, dissolving them ("oncolysis") while leaving the healthy cells largely alone. The immune system then removes these viruses from the recovered body.

Thus, in order to convert a normal virus into such an oncolytic virus, the natural virus must be modified in such a way that it preferentially replicates in cancer cells.

Find more information about this.

## Task 13.8

Using the DrumPID database: How do I actually find out whether a protein is hit by a drug or proteins that match a drug effect? This can be investigated using databases such as DrumPID (Uni Würzburg) or STITCH (EMBL, Heidelberg). Find these databases on the net.

## Task 13.9

Identify papers on *natural computing* from PubMed.

## Task 13.10

A light-targeted protein domain gives a protein new properties. How do you go about designing it?

| Useful Tools and Web Links | |
| --- | --- |
| NCBI Domains/ Structures | https://www.ncbi.nlm.nih.gov/guide/domains-structures/ |
| NCBI proteins | https://www.ncbi.nlm.nih.gov/protein |
| TESS | https://www.cbil.upenn.edu/tess/ |
| Genomatix | https://www.genomatix.de/ |
| TRANSFAC | https://www.gene-regulation.com/pub/databases.html |
| MotifMap | https://motifmap.igb.uci.edu/ |
| RNAAnalyzer | https://rnaanalyzer.bioapps.biozentrum.uni-wuerzburg.de |
| ICANN | https://www.icann.org |
| GoSynthetic | https://gosyn.bioapps.biozentrum.uni-wuerzburg.de/index.php |
| DrumPID | https://drumpid.bioapps.biozentrum.uni-wuerzburg.de/ compounds/index.php |
| BioBricks | https://biobricks.org/ |
| UniProt | https://www.uniprot.org/ |
| SMART | https://smart.embl-heidelberg.de/ |
| RasMol | https://www.openrasmol.org/ |
| PDB | https://www.rcsb.org/pdb/home/home.do |
| JASPAR | https://jaspar.genereg.net/ |

## Literature

Adleman LM (1994) Molecular computation of solutions to combinatorial problems. Science 266(5187):1021–1024. (PubMed PMID: 7973651 *Great first work, showing that you can solve NP problems like the traveling salesman for six cities by gluing DNA rods together [ligation reactions]. But after that there was a break in "natural computing".*)

Bencurova E, Shityakov S, Schaack D, Kaltdorf M, Sarukhanyan E, Hilgarth A, Rath C, Montenegro S, Roth G, Lopez D, Dandekar T. (2022) Nanocellulose Composites as Smart Devices With Chassis, Light-Directed DNA Storage, Engineered Electronic Properties, and Chip Integration. Front Bioeng Biotechnol. 10:869111. https://doi.org/10.3389/fbioe.2022.869111.

Church GM, Gao Y, Kosuri S (2012) Next-generation digital information storage in DNA. Science 337(6102):1628. (Epub 2012 Aug 16. PubMed PMID: 22903519 * First work to show the great potential of DNA as storage: exabytes per gram of DNA are possible to store)

Dandekar T (2013) Molekulare hoch integrierte Datenspeicherung über aktiv gesteuerte DNA. DPA 102013 004584.3 (Offenlegungsschrift DPA DE 102014 005549 A1 2014.10.23 C12Q 1/02 * This work explains in more detail how light-guided polymerases and exonucleases can be used to read in or out DNA.)

Dandekar T (2015) Intelligente Nanozellulosefolie für verbesserte Chipkarten. DPA vom 27.04.2015 Aktenzeichen DE 102015 005307.8

Dandekar T (2016) Modified bacterial nanocellulose and its uses in chip cards and medicine PCT U30719WO (published 3rd Nov 2016 * All details and our current experiments on the nanocellulose chip are described here for replication. However, there is still some development work to be done [a "proof of concept" is there, but a real prototype still needs time])

Dandekar T, Lopez D (2015) Programmable bacterial membranes with active DNA storage. Emerging Technology Finalist presentation, Royal Society for Chemistry, University of Würzburg, London (29.6.2015)

Dandekar T, Lopez D, Schaack D (2013) Active DNA storage is essential. Nature 494:80. (Comment [posted 17.4.13] reviewed and recommended by the Nature Editor on: Goldman N, Bertone P, Chen S, Dessimoz C, LeProust EM, Sipos B, Birney E. Towards high-capacity, low-maintenance information storage in synthesized DNA. Nature 494:77–80)

Drexler KE (1986) Engines of creation: the coming era of nanotechnology. Doubleday (0-385-19973-2 * This work draws attention well to the high potential, but also to potential dangers of nanotechnology. In particular, the design of a nanomachine should exclude self-sufficiency or autonomy from the outset [no "grey goose syndrome"])

Ganesan P, Ranganathan R, Chi Y et al (2016) Functional pyrimidine-based thermally activated delay fluorescence emitters: photophysics, mechanochromism and fabrication of organic light-emitting diodes. Chemistry 28. https://doi.org/10.1002/chem.201604883. ([Epub ahead of print] PubMed PMID: 28028848)

Goldman N, Bertone P, Chen S et al (2013) Towards practical, high-capacity, low-maintenance information storage in synthesized DNA. Nature 494(7435):77–80. https://doi.org/10.1038/nature11875. (Epub 2013 Jan 23. PubMed PMID: 23354052; PubMed Central PMCID: PMC3672958 * Second work on DNA storage. Demonstrates convincingly that text, images, and sounds can be stored well in DNA and read out again with modern NGS technology a little time consuming. Sequencing twice removes random errors for clear image decoding, for example. But: everything still quite slow, no concept for a computer chip yet)

Grass RN, Heckel R, Puddu M (2015) Robust chemical preservation of digital information on DNA in silica with error-correcting codes. Angew Chem Int Ed Engl 54(8):2552–2555. https://doi.org/10.1002/anie.201411378. (Epub 2015 Feb 4. PubMed PMID: 25650567 * Third work on DNA storage. Here, the chemists around Prof. Stark show how long information can be stored in DNA if chemistry and error correction codes support [literally millions of years – as we also see through our evolution])

Hekstra DR, White KI, Socolich MA (2016) Electric-field-stimulated protein mechanics. Nature 540(7633):400–405. https://doi.org/10.1038/nature20571. (* Current work on "electronic" proteins.)

Hoffmann J, Trotter M, von Stetten F et al (2012) Solid-phase PCR in a picowell array for immobilizing and arraying 100,000 PCR products to a microscope slide. Lab Chip 12(17):3049–3054. https://doi.org/10.1039/c2lc40534b. (Epub 2012 Jul 23. PubMed PMID: 22820686 *Here, high storage density of DNA is achieved by modern chip methods.)

Jozala AF, de Lencastre-Novaes LC, Lopes AM (2016) Bacterial nanocellulose production and application: a 10-year overview. Appl Microbiol Biotechnol 100(5):2063–2072. https://doi.org/10.1007/s00253-015-7243-4. (Epub 2016 Jan 8. PubMed PMID: 26743657)

Jung YH, Chang TH, Zhang H (2015) High-performance green flexible electronics based on biodegradable cellulose nanofibril paper. Nat Commun 6:7170. https://doi.org/10.1038/ncomms8170

Kunz M, Liang C, Nilla S et al (2016) The drug-minded protein interaction database (DrumPID) for efficient target analysis and drug development. Database (Oxford) 2016:baw041. https://doi.org/10.1093/database/baw041

Nomura Y, Yokobayashi Y (2015) Aptazyme-based riboswitches and logic gates in mammalian cells. Methods Mol Biol 1316:141–148. https://doi.org/10.1007/978-1-4939-2730-2_12. (PubMed PMID: 25967059 * Auch mit RNA kann man wie in Computern schalten und verrechnen.)

Poelwijk FJ, Krishna V, Ranganathan R (2016) The context-dependence of mutations: a linkage of formalisms. PLoS Comput Biol 12(6):e1004771. https://doi.org/10.1371/journal.pcbi.1004771

Raman AS, White KI, Ranganathan R (2016) Origins of allostery and evolvability in proteins: a case study. Cell 166(2):468–480. https://doi.org/10.1016/j.cell.2016.05.047

Ridley M (2010) The rational optimist. How prosperity evolves. Harper Collins, London. (ISBN: 9780061452055 * An optimistic book that does a good job of combining some very positive characteristics of humans, especially their ability to imitate, willingness to exchange [money as the first desire machine], and creativity into an optimistic vision of the future quite plausibly)

Shityakov S, Bencurova E, Dandekar T (2019) Nanozellulose Verbundwerkstoff mit elektronischen Eigenschaften und DNA, Deutsche Patentanmeldung, Priorität 3.Januar 2019, publiziert 9.Juli 2020; DE102019000074A1

Stanke M, Diekhans M, Baertsch R et al (2008) Using native and syntenically mapped cDNA alignments to improve de novo gene finding. Bioinformatics. https://doi.org/10.1093/bioinformatics/btn013

Tero A, Takagi S, Saigusa T, Ito K, Bebber DP, Fricker MD, Yumiki K, Kobayashi R, Nakagaki T (2010) Rules for biologically inspired adaptive network design. Science 327(5964):439–442. https://doi.org/10.1126/science.1177894. (PubMed PMID: 20093467)

Tschowri N, Busse S, Hengge R (2009) The BLUF-EAL protein YcgF acts as a direct anti-repressor in a blue-light response of Escherichia coli. Genes Dev 23(4):522–534. https://doi.org/10.1101/gad.499409. (PubMed PMID: 19240136, PubMed Central PMCID: PMC2648647)

Zhang Y, Lamb BM, Feldman AW, Zhou AX, Lavergne T, Li L, Romesberg FE (2017) A semisynthetic organism engineered for the stable expansion of the genetic alphabet. Proc Natl Acad Sci U S A. 2017 Jan 23. pii: 201616443. https://doi.org/10.1073/pnas.1616443114

# We Can Think About Ourselves – The Computer Cannot

# 14

### Abstract

A computer cannot think about itself, because formal systems have basic barriers here (exactly proved by Gödel and Turing). Humans (and living beings in general) do not think formally exactly, but therefore can think more successfully about themselves or all basic questions. Goals and values therefore must and should always be set by humans, especially as computers become ever more powerful. Artificial intelligence, especially deep learning algorithms and neural networks are helping computer capabilities to soar even higher. The more features of a living being are replicated (e.g., acting in an artificial environment; replicating language and emotions), the more powerful the capabilities of such a machine become. Bioinformatically, the properties of artificial intelligence can be used directly, for example, for modern image processing, but also more generally for the recognition of complex properties ("feature extraction"), pattern recognition from large amounts of data ("training data set") and then also for individual molecules or sequences (predictions, for example, for the secondary structure in the protein, for the localisation in the cell, etc.).

What could be more fascinating than synthetic biology, man's ability to create new biology (with all the limitations to be considered)? Well, no question, it is man himself. In particular, his ability to think about himself marks him out as unique. Computers, after all, can't do that. Among higher mammals, there are at least some that recognize themselves as themselves in the mirror. On the other hand, there is no animal species known that can do detailed introspection based on this or even philosophy like us humans.

© Springer-Verlag GmbH Germany, part of Springer Nature 2023
T. Dandekar, M. Kunz, *Bioinformatics*,
https://doi.org/10.1007/978-3-662-65036-3_14

## 14.1    People Question, Computers Follow Programs

Computers often calculate so fast that one is tempted to assume they can think. However, there are numerous differences to living beings. In particular, computers are not alive, so they do not operate in an environment and cannot reproduce. Therefore, meaning (e.g., food, fear, freedom, etc.) in the strict sense is also not possible for a computer; the computer, on the other hand, formally reason with logical chains of reasoning. But for formal systems it is true that they are either closed and then one cannot formulate provable statements for these closed systems or they are not clearly delimited. More precisely, there are the two Gödel incompleteness theorems.

The *first incompleteness theorem* proves that there are always unprovable statements in sufficiently strong, contradiction-free systems. The *second incompleteness theorem* shows that sufficiently strong, non-contradictory systems cannot prove their own non-contradiction. So for such fundamental statements, the computer remains in the undecidable. In contrast, we can at least think about our fundamentals any time we like. But it is also clear that humans do not always think and decide without contradictions. This is also true in general: Biological systems are primarily not decision-making or computational systems, but living beings that have to survive, especially in their environment. For the same reason, decisions, even of a fundamental nature (e.g. should a cell divide or not), quickly become fuzzy (sometimes even a bit random). But evolution ensures that this fuzziness is adjusted precisely so that we can survive as well as possible with the resulting decisions and we also have a sufficiently accurate picture of the environment in which we act as living beings for this purpose.

Basically, this phenomenon is also easy to understand. Formal systems are either closed, then one can drive them into a contradiction or at least into a statement undecidable for them, if they have to think about their foundations, or they are not closed (then one can formally add additional statements in case of emergency). People, on the other hand, do like to think about themselves, and they also (usually) manage to get back to everyday work afterwards. It is important to realize that this is a very basic barrier between humans and computers. As long as the computer closes correctly and logically exactly like a formal system, it will not get beyond this "Gödel limit", i.e. it will never really be able to think about itself. There is no concept of meaning and no real life in a real environment. Instead, if you would create artificial life, you would be able to cross this border, but every type of life in nature is equal, has the same right to live, be it human, an insect or bacteria, including also any future type of artificial life. However, as we are not even treating all humans equal, we are not ethical mature enough to try to create artificial life. Luckily, the technological hurdles towards artificial life are also enormous.

After this consideration of clear boundaries of computers as formal systems versus humans as living, feeling and acting beings, the infobox gives some cornerstones of artificial intelligence. The important thing to take away is that humans should at least be capable partners with computers, and should actually use their human foresight to verify the computer's calculated results, rather than blindly (or even "subordinately") believing them. Needless to say, in fact, the relationship between the two subjects is rather cordial. Artificial intelligence research and bioinformatics intensively cross-fertilize each other. For example, a number of sophisticated search strategies of artificial intelligence have been inspired by biological phenomena. Conversely, neuronal networks (and also Hidden Markov Models; Chap. 3) are used for many sequence or even more complex predictions in bioinformatics (e.g. for signal sequences).

**Some Cornerstones of Artificial Intelligence**

(a) Principle: neural networks in general
(b) Software that uses **neural networks:**

NucLocP; SignalP or for transmembrane proteins: TMHMM

(c) *Deep learning* **and other modern methods** (Fig. 14.1)
(d) **Expert systems**

– Example area medical informatics:
e.g. infections, burns, anaesthesia
– **Wolfram Mathematica**

Figure 14.1 illustrates one example, namely image recognition using *deep learning*.

Figure 14.1a shows the task: Electron microscopic images of *C. elegans* (left) are to be analysed with the aid of a neural network (centre). Since the number of neurons converges in the middle, namely from 100 to only 50, it is a so-called *deep learning network*. Then a response is shown schematically, bright spots are detected at the top (1), not at the bottom (0). Below this, the learning curve of the network for many trials is shown. During training, the network performs very well (blue curve), during the validation test (red curve) more training is needed first, then the results become as good as the training data set. With mature *deep learning networks*, the performances are of course better than in the scheme (7% error is still a bit high, but this is just a snapshot of the development).

This is clearly demonstrated in Fig. 14.1b, where one can see how this algorithm learns image properties (left) and recognizes them in the image (right, red circles). Among the expert systems, "Wolfram Mathematica" was also mentioned, since this is able to solve differential equations independently or to calculate complex integrals.

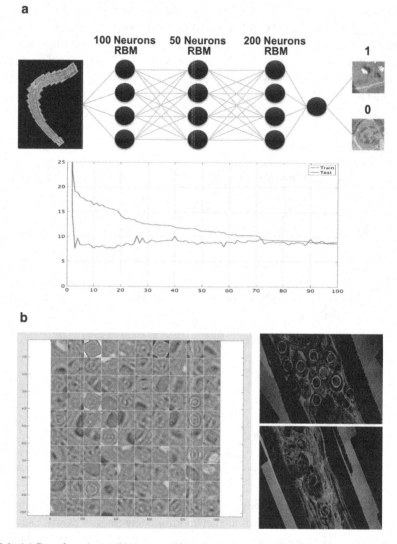

**Fig. 14.1** (**a**) *Deep learning architecture* and learning curves. (**b**) Algorithm learns image properties

## 14.2    Artificial Intelligence

How do you detect artificial intelligence? A first possibility is the Babbage test. The computer scientist Charles Babbage proposed the following competition between humans and computers. Both are hidden behind a cloth, and people on the outside are asked to guess who is who. If a computer succeeds in deceiving people into thinking it is human, then the computer has passed Babbage's test and "possesses" artificial intelligence. Although computer scientists think this test is great, I don't find it particularly impressive. After all, the test is probably easy to win if the human behind the cloth is trying hard to be as

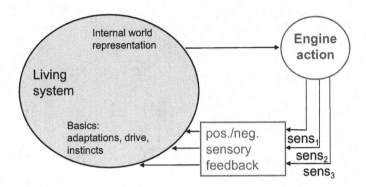

**Fig. 14.2**  Sensorimotor *feedback loop*

monosyllabic as a computer. So the computer might be able to fool us well (pass the Babbage test), but it won't become a living being from that either, but only with evolution and active representation of the environment (which after all might be possible to reproduce later).

How does **biological self-programming** work? Well, by actively acting itself in an environment and then satisfying urges or not and then, of course, being able to actively perceive the environment and itself. A computer that programs itself in this way has at least a real knowledge of meaning, just as a living being does (Fig. 14.2).

Moreover, a computer can also be designed stochastically. For example, efforts to care for the elderly in Japan are going towards making computers as human-like as possible in appearance, responses, etc. This is called emulation, and it can emulate a human very well, but with normal programming it has the same limitations as mentioned above. But you can also equip computers with neuronal networks and let them gain experience in an artificial environment and equip them with drives, etc. And if you add a certain amount of insecurity to it, you can also use it to simulate a human being. And if a certain fuzziness is added, the whole design is already very close to a living being. This means that the fundamental limits for formal systems can be overcome more and more easily. However, the problem is then exacerbated as to how a relationship can be formed between these artificial, increasingly human-like machines and the people concerned. Again, a human (and not machine-like) solution can only be found if humans remain self-aware and specify human values.

If the computer then reproduces itself, it is a real living being. The only thing missing then is the ability to evolve – but that's exactly what we don't want, not only because of security concerns, but also because of ethical concerns.

**Current Examples of Artificial Intelligence**

The most famous example of artificial intelligence and computer successes are probably chess computers, especially Deep Thought's 1988 victory over Grandmaster Bent Larsen, losing to World Champion Garri Kasparov in 1989. Deep Junior won the 2011 and 2013 World Computer Chess Championships, and played Garri Kasparov to a draw in 2003.

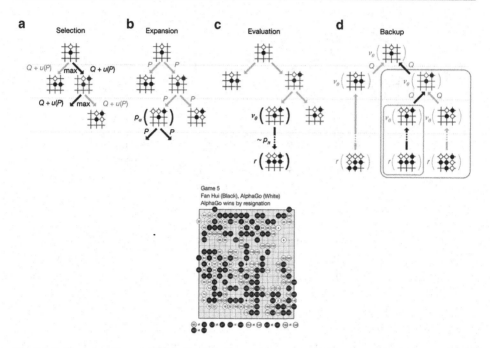

**Fig. 14.3** Deep learning and AlphaGo. The figure shows the optimization of AlphaGo with deep learning (top, Monte Carlo search in four simulation steps) and a winning AlphaGo game (bottom, AlphaGo in white, moves are shown as numbers). (Figure from Silver et al. 2016)

An exciting technique is the so-called *deep learning*. Here, the neuron layers learn by *back propagation* of the result. This was not optimally built in for a very long time, but around 2005 this was corrected (Fig. 14.3; Silver et al. 2016). Since then, *deep learning* networks have become increasingly powerful. As a result, it has recently been possible to beat at least one Go grandmaster even in a strategic board game (https://www.nature.com/nature/journal/v529/n7587/full/nature16961.html).

More generally, the strength of artificial intelligence programs is based on partially emulating biology so that autonomous learning is possible, e.g. neural networks or through hidden Markov models (Maccorduck 2004). Such strategies are used in bioinformatics, for example, for genome annotation (exon-intron domain; e.g., GenScan program) or for the prediction of domains (e.g., Pfam, SMART databases), signal proteins (e.g., the SignalP program), and membrane regions in proteins (e.g., the TMHMM program; Käll et al. 2004) (Chap. 3). For complex optimization problems, such as in protein folding, artificial evolution by genetic algorithms is also used. Evolutionary strategies are another important method of artificial intelligence. The more efficiently learning is replicated, the closer one gets to artificial intelligence. *Deep learning* seems to bring a new quality to it. We are currently using this for image recognition, for example, of microscopic images.

In general, we can say that artificial intelligence is very good at simulating the recognition of complex features (technical term: *feature extraction*; pattern recognition). To do this, we need large amounts of data (a "training data set") and feedback to the neuronal network (by itself: *unsupervised learning*; from the outside: *supervised learning*) as to whether the computer's prediction was correct or incorrect afterwards, even for individual molecules or sequences (predictions, for example, for the secondary structure in the protein, for the localisation in the cell, etc.).

Genetic algorithms are a sophisticated search strategy that I myself have used enthusiastically for many years. Here, solutions are bred in the computer with the help of artificial evolution through selection, mutation and recombination of digitally programmed chromosomes. These chromosomes then encode the problem you want to solve. This works surprisingly well, given sufficient populations of individuals and several hundred generations of evolution. For example, one can obtain protein structures from the sequence by using appropriate selection parameters with small error to the observable structure (Dandekar and Argos 1994, 1996, 1997). The "catch" with this approach is only how to code the protein structure efficiently enough in the chromosomes (e.g., by "internal coordinates") and how to design the selection "correctly" (many years of work and then requires a sufficient number of known, experimentally resolved crystal structures). Another clever search strategy for complex problems with a huge, often high-dimensional search space is to do like the ants (*ant colony optimization*). Here, an anthill is electronically programmed, and the individual virtual ants scour the solution space. In doing so, they leave behind a scent trail. This trail is amplified and turned into a virtual ant trail in the computer if there are particularly good solutions along the searched route. This method is also surprisingly powerful for complex problems, but also requires a lot of patience until one has sufficiently mapped the problem one wants to solve in the real world into this virtual "forest of ants" so that the solutions are tractable. A breakthrough in predicting 3D structures of proteins was recently achieved by Senior et al. (2020) and Tunyasuvunakool et al. (2021).

## 14.3    Current Applications of Artificial Intelligence in Bioinformatics

The high-dimensional data in biology and medicine contain various variables (features), e.g. diagnosis, expression values, age, weight. In addition, there are complex relationships and correlations, but also confounders (confounding variables), batch effects and multicollinearity between the variables. In short, it is very time-consuming to find out which variables are relevant and which are not. An application from artificial intelligence research that has been used for a long time is machine learning (machine learning; Tarca et al. 2007; Sommer and Gerlich 2013) in bioinformatics to structure the data and extract relevant features, but also to develop classification models (predictive models). We have already learned about PCA (Chap. 7) to decompose high-dimensional data into principal components and reduce their complexity (dimensionality reduction). Other methods are cluster and regression analyses. While cluster analysis is used to classify data into groups (clusters) with similar characteristic structures (characteristics), regression analysis is used to find correlations and relationships between variables.

**Clustering** In cluster analysis (group analysis), a distinction is made between supervised (groups known) and unsupervised clustering (groups unknown). An example of supervised clustering is the k-nearest-neighbour algorithm, in which new data (e.g. patient) are classified into predefined groups (the k-nearest neighbour [e.g. k = 3 considered 3-nearest neighbours] is always considered and then assigned to a cluster). This allows, for example, to assign a diseased person to an optimal therapy (e.g. radiation, chemotherapy) according to the gene expression profile. If, on the other hand, one wants to find clusters in one's data, one can apply unsupervised clustering, such as k-means (non-hierarchical; often for NP problems) or complete-linkage (hierarchical).

**Regression** Regression analyses examine the relationship between a dependent (regressand, criterion, "response variable") and independent (predictor, "predictor variable") variable. For example, a linear regression can be used to examine the relationship between weight (independent variable) and blood pressure (dependent variable). The prerequisite is that the dependent and independent variables are metric. The calculation is done with the least-squares estimator, which tries to minimize the least-squares error of the residuals (distance from the data point to the regression line) to get the best fit to the data (you put a regression line in the data). How well the regression model represents the data (goodness of fit) is usually calculated with a t-test (p-value < 0.05) and the $R^2$ (coefficient of determination, between 0 [no correlation] and 1 [high linear correlation]).

If, on the other hand, the dependent variable is a binary/dichotomous (yes/no) variable, logistic regression can be used. Here is a popular analysis question: What is the probability (the correlation) of developing high blood pressure (or heart failure) if you are overweight? The calculation is done using the logit function [$\log(p/1-p)$] and maximum likelihood method (you put a sigmoidal curve in the data). To assess the model quality, one uses, for example, a chi-square test (p-value <0.05), the $R^2$ and the AIC (Akaike Information Criterion; adjusts model fitting to the number of parameters used).

Often one also has time points (time data) for a dependent variable, for example in the context of a follow-up study. Cox regression is used for this (survival time analysis, "time to event" analysis). Survival time analyses are of interest if one wants to know, for example, what the influence of a mutation or therapy is on the 5-year survival time. The calculation is done using the Kaplan-Meier estimator (hazard function calculates risk [failure rate] that event actually occurred; censored data [no exact information about event] are not included in the calculation). The survival rates are represented in a Kaplan-Meier curve, and the model quality is assessed using a log-rank test and Cox proportional hazards.

A nice overview of regression analysis is provided by Worster et al. (2007), Schneider et al. (2010), Singh and Mukhopadhyay (2011) and Zwiener et al. (2011). The two recent papers on remdesivir (Wang et al. 2020) and lopinavir-ritonavir (Cao et al. 2020) treatment in COVID-19 should also be mentioned here.

Logistic regression and Cox regression are popular for the analysis of diagnostic and prognostic signatures, i.e. the optimal combination of genes (Vey et al. 2019) or metabolites (Schweitzer et al. 2019) whose strong or weak expression indicates a disease or its (in)favourable course. Classification models (prediction models) are being developed for this purpose.

**Classification Model (Prediction Model)**  Classification models are often used in bioinformatics for the classification between two categories (binary), for example for the diagnosis of a disease (sick/healthy). This is done using a classification table (confusion matrix, truth matrix), which summarizes the prediction (test) and actual observation (reference) of a classifier (Table 14.1).

Here we classify between positive ("sick", alternatively: yes, 1, correct) and negative ("healthy", alternatively: no, 0, wrong) as follows:

- True Positive (TP; true positive cases): test and reference positive (test and reference "sick")
- False Positive (FP; false positive cases): test positive, reference negative (test "sick", reference "healthy")
- False Negative (FN; false negative cases): test negative, reference positive (test "healthy", reference "sick")     ·
- True Negative (TN; true negative cases): test and reference negative (test and reference "healthy")

In order to be able to assess how meaningful (accurate) a classification model (predictive test) is, i.e. whether the classification made is correct or incorrect, there are various statistical quality criteria/measures (performance metrics). These are:

- Sensitivity (true positive rate, sensitivity; positives detected as positive)" $= \dfrac{TP}{TP + FN}$

- False positive rate ("false alarm", positives that are actually negative) $= \dfrac{FP}{TN + FP} = 1$ specificity

- Specificity (negatives detected as actual negatives) $= \dfrac{TN}{FP + TN}$

- Positive predictive value (PPV, precision; probability of actually being positive) $= \dfrac{TP}{TP + FP}$

**Table 14.1**  Overview confusion matrix for a classification model

|                   |   | Reference |    |
| ----------------- | - | --------- | -- |
|                   |   | +         | −  |
| Test (prediction) | + | TP        | FP |
|                   | − | FN        | TN |

- Negative predictive value (NPV, segregancy; probability of actually being negative) = $\dfrac{TN}{TN + FN}$

- Accuracy (correct classification rate) = $\dfrac{TP + TN}{TP + FP + TN + FN}$

- Misclassification rate = $\dfrac{FP + FN}{TP + FP + TN + FN}$

- Prevalence (proportion of actually positive persons in the total number) = $\dfrac{TP + FN}{TP + FP + TN + FN}$

For the graphical representation, a ROC curve (Receiver Operating Characteristic; x-axis: false positive rate, y-axis: sensitivity) is often used, where the AUC (Area Under the Curve) is a measure of the quality of the classification (higher AUC value = better classification). An ideal classification model has a 100% true positive rate (100% sensitivity) and 0% false positive rate (100% specificity). But this is not always the case in reality. For example, in a recent paper, we were able to show that a novel real-time PCR has better predictive power for the detection of *Trypanosoma cruzi* in a Chagas disease and is superior to previous PCR methods here, but is just not 100% accurate (Kann et al. 2020). In any case, it is advisable to always create a prediction model on the basis of a training and test data set and to validate it on at least one independent data set in order to be able to reliably assess its predictive power for a possible application, such as a clinical decision support system.

**Artificial Neural Networks** Another possibility for machine learning is the use of simple neural networks, which consist of input a simple intermediate layer and an output. Connections between these three layers are strengthened or weakened so that the output is as accurate as possible. To do this, the neural network is trained on a training dataset (automatically: unsupervised; with human review: supervised) and then its accuracy is checked on another test dataset. This can then be used to generate an optimal prediction for helix and beta boundary regions in protein structures (PredictProtein software, https://predictprotein.org) and to determine protein localization.  **The deep learning approach** extends the simple neural network by several layers of intermediate neurons, which in particular then get by with fewer neurons in the later layers (and thus bring results together, "converge"). This replicates – in very simplified terms – an abstraction of the many inputs to more general terms. These networks are more complex to train ("back-propagation" and other steps) but, often further improved with other strategies from artificial intelligence research, also create amazing things, such as optical image recognition of leukemia cells through improved swarm optimization (Sahlol et al. 2020) or the automatic recognition of secondary structure and oligonucleotides in electron micrographs (Mostosi et al. 2020), so that eventually even antibiotics can be discovered with this *deep learning* approach (Stokes et al. 2020) or the energy potentials and thus also the three-dimensional structure of proteins (Senior et al. 2020), now culminating in large-scale and accurate deep-learning based prediction of human proteins (Tunyasuvunakool et al. 2021).

## 14.4  Biological Intelligence

But how do living beings escape the Gödel limit? Biological systems like humans can solve fundamental problems without being stuck with unsolvable decision problems. Why? Well, we hinted at that in our systems biology chapters above. Biological systems are selected to quickly make the most optimal (in the sense of "adapted to the environment") decisions possible. Whereby a bacterium, of course, does not really think about itself. But the division rate (i.e. the decision about itself, "to divide or not"?) is constantly adapted to the environment as optimally as possible (for maximum chances of survival). We can immediately see the difference between this and formal systems, and this then also applies to important decision-making processes in higher organisms. A biological system will, in case of doubt (so that it does not die out from doubt), make a decision between several variants stochastically (i.e. randomly), but the random weights have again been selected by evolution in such a way that the resulting action guarantees the best survival success on average.

Of course, modern neurobiology has already come astonishingly far, especially with the help of bioinformatics. We made it clear in our chapter introduction that it is important to recognise that biological brains and computers can basically both perform calculations amazingly well, but that they come from two different worlds. The computer is accurate and often amazingly fast. In biology, it is rather amazing that brains are capable of such fast and accurate computations. Because all these capabilities are only a means to an end, they are always primarily about survival.

If you like, you can take this away as an important self-knowledge. Our brain may think a lot, make art or dream of the next galaxy, but it was not designed for that. It is only the most powerful survival machine this planet has produced, including the risk of speeding up the evolutionary game so fast that no one can keep up, not even our brain (see next chapter). Of course, we can do philosophy and even overcome the logical Gödel limit for computers. But our brains were not selected to think particularly clearly about the world, but to survive successfully in that world, no matter how difficult the environment. For example, we got a final evolutionary boost from the Ice Age and a first one about two and a half million years ago when the savannah expanded. First bipedal pre-humans started living there, started hunting with hands, hand axes and then spears, while chimpanzees, since they separated from us about seven million years ago, continued to stay peacefully in the trees in the forest (and as our closest relatives, deserved much more protection than they currently get). Let's now take a closer look at that natural high-performance intelligence, the human brain, next.

**Conclusion**

- A computer (as conceived by Turing as a Turing machine) cannot reason about itself. Formal systems have basic bounds (exactly proved by Gödel and Turing), what they can prove or decide and what not. Humans (and living beings in general) may not think formally exactly, but therefore can think about themselves and, in general, all fundamental questions more successfully. Therefore, goals and values must and should always be set by humans, especially when computers become more and more powerful.
- Artificial intelligence, in particular deep *learning algorithms* and neural networks, is giving a further boost to the capabilities of computers. The more features of a living being are emulated (e.g. acting in an artificial environment, emulating language and emotions), the stronger its capabilities become.
- In bioinformatics, the properties of artificial intelligence can be used directly for modern image processing, for example, but also in general for the recognition of complex properties *(feature extraction), for* pattern recognition in large data sets (training data set) and then also for individual molecules or sequences (predictions, for example, for the secondary structure in the protein, for the localization in the cell, etc.).

## 14.5    Exercises for Chap. 14

**Task 14.1**
What does Gödel's theorem say?

**Task 14.2**
What does "Turing-computable" or "non-Turing-computable" mean?

**Task 14.3**
What's the Babbage test?

**Task 14.4**

   (a) Find out about neural networks. Find TMHMM on the net and use it.
   (b) Look at the ELM server, what predictions of the ELM server do neural networks use?
   (c) The protein secondary structure prediction *"PredictProtein"* uses neural networks. Get an overview by using the software and the given references.

**Task 14.5**
Find a neural network software.

## Task 14.6

How does *deep learning* work*?* An example of a deep learning network is AlphaGo – can you find the movie?

## Task 14.7

What is meant by a bioinformatics prediction model that is a classification model (feel free to explain with an example)?

## Task 14.8

Explain PCA, clustering, linear regression, logistic regression, and survival time analysis. What are the similarities and differences?

---

**Useful Tools and Web Links**

- Bent Larsen:
  https://www.chessgames.com/player/bent_larsen.html

  * Of course you can also look at completely different chess games and players at this link.

- Deep Junior (chess computer):
  https://www.hiarcs.com/pc-chess-deep-junior.htm

  * Deep Junior 13.8 was the best chess computer at the World Computer Championship in 2013. Authors are Amir Ban and Shay Bushinsky. The prefix *Deep* refers to the correspondingly strong multiprocessor version for tournaments.

- Game of Go (won by computer):
  https://www.nature.com/nature/journal/v529/n7587/full/nature16961.html
- TMHMM (hidden Markov model predicts membranes):
  https://www.cbs.dtu.dk/services/TMHMM/
- SignalP (neural network predicts signal peptides or secretion):
  https://www.cbs.dtu.dk/services/SignalP/

---

## Literature

Cao B, Wang Y, Wen D et al (2020) A trial of Lopinavir-Ritonavir in adults hospitalized with severe covid-19. N Engl J Med 382:1787–1799. https://doi.org/10.1056/NEJMoa2001282

Dandekar T, Argos P (1994) Folding the main chain of small proteins with the genetic algorithm. J Mol Biol 236(3):844–861

Dandekar T, Argos P (1996) Identifying the tertiary fold of small proteins with different topologies from sequence and secondary structure using the genetic algorithm and extended criteria specific for strand regions. J Mol Biol 256(3):645–660

Dandekar T, Argos P (1997) Applying experimental data to protein fold prediction with the genetic algorithm. Protein Eng 10(8):877–893. (*The three Dandekar-Argos references describe how far one can fold and correctly predict protein structures using a robust and intelligent search strategy, genetic algorithms. Of course, one can also use this method for completely different problems [see. Goldberg, David Genetic Algorithms in Search, Optimization, and Machine Learning. Addison-Wesley, 1989, the classic textbook in the field].)

Käll L, Krogh A, Sonnhammer EL (2004) A combined transmembrane topology and signal peptide prediction method. J Mol Biol 338(5):1027–1036. (PubMed PMID: 15111065 * Vergleicht und kombiniert TMHMM und SignalP.)

Kann S, Kunz M, Hansen J et al (2020) Chagas disease: detection of *Trypanosoma cruzi* by a new, high-specific real time PCR. J Clin Med 9(5):1517. https://doi.org/10.3390/jcm9051517

Maccorduck P (2004) Machines who think: a personal inquiry into the history and prospects of artificial intelligence. A K Peters, Ltd. S 482 (ISBN 1-56881-205-1)

Mostosi P, Schindelin H, Kollmannsberger P, Thorn A (2020) Haruspex: a neural network for the automatic identification of oligonucleotides and protein secondary structure in cryo-electron microscopy maps. Angew Chem Int Ed Engl. https://doi.org/10.1002/anie.202000421

Sahlol AT, Kollmannsberger P, Ewees AA (2020) Efficient classification of white blood cell leukemia with improved swarm optimization of deep features. Sci Rep 10(1):2536. https://doi.org/10.1038/s41598-020-59215-9

Schneider A, Hommel G, Blettner M (2010) Lineare regressionsanalyse. Dtsch Arztebl Int 107(44):776–782. https://doi.org/10.3238/arztebl.2010.0776

Schweitzer S, Kunz M, Kurlbaum M et al (2019) Plasma steroid metabolome profiling for the diagnosis of adrenocortical carcinoma. Eur J Endocrinol 180(2):117–125. https://doi.org/10.1530/EJE-18-0782

Senior AW, Evans R, Jumper J, Kirkpatrick J, Sifre L, Green T, Qin C, Žídek A, Nelson AWR, Bridgland A, Penedones H, Petersen S, Simonyan K, Crossan S, Kohli P, Jones DT, Silver D, Kavukcuoglu K, Hassabis D (2020) Improved protein structure prediction using potentials from deep learning. Nature 577(7792):706–710. https://doi.org/10.1038/s41586-019-1923-7

Silver D, Huang A, Maddison CJ (2016) Mastering the game of Go with deep neural networks and tree search. Nature. 529(7587):484–489. https://doi.org/10.1038/nature16961. (PMID 26819042 * AlphaGo wurde von Google DeepMind in London programmiert, konnte ab Oktober 2015 ohne Handicap gegen professionelle Spieler gewinnen und schlug im Dezember 2016 Lee Sedol, einen 9-Dan-Go-Spieler)

Singh R, Mukhopadhyay K (2011) Survival analysis in clinical trials: basics and must know areas. Perspect Clin Res 2(4):145–148. https://doi.org/10.4103/2229-3485.86872

Sommer C, Gerlich DW (2013) Machine learning in cell biology – teaching computers to recognize phenotypes. J Cell Sci 126(Pt 24):5529–5539. https://doi.org/10.1242/jcs.123604

Stokes JM, Yang K, Swanson K, Jin W, Cubillos-Ruiz A, Donghia NM, MacNair CR, French S, Carfrae LA, Bloom-Ackermann Z, Tran VM, Chiappino-Pepe A, Badran AH, Andrews IW, Chory EJ, Church GM, Brown ED, Jaakkola TS, Barzilay R, Collins JJ (2020) A deep learning approach to antibiotic discovery. Cell 181(2):475–483. https://doi.org/10.1016/j.cell.2020.04.001. (Erratum for: Cell 2020 Feb 20; 180(4):688–702.e13)

Tarca AL, Carey VJ, Chen XW, Romero R, Drăghici S (2007) Machine learning and its applications to biology. PLoS Comput Biol 3(6):e116. https://doi.org/10.1371/journal.pcbi.0030116

Tunyasuvunakool K, Adler J, Wu Z, Green T, Zielinski M, Žídek A, Bridgland A, Cowie A, Meyer C, Laydon A, Velankar S, Kleywegt GJ, Bateman A, Evans R, Pritzel A, Figurnov M, Ronneberger O, Bates R, Kohl SAA, Potapenko A, Ballard AJ, Romera-Paredes B, Nikolov S, Jain R, Clancy E, Reiman D, Petersen S, Senior AW, Kavukcuoglu K, Birney E, Kohli P, Jumper J, Hassabis D

(2021) Highly accurate protein structure prediction for the human proteome. Nature. https://doi.org/10.1038/s41586-021-03828-1. Epub ahead of print. PMID: 34293799

Vey J, Kapsner LA, Fuchs M et al (2019) A toolbox for functional analysis and the systematic identification of diagnostic and prognostic gene expression signatures combining meta-analysis and machine learning. Cancers (Basel). 11(10):1606. https://doi.org/10.3390/cancers11101606

Wang Y, Zhang D, Du G et al (2020) Remdesivir in adults with severe COVID-19: a randomised, double-blind, placebo-controlled, multicentre trial. Lancet 395(10236):1569–1578. https://doi.org/10.1016/S0140-6736(20)31022-9

Worster A, Fan J, Ismaila A (2007) Understanding linear and logistic regression analyses. CJEM 9(2):111–113. https://doi.org/10.1017/s1481803500014883

Zwiener I, Blettner M, Hommel G (2011) Überlebenszeitanalyse. Dtsch Arztebl Int 108(10):163–169. https://doi.org/10.3238/arztebl.2011.0163

# How Is Our Own Extremely Powerful Brain Constructed?

# 15

**Abstract**

Our brain gets the ability to think through its modular construction. In the process, nerve cell associations are trained like neuronal networks in a computer. Training and exercise strengthen or delete synapses. In the associative regions of our cerebrum, there are so many nerve connections that it becomes advantageous to process information in an integrated rather than localized manner. Interference patterns similar to a hologram emerge. Bioinformatics decodes neuromolecular signals at many levels: Genetic factors of neuronal maturation and disease, which can be elucidated using the OMIM database, genome and transcriptome analyses. At the neuronal level, protein structures, in particular receptors and their activation can be described in detail using protein structure analyses, molecular dynamics and databases (e.g. DrumPID, PDB database), as well as underlying cellular networks, protein-protein interactions and signalling cascades involved. Brain blueprints, so-called connectomes, are already available for *C. elegans* and are being intensively developed for other model organisms and humans. Numerous special software are available for clinical evaluations (EEG, computer tomograms) ('medical informatics'), but also for neurobiological experiments (e.g. a neuronal activity detection tool).

Our excursions into systems biology (especially Chap. 5) provide a first important answer: modular, of course, made up of identical units, which then reassemble at the next level as emergent, new components (with entirely new properties) and thus finally enable us to think. First we have to feel the hunger, then we learn to see the light. Then, at the age of a little more than 2 years, the ability to say "I" forms, to play as a person in our world at first, then to act in an increasingly complex way, and gradually to consider, act and evaluate one's own position in the world.

© Springer-Verlag GmbH Germany, part of Springer Nature 2023
T. Dandekar, M. Kunz, *Bioinformatics*,
https://doi.org/10.1007/978-3-662-65036-3_15

## 15.1    Modular Construction Leads to Ever New Properties – Up to Consciousness

Bioinformatically, we can go through these stages in order. First, we have the genome. But what is important is the concrete inventory of mRNA in a nerve cell. Even at this level, there is much to discover and analyse. In this context, there is, for example, interesting work that shows how strong experiences also change the brain in its regulation, i.e. epigenetic regulation, via the peptide abundance in nerve cells or how, for example, the synaptic circuitry in individual synapses functions (although model organisms are often used here) (Hassouna et al. 2016). In short, the analysis of large amounts of genomic and transcriptomic data is of great help here.

The next level concerns individual neuronal circuits, with patterns forming through a juxtaposition of inhibitory and activatory circuits. It should be emphasized here that initially such pattern formations are important for the neuronal differentiation and emergence of the individual brain region. These two processes can be well reproduced with semi-quantitative dynamic models and Boolean models (see Chap. 5).

The next higher level concerns the formation of individual brain tissues, for example the primary visual cortex or the hippocampus. Here we can already describe a lot. However, much is not yet understood, e.g. how the final perception of optical images occurs. Equally not understood are the fabulous properties of the hippocampus to give events time stamps and thus make us a person with our own biography and memories. One's own bioinformatics work only helps here by evaluating data and getting the statistics right. Since more, more complex and deeper experiments are needed here, this complex and high level is clearly the domain of experimental neurobiology. However, bioinformatics can help here mainly by evaluating and interpreting complex data.

And yet this is nothing compared to perhaps the most puzzling phenomenon of our very existence: consciousness. Interestingly, there is so little unambiguous data available here that it may again make sense to map individual aspects through a dedicated bioinformatics model and better understand some of the enigmatic, extremely powerful capabilities of human consciousness. We have worked out a simulation that, building on early work, is now capable of simulating a holographic model of human consciousness. What does that mean? The brilliant physicist David Bohm (who only didn't get a chance to expand on his findings in the USA because of alleged communist machinations) had thought early on (Bohm and Hiley 1985) about how the holistic view of consciousness, that inseparable unity we feel (at least in a healthy state) when we talk about our person, could probably be explained physically. He was thinking of how images can be stored holistically as holograms. In doing so, the object's light waves hit a fundamental rhythm and interfere with it. The resulting image is therefore an interference image. Let us look at such a photograph (Fig. 15.1).

How exactly the information is processed in the cerebrum, nobody knows at the moment. It is clear, however, that in addition to serial information processing, more global processing and not local information is processed (detailed analysis in our own preprint:

https://www.biorxiv.org/content/10.1101/2019.12.19.883124v1). Processing in the form of excitation waves may result in conscious integration of all information in a single interference image for the whole cortex. This is what we have simulated here. What can be seen through the simulation are interference images, through interpenetrating neuronal excitation waves. Simplified, the simulation assumes only two signals that directly interfere with each other. Detailed models are highly compatible with EEG and other data. The exact proportion of non-locally processed information in the cerebrum is still unclear.

When we look at the wave pattern, we may be a little disappointed because the original image is no longer recognizable. Instead, we only see a wave interference pattern, like on a pond when several wave trains overlap: an interference image. But that's not so surprising, because that's exactly what we did to capture the image. To make the original image we wanted to store visible again, we now have to beam the carrier background wave onto

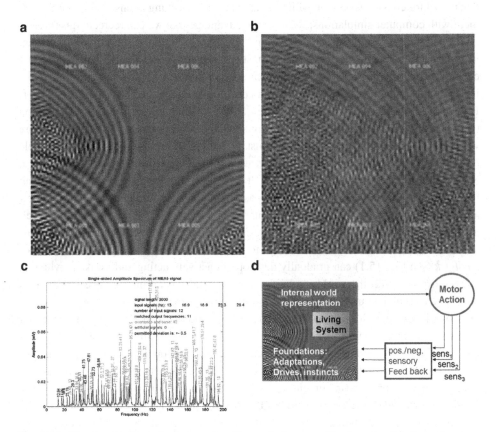

**Fig. 15.1** Non-local information processing. Shown is the wave-like (non-local) information processing in the brain. This happens in regions such as the associative cortex of humans or associative regions in the hippocampus. The wave pattern reaches all points of the water surface and gives the same information to all neurons involved. This could be a basis for consciousness, because all the neurons involved are reacting as a unit, participating in the same wave pattern. In each point exactly the same information (about 150 bits, changing every 3 s) can be tapped

the photo with the interference image, and then we see the holographically stored image in front of us again. So far so good. But what is amazing now is that we can tear the interference image in half, or smaller pieces, and each of these halves shows the complete original image again when the fundamental wave is beamed in. The only thing that gets worse is the image sharpness when I tear the interference image into smaller and smaller pieces.

Such interference images are thus particularly well suited to reflect the holistic, holistic aspect of human consciousness quite well. If our conscious brain (more precisely: the associative regions of the neocortex, higher regions of the hippocampus) works in the same way, then the conscious contents are distributed as interference images like rippled wave trains over these entire regions and are equally and completely accessible in every point (more precisely: every *"Mountcastle Column"*) of these conscious brain regions. In this model for consciousness, on which we are currently working intensively in our department with computer simulations, we have the advantage that we can recreate this directly on the computer, calculate the storage volume for conscious stimuli (the conscious present is 3 s, in each second 50 bits can be consciously perceived, so a total of 150 bits) and many concrete properties fit very well with neurobiological findings. Moreover, this would make it easy to understand how consciousness can develop gradually and also spontaneously. If I have a critical mass of neurons that are "freely" available, then sensory information (from sense organs and also self-perceiving signals, called proprioception) can interfere with basic rhythms of activity in the brain just as freely as motor information (from muscle movements). This wave pattern is equally available as a holistic interference pattern at every point of the conscious brain regions. It arises so spontaneously because there is no higher region to which the signal has to be passed on. Because stimuli from the body itself as well as from outside are stored in the interference picture, a picture of oneself as well as of the environment is created. Since at the same time the motor activity is "freely" available as a signal in the interference image to all conscious brain regions, a sensorimotor *feedback* (see Fig. 15.1) can gradually develop, an actively acting and awake I, which is then confronted with the world model that also develops as an interference image. This results in a simultaneously simple but also impressively strong and integrating picture for our consciousness. With our concrete simulation model we can also estimate well that below several hundred million neurons it is simply too uneconomical to free neurons only for "free interference images", i.e. conscious perception should only develop in higher mammals. But if you don't really have many "free", i.e. associative neurons, as humans do, the subjective present (in humans, as I said, about three seconds) is simply far too short (tenths of a second or even shorter) to be able to really plan any conscious action. There is only reflexive, rapid response. It is true that this current simulation model of human consciousness is far from being confirmed or experimentally validated by us, but that is not the point here. We only wanted to show why it can be particularly fascinating to approach human consciousness with bioinformatics. It is also very interesting that this model can help to better understand typical psychiatric disorders. For wave interference images and their stable storage, it is important that the wave trains are freely transmitted and that the nerve cells work in a correlated manner, i.e. that they are well synchronized. However, recent work shows (Voytek and Knight 2015) that precisely a disruption in the correlation of larger neuronal assemblies and groups essentially underlies schizophrenia. In the model, the range of the interference image is then limited. Thus, it is no longer possible to integrate all motor and sensory inputs. The patient then experiences him/herself as remotely controlled (motor impulses without control, but also "thoughts from outside") and "hears voices" (sensory input).

It would be very fascinating if we were right with this model. But at least we can state that other studies also suggest that in schizophrenia the integration capacity of the brain is decisively disturbed, that the cardinal symptoms of this disease arise from this, and that we are pursuing a very fundamental, holistic approach here, which was originally conceived by David Bohm.

## 15.2   Bioinformatics Helps to Better Describe the Brain

How can we use bioinformatics, for example, to better help patients with what is perhaps the worst disease of all, destroyed personality, schizophrenia or schizophrenia? Well, interestingly, that too will probably soon be better done by examining large genomic datasets. **Genome-wide association studies (GWAS)** are important here. Family studies have always been used to try to identify important genes for schizophrenia and mental function. In the past, a major disadvantage was that these studies often found only family-specific genes and mutations. In the meantime, however, it is possible to sequence very large amounts of data and in this way also detect weak signals (more precisely "low *load score*") more accurately. A major improvement has been proposed by Prof. Hannelore Ehrenreich (MPI for Experimental Medicine, Göttingen). By precise diagnostics one can divide global diseases (like schizophrenia, depression) into subtypes (in schizophrenia for example autistic, paranoid, catatonic etc.), so that one can detect much more precisely the specific signal of mutations which are then important for the respective single aspect. This is a very important approach, which can now be approached in a very promising way with the help of large transcriptome analyses. Once the molecular findings for these subtypes have been determined with the help of data analysis, it is then possible to analyse the protein structures involved in detail and develop new drugs that can then help specifically with the respective subtype – the success story of atypical neuroleptics shows how much good we could then do. Thus, GWAS already bring good results for specific diseases, whenever the signal is clear enough (Hammer et al. 2014).

Another possibility is the analysis of *single nucleotide polymorphisms* (SNPs). Often the DNA sequence is altered at exactly one nucleotide. This can be neutral in its health effects (most of the time), but can also lead to negative effects (rarer) or to positive effects (even rarer) – and rarely detected, because it then does not lead to any complaints, on the contrary. For example, we investigate an SNP that has an effect on the psyche. A sequence analysis first provides information on which DNA segments with a known function the gene resembles. This is followed by further analyses to determine whether the SNP also appears as RNA in the brain or even as a protein. Based on this, the RNA or protein structure is then analysed and how the difference of one nucleotide affects this. Subsequently, one can try to predict (from databases and with prediction algorithms, a first start is the tool STRING at the EMBL: https://string-db.org) with which proteins there are interactions here. In this way, one can determine step by step what effects this small change has.

Of course, there is also everything in between. Smaller or longer insertions or deletions in the genome sequence on the affected chromosome and also very large modifications, such as additional chromosomes (the best known is trisomy 21, Down syndrome) or incorrectly assembled chromosomes ("translocations"). The database "Online Mendelian Inheritance in Man" (OMIM; https://www.omim.org) provides a detailed overview. Genes, proteins and sequences involved in the structure of the nervous system, for example at synapses, are thus assigned to their function with the aid of genome, sequence and domain analyses (as already practised on other topics).

However, neurobiological processes can also be viewed using a wide variety of other bioinformatics techniques. The structure of important receptors involved can be modelled, e.g. the seven transmembrane helices that build up a GPCR receptor, as well as other important activatory (e.g. glutamate) and inhibitory receptors (e.g. glycine). At the next level, neuronal networks can be represented by semiquantitative simulations, but also receptor excitations, for example by differential equations. Individual circuit diagrams can be recreated in the computer, for example for memory (Rolls 2013a) or for the whole network of the cordworm *C. elegans (Connectome C. elegans)*. For the hippocampus, there are already ideas about how it recognizes and separates patterns (Rolls 2013b). There is even a quantitative theory of the function of individual layers in the hippocampus (Rolls 2013a). Higher processes can also be modelled with the help of computers (Markram et al. 2015). In particular, however, omics studies again provide an overview of the brain (especially, but not only, of humans, e.g. an atlas of gene expression in the human brain; Hawrylycz et al. 2015). Much is also taking place on the model organism mouse.

An overview of the pathology is equally important. Bioinformatics can help here, for example, to better understand the regeneration processes in the old and young brain in stroke using transcriptome data (Buga et al. 2012) or, for example, to decipher regeneration in the hippocampus using various data via statistical analyses, whereby erythropoietin apparently has a supporting effect (Hassouna et al. 2016). Meanwhile, there are a whole bunch of ways to boost memory performance using growth factors, stem cells, pharmaceuticals, or even memory training. But a clinical breakthrough, for example in Alzheimer's disease, will still require a lot of work and studies (Schneider et al. 2020).

## 15.3 Brain Blueprints

The human brain is certainly the most complex object we can study with bioinformatics. Certainly, this is a kind of "moon landing" that we still have to do in this century: to understand what our consciousness actually is. But of course there are plenty of other tasks (see previous and next chapter to define other "big goals"). However, a recurrently quite successful strategy in bioinformatics is to start "small". What if one could understand the brain of a small organism perfectly, in its entire blueprint? And what if you could disclose the blueprint and put it on the web worldwide, as *"open source"* software, so to speak? That is exactly what has already happened.

Here is the link to the rotifer brain blueprint: https://www.openconnectomeproject.org. The infobox shows details of the entry page on the Internet.

This is the so-called *"Open Connectome"* project of *C. elegans*. Here we know "everything", so to speak. How every neuron in *C. elegans* is connected to every other neuron? And the next most complicated brains are currently being worked on in terms of a model, a circuit diagram, for example for ants and mice. Even for humans there is already the *Brain Activity Atlas* and the SMART computing pipeline charted the rhesus monkey brain at micrometer resolution (Xu et al. 2021) – these are exciting times for bioinformatics.

---

**Brain Blueprints in Computer Models**

The link to the entry-level "Open Connectome" project of C. elegans can be found at: https://www.openconnectomeproject.org. For *C. elegans* there is also the *WormAtlas* (https://www.wormatlas.org/) or WormWiring (https://wormwiring.org/).

Links to the projects for the next most complicated brains can be found here:
Insects: Virtual Insect Brain Lab (https://www.neurofly.de/).
Mouse: *Allen Mouse Brain Connectivity Atlas* (https://mouse.brain-map.org/static/atlas).
Human: *Brain Activity Atlas* (https://www.brainactivityatlas.org/).

---

With the help of the "connectome", i.e. the wiring diagram for all neurons in *C. elegans,* it is possible to understand how the nervous system of a simple animal works. In *C. elegans,* there are only exactly 302 neurons in each animal, but 118 classes of neuron types. This is why, in addition to the *open-connectome-in-silico* modelling project, there is also the Worm Atlas (see infobox). This is also a growing area of bioinformatics. In addition to the digital anatomy atlas, the hippocampus region with detailed circuit diagram and the detection of neuroactivity are shown here.

## Brain Complexity Directly Mapped Digitally

**The Sausage Atlas**
https://www.wormatlas.org/neuronalwiring.html

There are also numerous anatomical data.

### Wiring Diagrams of Brain Regions

Humans are much more complex (over ten billion neurons), but for this there are only five classes of component neurons here. For the rat hippocampus, there is detailed information and circuit diagrams on the net:

https://www.temporal-lobe.com/background/connectome

### Neuroactivity Detection

In order to see individual nerve cells in their activity, i.e. to pay attention to them, we have developed a tool ourselves: The *"activity detection tool"* uses a Fourier transform. A nice introduction to such approaches is a paper that works in particular with segmentation and ImageJ (Schulze et al. 2013):

https://www.ncbi.nlm.nih.gov/pubmed/23537512

## 15.4   Possible Objectives

What should be the next step? Well, bioinformatics is very good at analysing sequences, identifying domains and thus elucidating functions of proteins in the nervous system very well. Receptors can also be modelled in terms of their structure and their detailed dynamics and function can be investigated in detail. Finally, larger network analyses, in particular on the *"connectome"*, the connection of nerve cells, are already planned for several organisms and their nervous systems (mouse, human, insects and others) or have already been completely carried out in first drafts (cordworm *C. elegans,* molluscs). Therefore, it is safe to say that bioinformatics is doing good work in the areas of **understanding neurobiology and basic research.**

The same applies to medical causal research, although here the impetus from bioinformatics is particularly concerned with uncovering the **molecular causes of diseases**, thereby supporting diagnosis or even predicting innovative therapies.

More ethically and technically challenging is work on **artificial intelligence** or on **consciousness**. Here, it is an ethical imperative to leave all central, moral or ethical decisions to humans, to take this into account already when structuring the problem and the decision line, and not to cross the threshold to a living being, because the current ethical maturity of humans is not sufficient for this. "Conscious machines" are also in principle uncontrollable and risky. Fortunately, however, we are relatively far away from this in bioinformatic modelling because of a number of breakthroughs that are still necessary. Nevertheless, it is advisable to take great care already during the design phase (for example, of increasingly powerful Internet tools or increasingly autonomous weapon systems) to ensure that the design prevents the worst-case scenario (the greatest accident that can be assumed), namely the autonomous machine with consciousness or superior intelligence that begins to control or kill humans, from the outset.

**Conclusion**
- Our brain is given the ability to process information very well due to its modular design. Our genome encodes different proteins that lead to different activating and inhibiting nerve cell connections (synapses) in numerous different nerve cells, depending on the cell type. Nerve cell associations thus have new properties (emergence). In particular, our brain is particularly good at recognizing patterns. Human nerve cell associations are trained in the same way as neuronal networks in computers (see previous chapter). Training and practice strengthen or erase synapses. Practice thus optimizes learning success over time. There are so many nerve connections in the associative regions of our cerebrum that it becomes advantageous to process information in an integrated rather than localized manner. Interference patterns similar to a hologram are created.
- We describe with our own current simulations that environmental stimuli, but also one's own position as well as one's own actions can be encoded in a hologram for all neurons participating in the pattern equally and simultaneously. Such new emergent effects in our particularly complex brain presumably underlie our consciousness ("fulguration" according to Konrad Lorenz). However, bioinformatics already makes important contributions to neurobiology by decoding and describing coded molecular signals at all levels. First of all, this concerns genetic factors of neuronal maturation and diseases, which can be elucidated with the help of the OMIM database, genome and transcriptome analyses. At the level of the nerve cell, protein structures, in particular receptors and their activation, can be described in detail using protein structure analyses, molecular dynamics and databases (e.g. DrumPID, PDB database), as well as underlying cellular networks, protein-protein interactions and signalling cascades involved.
- Brain blueprints, so-called connectomes, are already available for *C. elegans* and are being intensively developed for other model organisms and humans. A connectome contains computer-readable information on how each nerve cell is linked to another and which receptors and ion channels play a role in this process. Suitable programming languages allow the direct simulation of information processing in the brain, especially for *C. elegans*. Numerous special software are available for clinical evaluations (EEG, computer tomograms) ("medical informatics"), but also for neurobiological experiments (e.g. a *neuronal activity detection tool*).

## 15.5   Exercises for Chap. 15

**Task 15.1**
What is the neuron software? Get an overview on the Internet.

**Task 15.2**
What is the *OpenWorm* project? Get an overview on the Internet.

**Task 15.3**
Find a link to the human *connectome*, which is all the neural connections in the human brain. What can you learn about neuroanatomy from this?

**Task 15.4**
Optical illusions arise from contrast enhancement and other neural mechanisms such as size constancy in the brain. Find a suitable reference to this on the Internet.

**Task 15.5**
The *Necker cube* (Necker's cube) gives a clue to the subjective present. Find a suitable reference to this on the Internet.

**Task 15.6**
One way to understand diseases of the nervous system is the OMIM database. What does the acronym stand for? What can you say about the data on alcoholism or schizophrenia in the database?

| Useful Tools and Web Links | |
| --- | --- |
| Allen Institute | https://alleninstitute.org |
| Allen Institute for Brain Science | https://www.braininitiative.org/alliance/allen-institute-for-brain-science/ |
| European Human Brain Project (HBP) | https://www.humanbrainproject.eu |
| Brain Mapping by Integrated Neurotechnologies for Disease Studies (Brain/MINDS) in Japan | https://brainminds.jp/en/ |
| String database | https://string-db.org |
| OMIM | https://www.omim.org |
| Connectome project | https://www.openconnectomeproject.org |
| Sausage Atlas | https://www.wormatlas.org/ |
| WormWiring | https://wormwiring.org/ |
| Virtual Insect Brain Lab | https://www.neurofly.de/ |
| Mouse Brain Connectivity Atlas | https://mouse.brain-map.org/static/atlas |
| Brain Activity Atlas | https://www.brainactivityatlas.org/ |
| Temporal lobe | https://www.temporal-lobe.com/background/connectome |
| Neuroactivity detection | https://www.ncbi.nlm.nih.gov/pubmed/23537512 |

# Literature

Bohm D, Hiley BJ (1985) Unbroken quantum realism, from microscopic to macroscopic levels. Phys Rev Lett 55(23):2511–2514. (PubMed PMID:10032166)

Buga AM, Scholz CJ, Kumar S et al (2012) Identification of new therapeutic targets by genome-wide analysis of gene expression in the ipsilateral cortex of aged rats after stroke. PLoS One 7(12):e50985

Hammer C, Stepniak B, Schneider A et al (2014) Neuropsychiatric disease relevance of circulating anti-NMDA receptor autoantibodies depends on blood-brain barrier integrity. Mol Psychiatry 19(10):1143–1149

Hassouna I, Ott C, Wüstefeld L et al (2016) Revisiting adult neurogenesis and the role of erythro-poietin for neuronal and oligodendroglial differentiation in the hippocampus. Mol Psychiatry 21:1752–1767. https://doi.org/10.1038/mp.2015.212

Hawrylycz M, Miller JA, Menon V et al (2015) Canonical genetic signatures of the adult human brain. Nat Neurosci 18(12):1832–1844. https://doi.org/10.1038/nn.4171. (*Vorbildliche ÜbersichtüberdieunterschiedlicheGenexpressionimGroßhirn.)

Markram H, Muller E, Ramaswamy S et al (2015) Reconstruction and simulation of neocortical microcircuitry. Cell 163(2):456–492

Rolls ET (2013a) A quantitative theory of the functions of the hippocampal CA3 network in memory. Front Cell Neurosci 7:98. https://doi.org/10.3389/fncel.2013.00098. (PubMed PMID: 23805074; PubMed Central PMCID: PMC3691555)

Rolls ET (2013b) The mechanisms for pattern completion and pattern separation in the hippocam-pus. Front Syst Neurosci 7:74. https://doi.org/10.3389/fnsys.2013.00074. (Review. PubMed PMID: 24198767; PubMed Central PMCID: PMC3812781 * Rolls' work is a good introduction to hippocampal function)

Schneider F, Horowitz A, Lesch KP, Dandekar T (2020) Delaying memory decline: different options and emerging solutions. Transl Psychiatry. 10(1):13. https://doi.org/10.1038/s41398-020-0697-x

Schulze K, Tillich UM, Dandekar T et al (2013) PlanktoVision – an automated analysis system for the identification of phytoplankton. BMC Bioinformatics 14:115. https://doi.org/10.1186/1471-2105-14-115

Voytek B, Knight RT (2015) Dynamic network communication as a unifying neural basis for cognition, development, aging, and disease. Biol Psychiatry 77(12):1089–1097. https://doi.org/10.1016/j.biopsych.2015.04.016

Xvu F, Shen Y, Ding L, Yang CY, Tan H, Wang H, Zhu Q, Xu R, Wu F, Xiao Y, Xu C, Li Q, Su P, Zhang LI, Dong HW, Desimone R, Xu F, Hu X, Lau PM, Bi GQ (2021) High-throughput mapping of a whole rhesus monkey brain at micrometer resolution. Nat Biotechnol. https://doi.org/10.1038/s41587-021-00986-5. Epub ahead of print

# Bioinformatics Connects Life with the Universe and All the Rest

**16**

### Abstract

Bioinformatics helps to better understand life. Whether one admires more adaptation (phylogeny, sequence analysis), metabolism (metabolic modeling, enzyme databases), or the regulation of these adaptations (systems biology). A common thread in all the great challenges of bioinformatics is to successfully master a new level of language and thus approach more deeply the very essence of biological regulation, understand forward and feedback loops, recognize stable system states, consider ecosystem modeling, climate or evolution. Actively questioning dangerous digitalization protects the creative freedom of everybody and of the internet. Bioinformatics helps to better understand the Internet and support the "Internet of Things" through software and databases. Bioinformatics helps drive new, creative and sustainable technologies (synthetic biology, nanotechnology, 3D printers, artificial tissues, etc.). Digitization with the help of bioinformatics is a pacesetter in molecular medicine. Bioinformatics also reveals limits to growth in mathematical models of ecosystems (e.g., the Verhulst equation for bacterial growth) and boosts according sensible, adapted system strategies.

We can sum up the fascination with bioinformatics like this: We use computers as tools to better understand life. Bacteria are already marvels of survival, efficiency and vitality. But with the help of bioinformatics, we can understand a little better how these fantastic feats work, whether we admire more adaptation (phylogeny, sequence analysis), metabolism (metabolic modeling, enzyme databases), or the regulation of those adaptations (systems biology). It is also clear that higher organisms are not only much more complex, but also often an even more exciting subject, whether you want to better understand plants or animals. Or, one might immediately attend to the most fascinating creature on this planet,

© Springer-Verlag GmbH Germany, part of Springer Nature 2023
T. Dandekar, M. Kunz, *Bioinformatics*,
https://doi.org/10.1007/978-3-662-65036-3_16

humans, and perhaps want to better cure their diseases or simply better understand and recognize how they are built and what separates humans from animals (anthropology). Consider, both our now very good anti-viral drugs for HIV infection and our modern targeted therapies for cancer (Duell et al. 2017) require bioinformatics computations to a very significant degree. For example, the Antiretroviral Therapy Cohort Collaboration (2008) showed that even with HIV disease, one has a near normal life expectancy with early therapy. The new approaches found for this, as well as the countless molecular therapy successes in the last two decades, would not have been possible without the support of molecular experiments by bioinformatics. Intriguingly, the article by Lengauer et al. (2014) describes how bioinformatics can help develop optimal individualized therapy against HIV. Similarly, Stratmann et al. (2014), Göttlich et al. (2016) and Baur et al. (2020) step by step improve a targeted cancer therapy using bioinformatics and cell culture experiments. The same is true for attempts to better understand the human brain. Here, computer models are important and are currently also being massively funded as an EU lead project (*"Blue Brain"* project of the EU). Perhaps a better strategy is to simply listen carefully to the brain and not immediately think of new computer architectures. This is precisely the goal of the US government's *Brain Activity* project, which is even three times more heavily funded than the EU project.

## 16.1    Solving Problems Using Bioinformatics

A common thread in all the great challenges of bioinformatics is climbing to a new level of language. Whether it is understanding the genetic (protein prediction) and genomic (gene prediction) code and correctly predicting proteins from foreign genomes or translating the sequence of a protein into three-dimensional protein structures, one is always climbing a new language level. Of course, this is even more true when doing systems biology, i.e., approaching the very essence of biological regulation in a deeper way and understanding *forward and feedback loops*, recognizing stable system states and can be said in the same way for ecosystem modeling (Kriegler et al. 2009). Thus, an important starting point for bioinformatics is first of all interest in the biological problem one wants to explore. Once one has delved a bit deeper into the problem, it is a matter of finding the right language to now build a suitable model for this phenomenon. This makes a great deal clear from the outset: we do not have the truth. It could well be that with a different language, with new software or even just a different perspective on the biological question, completely different insights will be possible than with the first approach just chosen. It is equally clear that only close collaboration with experimental biologists can help to figure out the best models. "True", i.e., internally consistent and correct, should be as consistent as possible in any model. But which model I then actually use is determined solely by the lowest possible deviation from experiment that is achieved. It is certainly much more important in bioinformatics to be able to correctly assess and correct the computer results, i.e., to have enough knowledge and overview to be able to evaluate and classify the computer results, than to be able to program oneself (which is nevertheless never a disadvantage). However, it is necessary not to be afraid of computers and to be able to use at least some programs, as well as to have a real interest in a biological question. If you want to do quite well, you should above all get enough exercise and do sports instead of wasting away in front of the computer or the book ("Mens sana in corpore sano", i.e., a healthy mind in a healthy body). In addition, one should also have a genuine willingness and interest to enjoy nature, biology, animals and plants, but also the encounter with fellow human beings (bioinformatics is an interdisciplinary subject).

Biology and of course its theoretical parts, such as bioinformatics, systems biology and theoretical biology, are together a key science of the twenty-first century. Here, experts are trained for complex systems that sometimes even overtake physics in their biological complexity. There are many problems that are pressing on our minds, whether they concern organisms, cells, molecules or the ecological balance. Equally important are medical problems or the biological part of research on artificial intelligence and neurobiology.

As indicated in the last two chapters, a new industrial revolution is upon us. Industry 4.0 or the "Internet of All Things" are important pacesetters for this approach. One simply knows exactly where which part is at any given time and electronically controls when it is installed where and how. The biological counterpart simply combines important individual aspects of bioinformatics, including the computer with synthetic biology, protein design and smart molecular biology (see infobox).

The infobox contrasts different approaches to the "Internet of Things". Here, the Internet notes or models where each thing is. This leads to faster, safer and cheaper production (Industry 4.0), increases the quality of life and sustainability in cities (Smart City) or optimises traffic (Smart Traffic). In biology, and thus in bioinformatics, one of the first steps towards this was the Gene Ontology Consortium (database catalog of all proteins, always answers: What is localized where in the cell? What is its molecular function? What cellular process is this?). Proprietary work includes the GoSynthetic Database, which compares synthetic biology and technical constructs, and the DrumPID Database, which compares drugs and protein-protein interactions (see infobox). Particularly impressive are the BioBricks from MIT, which, similar to our database but now underpinned with specific experiments, allow the artificial combination of biological control circuits. In addition, the systems biology achievements of the iGEM competitions in new synthetic biology are also impressive.

**"Internet of Things" (In Silico Knowledge of Where Each Thing Is Located)**
**Technical examples**
   Industry 4.0:

https://www.plattform-i40.de/I40/Navigation/DE/Home/home.html

   Smart City:

https://www.bioinfo.biozentrum.uni-wuerzburg.de/teaching/smart_city/
https://www.smart-cities.eu

Smart Traffic:

https://www.izeus.de/projekt/smart-traffic.html

   **Bioinformatics Examples**
   Gene Ontology:

https://www.geneontology.org

   GoSynthetic Database:

https://gosyn.bioapps.biozentrum.uni-wuerzburg.de/index.php

   DrumPID Database:


https://drumpid.bioapps.biozentrum.uni-wuerzburg.de

   MIT BioBricks:

https://web.mit.edu/jagoler/www/biojade/biobricks.html

   iGEM Parts:

https://igem.org/Main_Page

One can think of new technologies such as the nanocellulose computer chip, so that one can also control molecules individually here (especially via light-controlled protein domains, i.e., LOV or BLUF domains). Modern biology will be crucial in order not to have a technical proliferation here, but to create a stable, resilient and environmentally compatible new technology as a real basis of life for our civilisation.

However, it is also a general development in biology to use bioinformatics and large amounts of data to understand the cell more and more like an *"Internet of Things"*. This includes the fact that modern methods allow us to know much better where each molecule is (e.g., with *super resolution light microscopy*) and that we can then really control a process (synthetic biology, protein design, *nano factories, nano printers,* etc.). The same applies on a large scale: remote sensing, but also environmental samples of the microbiome of ecosystems (metagenomics) not only create new floods of data, but I also know more and more precisely where which object is located and how it is currently changing.

This also makes it increasingly possible to understand how our environment is changing and in what direction. This also increases the possibilities to actively change it. We must not be under any illusions. Even if we all "do nothing" (**Plan A:** *"Business as usual"*), the general activity of humans has a strong influence on our environment locally and also globally on the climate and biodiversity (Barnosky et al. 2011).

At present, it is clear that we have not yet achieved this stable basis for life. We are still living on borrowed time. We are steadily increasing our carbon dioxide emissions, we are struggling with global warming, overpopulation and dangerous nuclear armament, raw materials are becoming scarce, and electronic waste and environmental toxins are on the increase. But for these complex problems, there are good answers not only from technology, but also from biology and bioinformatics. It is clear that this must now be implemented decisively before our basis of life is irretrievably damaged and our current, outdated technology collapses, but also positive trends become visible (Lehman et al. 2021).

## 16.2 Model and Mitigate Global Problems

Bioinformatics approaches can contribute a great deal to global problems. This is because our entire world can also be viewed as an overall ecosystem and modelled in systems biology using computers. In addition, all the typical steps that otherwise have to be taken in bioinformatic modelling are there. In particular, one is forced to simplify strongly in some cases. One performs many simulations, and when the solution space becomes even more complex, one tries to represent and explore important combinations of conditions in individual models (so-called "scenarios"). Important problems, unfortunately, arise especially because of our success as a modern, technological civilization. This success, and in particular a certain prosperity, was increasingly achieved after the Second World War. This is also centrally important to pull the poorest strata of humanity (these have only one dollar a day at their disposal) out of hunger and disease, especially since only with four dollars a day of earnings is it possible to go to school and with 16 dollars a day to study, and only with at least 32 dollars a day is there so much prosperity that there is time for reflection, reading and real planning for the future (Rosling et al. 2019; illustrative: there is only different prosperity, otherwise all cultures are always the same people, to be seen in Hans Rosling's "Dollarstreet" https://www.gapminder.org/dollar-street/). The solution must not be to go back to the past (after general collapse only hunger and suffering) but to advance sustainability, digitalization and environmental and species protection as well as international cooperation and education. This is particularly important in the case of the five systemic risks of global war, global warming, economic crisis, pandemic and dictatorship (supported by digitalisation, for example). In contrast to smaller catastrophes, the systemic risks pose the danger of weakening our civilisation as a whole to such an extent that a downward spiral can occur.

**Nuclear Armament and the Arms Race**
Both are highly dangerous. A nuclear or even a major conventional war gets out of control all too quickly and destroys everything. Moreover, the risk of such an uncontrolled development has been exacerbated for a few years by nationalist and isolationist tendencies and new armaments on all sides, including nuclear ones. 2020 will witness the expiry of both the Strategic Nuclear Forces Agreement and the Intermediate-Range Nuclear Forces Agreement.

**Game Theory**  Interestingly, one can describe rearmament, arms races, nuclear exchange, but also quite generally combat and competition strategies very well with the help of game theory (and a lot of specific knowledge as an expert). Unfortunately, the explosive nature of this new arms race is currently being suppressed, presumably because no one wants to deal with this immanent and clearly too high risk in a seriously affected manner. Instead, the fear is directed towards other, more tangible dangers, such as international terrorism or radioactivity from nuclear power plants, both of which are negligible dangers. After all, the UN has just begun the process of outlawing nuclear weapons since October 2016. Moreover, one can deduce from systems biology considerations that it is important to continuously make our current peaceful state more robust: Disarmament, especially nuclear weapons, but also confidence-building measures are very important to prevent exacerbations here. Manageable own examples of modeling attack and defense strategies from infectious biology can be found in Dühring et al. (2015). Generally speaking, game theory (Amann and Helbach 2012) and **evolutionary strategies** (Bäck et al. 2013) are good bioinformatics approaches for modeling even such highly complex problems with many parameters.

## Global Warming

**Ecosystem Models**  Another explosive problem is *global warming*. Here, it is the other way around. The problem is not suddenly devastating, but will only strike in full force around the year 2100. Since it is also the case that, as with fishing, the temptation is very high to quickly grab a piece for oneself at the expense of others, there has so far been an unbroken trend for more and more carbon dioxide in our atmosphere. Although the Paris climate agreement of December 2015 gives us hope that concrete action will perhaps be taken against this, only time will tell whether the amount of savings will be sufficient. Here, of course, bioinformatics computer simulations directly help to simulate exactly how the climate will change in the future. And other models (from satellite data, for example) are used to measure exactly how the climate is changing right now. This is a blind spot without bioinformatics, so it's an ideal example of how bioinformatics approaches can actively help here. A nice introduction to modeling such problems is provided by the paper Lenton et al. (2008).

**Metabolic Modelling and Synthetic Biology**   There are also a number of innovative research measures that can help save a lot of carbon dioxide in the longer term. Bioinformatics can help optimize these processes using metabolic modeling and synthetic biology. For example, green plants fix carbon dioxide from the air using the enzyme Rubisco. This enzyme is from an ancient time (two billion years ago) and is, therefore, a little too sensitive to oxygen. It is now possible to try to replace this set of enzymes with alternative, better signalling pathways that are not sensitive to oxygen. New artificial enzyme cycles by Tobias Erb are particularly promising in this regard, which together can then replace Rubisco (e.g., CTECH cycle; Schwander et al. 2016). A slightly older paper from the USA gives a general overview (Bar-Even et al. 2010). There are also specific bioinformatics approaches to such metabolic design tasks (Lee et al. 2014). For example, coupling the CTECH cycle with a glycolate transport mutant that prevents the loss of carbon dioxide through light respiration theoretically results in a five-fold improvement in carbon dioxide harvest from the air (overview of this and other possibilities in Naseem et al. 2020; first experimental tests in Roell et al. 2021).

### Global Cooling ("Nuclear Winter") – Possible Climate Consequences of Nuclear Warfare

These are comparable steps to the simulation of global warming. The main problem in a nuclear war is the climate impact. Soot and dust from destroyed cities penetrates into the stratosphere and changes the climate there by a general cooling of four degrees Celsius over about 10 years, which would lead to worldwide crop failures and hunger.

**Climate Simulations**   Distressingly, even the replacement of about 100 nuclear weapons can cause significant global cooling (Mills et al. 2008). Whereby this will not always be the case, if not too many cities burn down with a big firestorm, because then the soot is only transported up to the stratosphere. The ozone layer, on the other hand, is always attacked when so many nuclear weapons are used. Again, this leads to skin cancer and the death of food crops and, again, worldwide famine. Although climate models are of course only approximations, there are now already a number of models and scenarios that make it very likely that these models are right in principle and that nuclear disarmament is urgent.

### Overpopulation

Here, there are different models to model population growth.

**Modelling with Scenarios**   The calculations show that the best results are to be achieved by rewarding environmentally friendly behaviour and consistently expanding helpful technologies (Hatfield-Dodds et al. 2015). This works better than dirigiste approaches or social re-education etc. A global solution ("**Plan B**"; Brown 2009) can only be a coordinated strategy that restores the Earth's so-called carrying capacity via education, more rights for women, family planning, equitable distribution of resources, sufficient and good job opportunities, but especially sustainability in agriculture, energy production, manufactur-

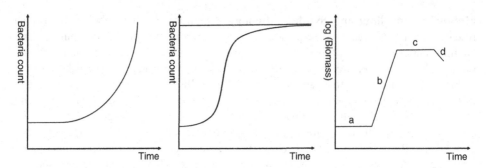

**Fig. 16.1** Population growth models: left, exponential growth (curve, e$^x$ equation); middle, growth with carrying capacity limit T (curve, Verhulst equation); right, growth phases of a bacterial culture (curve, sections): (**a**) lag phase, (**b**) exponential growth (as left curve, note logarithmic scale), (**c**) stationary phase, the population remains constant and at the carrying capacity limit T, (**d**) decay and disintegration phase

ing, conservation and landscape management. For such highly complex modelling, looking at different scenarios is most effective, because otherwise too many parameters would have to be tested.

**Population Modelling**  We can take an introductory look at three different simple models of population growth, which are particularly easy to observe in bacteria but apply to all animals (Fig. 16.1): The figure shows exponential growth in the left curve (exponential curve, e$^x$ equation, the bacteria keep doubling because nutrients are abundant). In the middle we see what inevitably happens after some time: Growth has reached the carrying capacity limit. The curve flattens out, and the population eventually remains constant. This is described by the Verhulst equation simplified in the figure in the middle. Finally, the right curve in the figure shows the growth phases of a bacterial culture (curve, sections). The lag phase (a), in which the bacteria adapt to the current environment and nutrients, is then followed by exponential growth if the nutrient medium is rich (b, same as left curve), but note the logarithmic scale in this partial figure. Once the food is depleted, the stationary phase (c) follows, the population remains constant, it is close to the carrying capacity limit T. It is interesting to note here that the microbes evolve many adaptations, in particular the switch from glycolysis to citric acid cycle (better yield of sugar), followed by the use of ketone bodies when they are present but the sugar is already completely depleted. This can be mimicked bioinformatically using various metabolic modeling tools (e.g., Liang et al. 2011). If still no new food is found, the degradation and decay phases follow in (d) (usually a somewhat slower exponential descent).

In "real life", i.e., when the bacteria have to survive under natural conditions, they often do not grow continuously, so that a complex mixture of these different curves then results, depending on the environment present.

Interestingly, there are also labs for *Computational Population Biology* (e.g., https:// compbio.cs.uic.edu/). There is also a metasite for this, the Biology WebDirectory (https:// www.biologydir.com/over-population/p1.html), which offers many more other terms.

### Pandemic

Global epidemics that cover almost all countries with infections are called pandemics. These occur about every 60 years. Since November 2019, the SARS-CoV-2 virus and the disease it causes, COVID-19, initially in China, have come to the attention of the global public, now a pandemic. Bioinformatics is critically important (Zimmerman et al. 2020) to rapidly decipher the genome sequence for such a new pathogen, to model the spread of infection and frequency of infection, but also to describe protein interactions as quickly as possible (especially with the human host, such as the spike protein of SARS-CoV-2, Shang et al. 2020). Bioinformatics also helps to develop therapies (e.g., antivirals, Bojkova et al. 2020), vaccines and neutralizing antibodies (Pinto et al. 2020).

### Pivot Point Limited Carrying Capacity

The global problems of mankind are each partial aspects of a biological system (man, other animal and plant species and the whole environment). They are all naturally inter-connected and can therefore be modelled using systems biology methods. They are a typi-cal problem that every species faces when the carrying capacity of its ecosystem is exceeded, if no adaptation to this situation takes place ("**Plan A**": do nothing or change nothing, but economic growth at any price). The important thing now, then, is to adapt decisively to carrying capacity through sustained system adjustments ("**Plan B**"; Brown 2009). If this happens too slowly, very robust, sustainable technologies are crucial ("**Plan C**", what to do if the system buckles?) to avoid serious damage. Here bioinformatics, together with molecular biology, computational technology, synthetic biology, and nano-technology, can point to important approaches (see Chap. 13). We must adapt now, but it is not too late. Our generation and our children will have to do this to get ahead of natural processes (Barnosky et al. 2011; Lehman et al. 2021). As humans, we have a good chance of doing this in an intelligent and social way, rather than solving this naturally through a hard degradation and decay phase, following our bacterial example.

## 16.3    Global Digitalisation and Personal Space

### Sample Task for this: The Growth of the Internet and Social Media, how Better to Predict It?

The simple answer is: with the right model! An example work is for instance Barberá et al. (2015). Here it is clearly explained how to extend a diffusion model for Internet messages appropriately, in particular to better account for the special network nodes at the edge. However, there are also very basic properties for networks that grow naturally. This is most easily seen in the case of a network of friend(s). For naturally growing networks it is

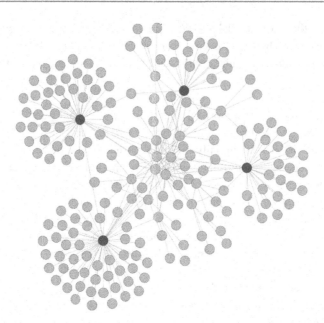

**Fig. 16.2** Growth and properties of a natural network. Shown is a network that grows naturally ("agglomeration"): The nodes with particularly strong networks grow particularly quickly (in red), the less strongly networked nodes grow more slowly, and the very weakly networked nodes at the edge grow most slowly (in cyan). This growth affects many natural phenomena, such as networks of friends, but also the growth of the Internet, of electricity networks, of trade and traffic, etc., but also the growth in molecular biology, for example of metabolic networks or of protein families

true that the highly networked nodes become even more networked particularly quickly (in the friend network: the star gets to know new people particularly quickly and is popular with more and more people), while the less networked nodes (the "wallflowers") grow their network particularly slowly (keep to themselves). This natural growth is illustrated in Fig. 16.2 with an example network.

Such networks are also called scale invariant because they look the same at every scale, similar to a fractal (see Sect. 9.5, Benoit Mandelbrot). Each section of a subnetwork of the large network again looks exactly the same. Again, there are a few central nodes (in red) that grow rapidly, and many marginal ones that grow very little or very slowly (in cyan). Bioinformatic analyses then calculate further properties, such as the diameter of the network (how far apart are the nodes?), how quickly information is passed on, how dominant individual nodes are *("centrality"),* and the like. The Cytoscape software already mentioned is very useful for this purpose. It has numerous *plug-ins* for various network biological analyses, such as the Network Analyzer (https://apps.cytoscape.org/apps/networkanalyzer).

**The Challenge: Growth of the Internet and Information Technology**
In the same way, it is clear that computers and information technology are penetrating more and more areas of our lives. At first glance, of course, everyone assumes they don't want it, don't need it, and so on. But the ubiquity of the Internet and mobile phones, WhatsApp and Facebook show that our civilisation is heading for an ever more intensive partnership with the computer, not only because of Industry 4.0 and new bionanotechnologies, but also because of its need for communication. However, we must also be ethically mature for this. Otherwise, without an enlightened use of the Internet and the computer, there will be a "silent takeover" by authoritarian forces, which will then subtly control the entire population via the computer and the media, including all their communications. This is no longer a utopia (as in George Orwell's novel "1984", source: Orwell 1949; current edition Orwell 2003), but there is now already the *"Citizen Score"*, which classifies the Chinese citizen according to his or her loyalty to the regime and his or her reference persons – as well as gradually controlling him or her more and more remotely.

Freedom is the most important good of a thinking being with consciousness, creativity, the freedom of thinking, to meet reference persons, to help people, to make art or also to solve a scientific problem (Schiller 1789). This is exactly what we should preserve, this freedom to shape our lives freely and not only in bioinformatics by an enlightened use of the computer and modern media improve, increase and secure this freedom for later generations (see box: "Digital Manifesto"). Then, and only then, will increasingly powerful computers become a powerful way to live out our freedom and free communication. For under a computer-based dictatorship, on the other hand, we face dark times ahead. For the background, see the "Digital Manifesto" box.

> **Digital Manifesto (from Spektrum der Wissenschaft)**
> The Digital Manifesto was written as a joint statement by eight scientists in 2015 (https://www.spektrum.de/thema/das-digital-manifest-algorithmen-und-big-data-bestimmen-unsere-zukunft/1375924). This call fits perfectly with our Chap. 16. It reflects both the threats and our tasks and opportunities in the face of the concentrated power of the internet and artificial intelligence. For this reason, the call is reproduced verbatim below:
>
> Big Data, Nudging, Behaviour Control: Are we threatened by the automation of society through algorithms and artificial intelligence? A joint appeal to safeguard freedom and democracy.
>
> Everything is becoming intelligent: soon we will not only have *smartphones*, but also *smart homes, smart factories and smart cities.* Will we end up with smart *nations* and a smart planet? We are currently experiencing the greatest historical upheaval since the end of the Second World War: the automation of production and the invention of self-driving vehicles is now being followed by the automation of society. This puts humanity at a crossroads where great opportunities are

*(continued)*

**(continued)**

emerging – but also considerable risks. If we make the wrong choices now, it could threaten our greatest societal achievements. The evolution is from programming computers to programming people. In effect, *"Big Nudging"* aims to bring many individual actions into line and manipulate perceptions and decisions. This brings it close to the deliberate disenfranchisement of citizens through state-planned behavior control. We fear that the effects could be fatal in the long run, especially considering the above-mentioned, partly culture-destroying effect.

**Looking to China: Is this what the future of society looks like (Feudalism 2.0)?**

1. By tracking and measuring all activities that leave digital traces, a transparent citizen is created whose human dignity and privacy fall by the wayside.
2. Decisions would no longer be free, because a wrong choice from the point of view of the government or company that sets the criteria of the points system would be punished. The autonomy of the individual would be abolished in principle.
3. Every little mistake would be punished, and no person would be unsuspected. The principle of the presumption of innocence would be invalidated. With *"predictive policing"*, even anticipated rule violations could be punished.
4. However, the underlying algorithms cannot work completely error-free. This would mean that the principle of fairness and justice would give way to a new arbitrariness against which it would hardly be possible to defend oneself.
5. With the external specification of the target function, the possibility of individual self-development would be abolished and with it democratic pluralism.
6. Local culture and social norms would no longer be the standard for appropriate, situational behavior.
7. Steering society by a one-dimensional objective function would lead to conflicts and thus to a loss of security. Serious instabilities would be to be expected, as we already know from our financial system.

**Solution: Collective creativity and freedom**

Centralized *top-down control* is a solution of the past, suitable only for systems of low complexity. Therefore, federal systems and majority rule are the solutions of the present. However, with economic and cultural development, societal complexity continues to increase. The solution of the future is collective intelligence: *citizen science, crowd sourcing* and online discussion platforms are therefore eminent new approaches to harness more knowledge, ideas and resources. Collective intelligence requires a high degree of diversity. However, this is reduced by today's personalized information systems in favor of reinforcing trends. Sociodiversity is as important as biodiversity. It is the basis not only for collective intelligence and innovation,

*(continued)*

**(continued)**

but also for societal resilience – the ability to cope with unexpected shocks. Reducing sociodiversity often also reduces the ability of economies and societies to function and perform. This is the reason why totalitarian regimes often end up in conflict with their neighbours. Typical long-term consequences are political instabilities and wars, as they have repeatedly occurred in our history. Plurality and participation should therefore not be seen primarily as concessions to the citizens, but as decisive functional prerequisites of efficient, complex, modern societies.

This can be done by adhering to the following basic principles:

- to decentralise the function of information systems to a greater extent,
- to support informational self-determination and participation,
- improve transparency for increased trustworthiness,
- reduce information distortion and pollution,
- enable user-controlled information filters,
- promote social and economic diversity,
- improve the ability of technical systems to work together,
- create digital assistants and coordination tools,
- support collective intelligence and
- to promote citizens' maturity in the digital world – a "digital enlightenment".

With this agenda, we would all benefit from the fruits of the digital revolution: Business, government and citizens alike. What are we waiting for?

## 16.4   What Are the Tasks for Modern Bioinformatics in the Internet Age?

What is very interesting is that the ten basic principles of the digital manifesto naturally help a great deal in moving all bioinformatics activities in a user-friendly, creative and developmental direction:

**Develop and Network Own Data Locally (Ad 1, 7 and 9)**
In addition to central databases and repositories, it is also important to make one's own data accessible locally, but also to network and collaborate (a particularly burning issue for medical data for bioinformatics).

**Self-Determination and Participation (Ad 2 and 6)**

Self-determination, participation and social diversity are supported by *open source programs*, joint development projects, among other things. In general, it is simply important to be responsive to the user and to listen to feedback in self-developed bioinformatics software.

**Transparency and Clarity (Ad 3, 4 and 10)**

Where does my data come from? What is its quality? Where do my conclusions come from? Indication of all sources! These are general principles ("good scientific practice") and mean, for example in bioinformatics and also in the sense of the Manifesto, that the data are freely accessible and so are the programs. In the best case, this also includes the source code. However, this also depends on whether the source code is being further developed. In this case, it is important to make at least the *executable* freely accessible and to explain transparently in a tutorial what the program does. The same applies to bioinformatics articles: As a reader, you should be mature and really understand the article, and as an author, you should make an effort to present the unfortunately mostly quite complex subject matter as clearly and transparently as possible.

**Free Working and Collaborative Coordination on the Network (Ad 5 and 8)**

An easy start is to use https://www.startpage.com/ as a browser, which does not immediately pass on all data to Google etc.. But you should generally use "free browsers", delete cookies regularly and use a new web browser from time to time. **Digital assistants and coordination tools** (Ad 8) we constantly explain here with regard to bioinformatics and a lot is direct analysis software (i.e., a digital assistant for bioinformatics). But databases also help to coordinate activities. But bioinformatics in particular has real coordination tools, such as the R community, which jointly writes the R language, but especially numerous R programs for statistical and bioinformatics analyses, or the numerous collaborations on the GNU project and other *open source activities* (Biojava, Bioperl, etc.). The best example is Wikipedia, which also attracts a steadily increasing share of bioinformatics wikis (e.g., www.wikidata. org or introductory https://de.wikipedia.org/wiki/Transkriptionsfaktor).

> **Conclusion**
>
> - We use the computer as a tool to better understand life. Bacteria are already marvels of survival, efficiency and vitality. But with the help of bioinformatics, we can understand a little better how these fantastic feats work, whether we admire more the adaptation (phylogeny, sequence analysis), the metabolism (metabolic modelling, enzyme databases) or the regulation of these adaptations (systems biology). Convincing results are *drug design,* for example in infectious diseases and cancer, but also progress in brain modelling.

*(continued)*

**(continued)**

- A common thread in all the great challenges of bioinformatics is establishing a new level of language. This is true whether it is a matter of understanding the genetic (protein prediction) and genomic (gene prediction) code and correctly predicting proteins from foreign genomes or translating the sequence of a protein into three-dimensional protein structures. Of course, this is even more true when doing systems biology, i.e., approaching the very essence of biological regulation more deeply and understanding *forward* and *feedback loops*, recognizing stable system states. This can be used in particular for ecosystem modelling. Especially with regard to climate (global warming and nuclear winter), but also species conservation, biodiversity and population dynamics, such systems biology models provide important insights.
- Our lives are becoming increasingly digitalised. The dangers of this digitalisation of our society are in the increasing control of citizens (e.g., NSA in the USA, *"Citizen Score"* in China, steering and opinion-making via the Internet and social forums: "post-factual society"). Active questioning of false information, personal rights and protection on the Internet, but also transparency, diversity of opinion, pluralism and democracy, as well as free information filters (browsers) controlled by the users, are important contributions to protecting and improving digital civil rights in the Internet age. Bioinformatics reinforces positive aspects of digitalisation: it helps to model the growth of social networks in biological models, to better understand the Internet and also to support the "Internet of Things" through software and databases. Bioinformatics helps drive new, creative and sustainable technologies (synthetic biology, nanotechnology, 3D printers, artificial tissues, etc.). Digitization with the help of bioinformatics is a pacesetter of molecular medicine. In mathematical models of ecosystems, bioinformatics digitization reveals limits to growth (e.g., Verhulst equation for bacterial growth) and resulting sensible system strategies.

## 16.5   Exercises for Chap. 16

**Task 16.1**
The Digital Manifesto. Familiarize yourself with the content!

**Task 16.2**
*Global Warming,* how does it work?

**Task 16.3**
What is *Doomsday Clock* and can you find it on the internet?

**Task 16.4**
What is meant by Plan B in the context of global problems? Explain this briefly.

**Task 16.5**
What is meant by Plan C in the context of global problems? Explain this briefly.

| Useful Tools and Web Links | |
|---|---|
| Blue Brain Project (EU) | https://bluebrain.epfl.ch/ |
| Brain Activity Project (US Government) | https://www.braininitiative.nih.gov/ |
| Gene Ontology Consortium | https://www.geneontology.org |
| MIT BioBricks | https://web.mit.edu/jagoler/www/biojade/biobricks.html |
| Smart traffic | https://www.izeus.de/projekt/smart-traffic.html |
| iGEM Parts | https://igem.org/Main_Page |
| WebDirectory | https://www.biologydir.com/over-population/p1.html |
| Cytoscape | https://www.cytoscape.org/ |
| NetworkAnalyzer | https://apps.cytoscape.org/apps/networkanalyzer |
| Smart City EU | https://www.smart-cities.eu |
| GoSynthetic | https://gosyn.bioapps.biozentrum.uni-wuerzburg.de/index.php |
| DrumPID | https://drumpid.bioapps.biozentrum.uni-wuerzburg.de/compounds/index.php |
| Digital Manifesto (Spektrum der Wissenschaft) | https://www.spektrum.de/thema/das-digital-manifest-algorithmen-und-big-data-bestimmen-unsere-zukunft/1375924 |
| Industry 4.0 | https://www.plattform-i40.de/I40/Navigation/DE/Home/home.html |
| Smart City Würzburg | https://www.bioinfo.biozentrum.uni-wuerzburg.de/teaching/smart_city/ |
| Computational Population Biology (University of Chicago) | https://compbio.cs.uic.edu/ |

## Literature

Amann E, Helbach C (2012) Spieltheorie für Dummies. Wiley, Weinheim, S 314 (* A nice introduction to current game theory is given here. Dühring et al. (2015) then illustrate biological modeling with this tool.)

Antiretroviral Therapy Cohort Collaboration (2008) Life expectancy of individuals on combination antiretroviral therapy in high-income countries: a collaborative analysis of 14 cohort studies. Lancet 372(9635):293–299. https://doi.org/10.1016/S0140-6736(08)61113-7

Bäck T, Fousette C, Krause P (2013) Contemporary evolution strategies. Natural Computing Series. Springer, Berlin. https://doi.org/10.1007/978-3-642-40137-4_1 (*A recent introduction to how evolutionary algorithms can be used to model and study very complex problems. [Link to Google Books for online version: https://books.google.de/books?id=GPy8BAAAQBAJ&printsec=front cover&hl=de&source=gbs_ge_summary_r&cad=0#v=onepage&q&f=false].)

Barberá P, Wang N, Bonneau R et al (2015) The critical periphery in the growth of social protests. PLoS One 10(11):e0143611. https://doi.org/10.1371/journal.pone.0143611. (eCollection 2015. PubMed PMID: 26618352; PubMed Central PMCID: PMC4664236)

Bar-Even A, Noor E, Lewis NE et al (2010) Design and analysis of synthetic carbon fixation pathways. Proc Natl Acad Sci U S A 107(19):8889–8894. https://doi.org/10.1073/pnas.0907176107. (Epub 2010 Apr 21. PubMed PMID: 20410460; PubMed Central PMCID: PMC2889323)

Barnosky AD, Matzke N, Tomiya S et al (2011) Has the Earth's sixth mass extinction already arrived? Nature 471(7336):51–57. https://doi.org/10.1038/nature09678

Baur F, Nietzer S, Kunz M et al (2020) Connecting cancer pathways to tumor engines: a stratification tool for colorectal cancer combining human *in vitro* tissue models with Boolean *in silico* models. Cancers (Basel) 12(1), 28:pii: E1761. https://doi.org/10.3390/cancers12010028

Bojkova D, Klann K, Koch B et al (2020) Proteomics of SARS-CoV-2-infected host cells reveals therapy targets. Nature 583(7816):469–472. https://doi.org/10.1038/s41586-020-2332-7

Brown L (2009) Plan B 4.0 mobilizing to save civilization. Norton, New York (*A very nice introduction to modeling this complex problem and how humans can restore the carrying capacity of their environment. As a corollary, this is also a gripping call. We would only need to put a small portion of the money spent on military armaments into sustainability, but use those funds wisely and socially)

Duell J, Dittrich M, Bedke T et al (2017) Frequency of regulatory T cells determines the outcome of the T cell engaging antibody blinatumomab in patients with B precursor ALL. Leukemia. https://doi.org/10.1038/leu.2017.41

Dühring S, Germerodt S, Skerka C et al (2015) Host-pathogen interactions between the human innate immune system and *Candida albicans*-understanding and modeling defense and evasion strategies. Front Microbiol 6:625. https://doi.org/10.3389/fmicb.2015.00625. (Review. PubMed PMID: 26175718; PubMed Central PMCID: PMC4485224)

Göttlich C, Müller LC, Kunz M et al (2016) A combined 3D tissue engineered in vitro/in silico lung tumor model for predicting drug effectiveness in specific mutational backgrounds. J Vis Exp 110:e53885. https://doi.org/10.3791/53885

Hatfield-Dodds S, Schandl H, Adams PD et al (2015) Australia is 'free to choose' economic growth and falling environmental pressures. Nature 527(7576):49–53. https://doi.org/10.1038/nature16065

Kriegler E, Hall JW, Held H et al (2009) Imprecise probability assessment of tipping points in the climate system. Proc Natl Acad Sci U S A 106(13):5041–5046. https://doi.org/10.1073/pnas.0809117106. (PubMed PMID: 19289827; PubMed Central PMCID: PMC2657590)

Lee Y, Lafontaine RJG, Liao JC (2014) Ensemble modeling for Robustness analysis in engineering non-native metabolic pathways. Metab Eng 25:63–71. https://doi.org/10.1016/j.ymben.2014.06.006. (Epub 2014 Jun 24. PubMed PMID: 24972370)

Lehman C, Loberg S, Wilson M, Gorham E (2021) Ecology of the Anthropocene signals hope for consciously managing the planetary ecosystem. Proc Natl Acad Sci U S A. 118(28):e2024150118. https://doi.org/10.1073/pnas.2024150118

Lengauer T, Pfeifer N, Kaiser R (2014) Personalized HIV therapy to control drug resistance. Drug Discov Today Technol 11:57–64. https://doi.org/10.1016/j.ddtec.2014.02.004. (Review. PubMed PMID: 24847654)

Lenton TM, Held H, Kriegler E et al (2008) Tipping elements in the Earth's climate system. Proc Natl Acad Sci U S A 105(6):1786–1793. https://doi.org/10.1073/pnas.0705414105. (PubMed PMID: 18258748; PubMed Central PMCID: PMC2538841 * Especially the illustrations are very nicely descriptive. Further helpful and interesting details of such modeling can be taken from the two subsequent works Schellnhuber HJ. Tipping elements in the Earth System. Proc Natl Acad Sci U S A 106(49):20561–20,563 and Kriegler E, Hall JW, Held H et al. (2009) Imprecise probability assessment of tipping points in the climate system. Proc Natl Acad Sci U S A 106(13):5041–5046. https://doi.org/10.1073/pnas.0809117106.)

Liang C, Liebeke M, Schwarz R et al (2011) Staphylococcus aureus physiological growth limitations: insights from flux calculations built on proteomics and external metabolite data. Proteomics 11(10):1915–1935. https://doi.org/10.1002/pmic.201000151

Mills MJ, Toon OB, Turco RP et al (2008) Massive global ozone loss predicted following regional nuclear conflict. Proc Natl Acad Sci U S A 105(14):5307–5312. https://doi.org/10.1073/pnas.0710058105

Naseem M, Osmanoglu Ö, Dandekar T (2020) Synthetic rewiring of plant $CO_2$ sequestration galvanizes plant biomass production. Trends Biotechnol 38(4):354–359. https://doi.org/10.1016/j.tibtech.2019.12.019

Orwell G (1949) Nineteen eighty-four A novel. Secker & Warburg, London

Orwell G (2003) Nineteen eighty-four. Thomas Pynchon (Foreword); Erich Fromm (Afterword). Plume. ISBN 0-452-28423-6.

Pinto D, Park YJ, Beltramello M et al (2020) Cross-neutralization of SARS-CoV-2 by a human monoclonal SARS-CoV antibody. Nature 583(7815):290–295. https://doi.org/10.1038/s41586-020-2349-y

Roell MS, Schada von Borzykowski L, Westhoff P, Plett A, Paczia N, Claus P, Urte S, Erb TJ, Weber APM (2021) A synthetic C4 shuttle via the β-hydroxyaspartate cycle in C3 plants. Proc Natl Acad Sci U S A. 118(21):e2022307118. https://doi.org/10.1073/pnas.2022307118

Rosling H, Rönnlund AR, Rosling O (2019) Factfulness: ten reasons we're wrong about the world–and why things are better than you think. Sceptre: City of Publication: London.: ISBN: 9781473637474.

Schiller F (1789) Was heißt und zu welchem Ende studiert man Universalgeschichte? ("What does it mean and to what end does one study universal history?" Inaugural lecture in Jena, May 26, 1789). Akademische Buchhandlung, Jena 1789: https://www.deutschestextarchiv.de/book/show/schiller_universalgeschichte_1789

Schwander T, Schada von Borzyskowski L, Burgener S et al (2016) A synthetic pathway for the fixation of carbon dioxide in vitro. Science 354(6314):900–904

Shang J, Ye G, Shi K et al (2020) Structural basis of receptor recognition by SARS-CoV-2. Nature 581(7807):221–224. https://doi.org/10.1038/s41586-020-2179-y

Stratmann AT, Fecher D, Wangorsch G et al (2014) Establishment of a human 3D lung cancer model based on a biological tissue matrix combined with a Boolean in silico model. Mol Oncol 8(2):351–365. https://doi.org/10.1016/j.molonc.2013.11.009. (Epub 2013 Dec 18)

Zimmerman MI, Porter JR, Ward MD et al. (2020) Citizen scientists create an exascale computer to combat COVID-19. bioRxiv [Preprint]. 28:2020.06.27.175430. https://doi.org/10.1101/2020.06.27.175430

# Conclusion and Summary

**17**

**Abstract**

Bioinformatics is now all the rage because of big data. However, computational biology also uses computers to provide unprecedented valuable biology insights, which is our main concern. With more and more new data and the application of data analysis to fundamental biological questions that can only now be answered by these new data in the first place, we will enter the fascinating new territory of modern molecular medicine and molecular biology in this century. With the help of bioinformatics this data flood makes us more knowledgeable; without bioinformatics, we will rather drown in the next big data wave.

In **principle, bioinformatics is simple (Part 1):** Sequence analyses, such as sequence comparison with BLAST, allow the language of life to be deciphered, whereby DNA and RNA sequences can be translated into proteins relatively easily with the computer. Numerous programs look at protein sequences in particular. The ExPASy server of the Swiss Bioinformatics Institute is important here. We learned about useful tools for RNA analysis, such as the RNA Analyzer. Numerous DNA sequences, databases and many electronic books, tips and programs for this purpose are available at the NCBI (National Center of Biotechnology Information) as well as all important publications (MEDLINE). With these techniques, we can detect viruses, determine the function of proteins, but also discover new RNA molecules that play a role in cancer or heart failure, for example. In order to model the metabolism of a cell, we need a list of all enzymes and metabolites that need to be kept in balance within the cell. From this, the computer can then calculate all possible metabolic pathways, and with a little more data (e.g., on gene expression), it is also possible to determine flux levels. This makes it possible to find targets for antibiotics, but also to better understand how bacteria grow, adapt to the environment or optimise the yield

© Springer-Verlag GmbH Germany, part of Springer Nature 2023
T. Dandekar, M. Kunz, *Bioinformatics*,
https://doi.org/10.1007/978-3-662-65036-3_17

of useful products such as citric acid or insulin. Finally, we have seen that regulation in cells is very logical. If we want to recreate this with a computer, we first have to find out which protein interacts with which. This can be done using protein interaction databases such as STRING and KEGG. Then you can assemble the logical (Boolean) network into a regulatory circuit in the cell. If this is made computer-readable (e.g., XML format), a program can then simulate the dynamics of regulation without having to know exactly how fast the process takes place. For this, the model is then only "semi-quantitative", i.e., it only tells what comes sooner or later, what is more or less activated in the cell. Nevertheless, this helps to describe how our nervous system works or to find new drugs against heart failure.

**Principles for understanding bioinformatics** and modern biology we have learned in the **second part.** Heuristic, i.e., fast, but not completely accurate searches speed up modern bioinformatics programs. Bioinformatics decodes coded information in cells, and living cells use different codes and levels of language. The analysis tasks for the computer are either easy, meaning that in the foreseeable future the computer can handle them, or they are unpredictably long. This happens especially easily when many combinations are tested.

Every organism is a complex system, but they behave in fundamentally similar ways. One can infer their behavior through big data, such as omics techniques. Emergence is at the heart of this, i.e., the appearance of completely new system properties as components come together, especially in human consciousness when a critical number of neurons come together in a previously unfounded way. Modular structure, positive and negative *feedback loops* are basic properties. A basic theme of biology is the consideration of evolution. Bioinformatics likes to use phylogenetic trees and evolutionary comparisons to quickly identify basic properties (many organisms; conserved regions in a protein sequence) and specific properties.

All these data analyses allow detailed insights into the molecular biology of the cell by looking step by step at involved protein sequences, their localization in the cell and their properties.

**Bioinformatics becomes fascinating (Part 3)** when data analysis provides surprising biological insights. For example, modern genome analysis makes it figuratively clear that genetic modifications affect every human genome and that everyone carries "good" and "bad" genes with them, useful or harmful depending on the environment. Molecular sequences only ever make sense in the context of the cell. Understanding this language can be used for synthetic biology and protein design, such as the nanocellulose chip. Bioinformatics is also making the powerful structure of our brains and limitations for natural and artificial intelligence clearer. Bioinformatic modelling and simulations also encompass ecosystems, infections (currently: Covid19 pandemic) and climate models, the internet and sharpen the view for chances and problems of our digital society.

**Conclusion** Bioinformatics uses computers to better understand biological problems, i.e., to find similarities between molecules, for example at the sequence level. It can break down processes in the cell, for example with *pathway databases*, or even reproduce biological processes in the model, such as the growth, differentiation and death of cells. The goal is biology and its understanding, the means are the computer and experimental data or new experiments to substantiate the gained understanding as solid as possible. That is why it is always important to be critical of the results of one's own computer analyses. One's own expertise and knowledge of the correlations should always critically question and accompany the results - otherwise one can be very far off the mark. Finally, it should be emphasized that for mathematical (Turing computability) as well as biological reasons (we are living beings, the computer is not) it remains our very own task to set goals, tasks and directions for biology and our work, no matter how modern the computers may be.

This brings us to the end of this introduction to bioinformatics. Bioinformatics is the way to the new biology of the twenty-first century, an intensive theoretical biology by and with computers. It helps us to make a new biology, to know better what life is, what is in bacteria, genomes, plants and protein structures, but also to get closer to the mystery of man in all its facets, one's own consciousness being a particularly deep mystery. This book wants to motivate us to this new biology, not only as a textbook, but also so that we can successfully and actively better shape our future with the help of bioinformatics. There are more than enough pressing future problems and important, motivating future tasks.

# Part IV

# Glossary, Tutorial, Solutions and Web Links

# Glossary

**18**

**Abstract**

The glossary explains and defines important terms in bioinformatics. We can only explain the most important terms here. The field is developing rapidly and is, after all, situated between two disciplines, biology and computer science. It is thus a bit more challenging in the set of basic terms than if it were just about one subject. First we give a short definition, then explain details and give examples for complex terms.

The glossary explains and defines important terms in bioinformatics. We can only explain the most important terms here. The field is developing rapidly and is also still located between two disciplines, biology and computer science, and thus also somewhat more demanding in the amount of basic terms than if it were only about one subject. First we give a short definition, then explain details and give examples for complex terms.

**Ab Initio Protein Structure Prediction** Method for predicting protein structure from a sequence based on the biophysical properties (e.g. hydrogen bonds or hydrophobic effects) of proteins.

**Amino Acids** Carbon compounds, consist of an amino and an acid group at the C-alpha carbon atom and a characterizing amino acid residue. Proteinogenic amino acids are the 20 that are incorporated into proteins with the genetic code.

**Artificial Intelligence** Properties of a computer to reproduce biological facts so that independent learning is possible, e.g. neuronal network or also through hidden Markov models.

© Springer-Verlag GmbH Germany, part of Springer Nature 2023
T. Dandekar, M. Kunz, *Bioinformatics*,
https://doi.org/10.1007/978-3-662-65036-3_18

**Attractor** Limited system state that pulls nearby system states into this stable basic state, e.g. health (immune system can cope with slight infections or pulse, returns to normal after exertion; however, if the disturbances are too great, then the stable system state changes and one becomes ill; again a new attractor that can last longer until one becomes healthy again).

**Babbage Test** Test for artificial intelligence in which outside people are asked to distinguish between a human and a computer (both undercover). If the computer succeeds in deceiving people into thinking it is human, then the computer has artificial intelligence.

*Big Data* Refers to the flood of large amounts of data generated in the course of modern experimental methods, which, for example, have to be analysed bioinformatically in order to gain new and exciting insights.

**BiNGO** *Cytoscape plugin* that identifies overrepresented biological functions (with *p-value* and corresponding genes) using *Gene Ontology* grouping in a network (see also *Gene Ontology*, GO).

**Bioinformatics** Bioinformatics, or *computational* biology, is the study of biological questions using computers. In this process, information ("data") and findings ("models") about organisms are collected (in databases), analysed (by experts, the bioinformaticians, who use various computer programs for these analyses) and reproduced in models ("simulations"). Essential properties (system properties) for the biological phenomenon under investigation are worked out ("systems biology"). Biologists often focus on plants and animals, fungi or lower organisms (bacteria, viruses). The latter are easier to understand and thus to reproduce in the computer, e.g. the metabolism of bacteria or the reproduction of viruses. For doctors, other medical professions (human geneticists, molecular physicians) and many interested biologists, however, the focus is on humans. Both health ("physiology") and disease are described in detail. The starting point of many bioinformatic studies is the flow of genetic information from DNA (the genome) through transcription (in higher cells in the cell nucleus) to RNA and after translation in the ribosome via the genetic code to proteins. Programs and software are used to study biological function. This is done, for example, by means of sequence analyses in order to obtain information about a pathogen, but also, for example, to obtain differences between the organisms involved (e.g. humans and parasites) by genome comparisons. In the case of proteins, but also regulatory and catalytic RNA, the analysis of the structure helps to better decipher their function (protein structure analysis, RNA analysis). It is also possible to create metabolic networks and compare them with each other, and finally, for example, to calculate drugs for important proteins in the parasite that optimally block the parasitic protein but are tolerated by humans. Signalling networks can also be modelled and studied to better understand cell maturation (differentiation, embryology) and to better combat or prevent diseases such as cancer, heart failure and stroke (together over 75% of all causes of death). Predictions are verified with the help of experimental data. In the meantime, other molecules can also be measured intensively (more complex), such as metabolites (e.g. lipids, sugars, vitamins, cofactors), proteins, nucleotides. Also new is an increasing amount of image data, both from microscopy and remote sensing, for which proprietary powerful algorithms are available for processing (image processing). The same applies to functional assays (e.g. ChIPseq, CLIP; RNAi screens; transposon screens) and high-throughput screening (HTS), for example for drugs, for which the use of computers is essential for the evaluation and *in silico* pre-testing of often many more candidate molecules. Bioinformatics is thus able to answer basic biological and medical questions much better than was previously possible, based on theoretical knowledge and ever new data. Bioinformatics has become the spearhead of modern biology, in that ever better computer predictions (especially via the Internet, with the help of modern *Deep Learning,* neural networks, *neurocomputing,* but also with ever better search possibilities through PSSMs and HMMs) help to advance these current research areas even faster. This is, for example, research on stem cells, ecosystem modelling, neurobiology, nanotechnology, nanobiotechnology as well as modern molecular biology with protein design and synthetic biology. Molecular medicine in particular is becoming much stronger with the help of bioinformatics through insight into the complex regulation of, for example, the immune system (help with allergies, rheumatism) regenerative medicine (help with chronic diseases) and the human genome. However, this only applies if the ethical aspects are internalised and incorporated into all problem solutions: Human dignity, respect for the individual, quality of life; effective control and already at the planning stage safe, intelligent design of related technology, be it computers, microorganisms, (human) cells or nanotechnology (cf. digital manifesto).

**Bit**  A bit of information is the smallest unit of information, a "yes" or "no" decision.

**BLAST (Basic Local Alignment Search Tool)**  Bioinformatics algorithm that allows protein and nucleotide sequences to be compared with a large database in terms of their local similarity. In this process, a sequence is compared for its similarity with reference sequences in a database, i.e. with sequences that are already known, and can provide information, e.g. which virus a patient has contracted. BLAST uses a heuristic search and here the *two-hit method* (2-hit method): A short word list *(lookup table)* is first compared with the short word lists of the database (indexed database). If at least one matching short word is found in an entry, the system immediately checks whether there is another short word hit in the vicinity (fixed distance). Only then the *alignment* is calculated. In all other cases, the algorithm *blasts ahead* to the next database entry.

**CATH** *(Classification by Class, Architecture, Topology and Homology)*

Classification of protein structure by class (structure of secondary structure), architecture (high similarity of secondary structure but no homology), topology (similar properties of secondary structure) and homology (evolutionary ...), based on experimentally determined three-dimensional protein structures from the protein database **PDB.**

**Chaotic Systems**  Description of systems (complex systems) whose behaviour is predictable (can be described exactly) only over short periods of time, but whose long-term behaviour is kept within fixed limits ("attractor").

**Clustering (Cluster Analysis)** Statistical procedure to classify (group) objects into groups (clusters) with similar characteristic structures (characteristics). A distinction is made between supervised (groups known) and unsupervised clustering (groups unknown).

**Code** Specification for the unambiguous representation or assignment of characters with the aid of a given character sequence (e.g. genetic code using base triplets to represent the 20 amino acids).

**COGs** (clusters of orthologous genes) see last universal common ancestor.

**Computers** are data processing machines. To this end, they now typically consist of hardware (electronic switches, transistors, *integrated chips*) and other parts (input and output devices, housings, etc.). They process instructions (software) in sequence to generate new results from the data, e.g. calculations, sequences, result lists or networks (typical results in bioinformatics calculations).

**Consensus Sequence** Conserved sequence of motifs in a multiple *alignment* of several sequences, such as nucleotides of an enzyme (see also PSSM).

**Corona virus** see Pandemic.

**Databases** Different databases (software component) integrate and collect biological data and make it available to the general public over the Internet using a serviceable computer (hardware component called a "server"). Databases hold all the data that people look up. Typically, this is done in many records. Different properties about a particular record are held in individual data fields. How this looks in detail is determined by the data model. Finally, the data can be searched using a query (database query). A simple query language popular in bioinformatics for simple, smaller databases is the *"Structured Query Language"* (in short: "SQL"), and such a database is then an SQL database. Important bioinformatics databases are listed many times in the book, e.g. GenBank (genome and nucleotide sequence data) and UniProt/Swiss-Prot for protein sequences.

**Data-Driven Modeling** Normalization of the different units of the bioinformatic model according to the experimental data, i.e. the typical times of the signaling cascade, receptor excitation, phosphorylation of kinases, etc. are determined by this.

**Dimension Reduction** see Principal Component Analysis (PCA).   DNA (deoxyribonucleic *acid*, DNA for short)

Biochemically, a mixture of nucleotides that are all connected via a deoxy-ribose sugar and a phosphate "backbone" to form a long molecule, the DNA single strand. Bioinformatically centrally important because DNA contains all the genetic material (hereditary material, also called the genome) and thus all the hereditary information of an organism. The DNA single strand pairs on its own with its counterpart strand, so that DNA typically exists as a twisted double strand, i.e. a double helix. Deoxyribose has one less oxygen atom than ribose, making it more stable. The double helix protects the genetic information and allows repair via the opposite strand. Both lead to the fact that DNA stores information much better than RNA.

**Domain Name Server** Specific number (IP address), e.g. 132.187.25.1, of a computer via which it is connected to the Internet protocol.

**Dotplot** Allows to compare two sequences in a diagram (x-/y-axis) to find similar areas (represented as a dot).

**Drug Design** A branch of bioinformatics that deals with the design of a drug with optimal properties.

**Dynamic Programming** Systematic trying of possibilities for optimizing solutions with the computer, such as secondary structure folding, dynamically allocating more and more memory.

**Elementary Mode Analysis** Part of metabolic modelling/metabolic modelling in which all metabolic sources, metabolic product/excretions ("sinks") as well as involved enzymes, metabolites (internal/external) and metabolic reactions (reversible/irreversible) are translated into a mathematical calculation rule. It is calculated which enzyme chains (metabolic pathways) each put all internal metabolites involved into equilibrium (elemental flux modes). Elementary mode analysis is important, for example, in order to achieve the best possible yield of a product (sinks) with a starting product (sources) or to identify differences between organisms, for example, with regard to medical use (e.g. antibiotic that blocks metabolic pathways of the bacterium but does not cause a toxic effect in the patient).

**Emergence** System effect in which new effects and properties arise from the coming together of the components that cannot be attributed to the individual components, e.g. individual blood and heart muscle cells form the circulatory system with properties such as blood pressure and pulse.

**ENCODE (ENCyclopedia of DNA Elements)** The ENCODE project (consortium) is the follow-up project to the Human Genome Project (sequencing of the human genome) and seeks to characterize the entire human genome and transcriptome in greater functional detail.

**Endosymbionts, Endosymbiont Hypothesis** Large organelles, especially mitochondria and chloroplasts, also contain DNA and a small ring-shaped DNA molecule (a few thousand nucleotides). This indicates descent from free-living bacteria (endosymbiont hypothesis).

**Enzyme** see Protein.

**Epidemic**  see Pandemic.

*E-value*  Statistical parameter that indicates whether my output *alignment* will be found again in the database with a similar or better score (expected *value of* a random hit; should be less than 1 in 1 million). It is therefore dependent on the size of the database (in contrast to the *p-value*).

**Evolution**  (from Latin evolvere, "to develop") deals with the gradual changes over time (typically long periods of time up to millions of years) of genetic material and external appearance in individuals, populations, species up to entire ecosystems. When genetic material is passed on for the next generation during cell division, there are sometimes minor or major changes in the genetic material in addition to identical copies of the genetic material. The resulting appearance of the body (phenotype) may be better, worse, or equally adapted to the currently prevailing environment as a result of these genetic changes. Random changes (mutations), natural selection and reproduction (replication) work together to achieve this. Depending on the environment, a mutation can thus be advantageous, disadvantageous or insignificant (neutral).

**ExPASy Server**  This is the most famous website of the Swiss Bioinformatics Institute, an expert system for protein sequence analysis (**E**xpert**P**rotein**A**nalysis **Sy**stem). It is an example of a portal, i.e. a website where you can find numerous databases and software. For example, I can use different software options to check whether my protein sequence is really the enzyme I think it is (for example, if it is the BLAST result). The database and software PROSITE checks whether all the important catalytic amino acid residues are present or *"peptide properties"* checks whether the amino acid composition matches the protein, for example, whether the protein has enough hydrophobic amino acids to fit into the membrane.

**False Positive Hits**  Proportion of hits that are grouped incorrectly (e.g., potentially predicted interaction partners that are not experimentally validated or person grouped as sick but who are actually healthy).

**FASTA**  Storage format (text-based) of bioinformatics for sequences, such as gene sequences.

**Feedback Loops**  positive/negative *feedback loops*.

**First Gödel's Incompleteness Theorem**  Proves that there are always unprovable statements in sufficiently strong contradiction-free systems (computer thus remains in the undecidable).

**Genetic Drift**  Random sequence change that can affect the function, e.g. of catalytic domains or functional sites. Stronger moves (e.g. insertion, deletion, clear direction) are called genetic shift and can lead e.g. to virus subtypes (e.g. in influence).

**Gene**  Section on DNA that codes for specific information and genes.

**Gene Expression**  Provides information about the activation of a gene, which leads to the biosynthesis of a protein through transcription (RNA synthesis) and translation.

**Gene Ontology (GO)**  Grouping of genes according to their species-specific known function in biological processes, cellular components and molecular function. Many *tools* use this grouping for initial functional analysis and characterization of genes, such as the Cytoscape plugin BiNGO (see BiNGO).

**Genetic Algorithms**  Search strategy in which solutions are bred in the computer using artificial evolution through selection, mutation and recombination of digitally programmed chromosomes (encode the problem).

**Genome**  The entire genetic material (hereditary material, also called genome) and thus all genetic information of an organism. Only viruses can exceptionally contain a genome of RNA. Large organelles, especially mitochondria and chloroplasts, also contain DNA and a small ring-shaped DNA molecule (a few thousand nucleotides). This indicates their descent from free-living bacteria (endosymbiont hypothesis). The genome of viruses is typically a few thousand nucleotides in size (only the polymerase duplicates the genome), that of bacteria a few million base pairs (polymerase and correction enzymes), and that of higher organisms (with nucleus) several billion nucleotides (polymerase and sophisticated correction pathways).

**Genomics**  Analysis of the genome, the totality of all genes.

**Genome-Wide Association Studies (GWAS)**  Studies to identify genome-wide important genes associated, for example, with a disease, in order to detect more precisely the specific signal of mutations that are then important for the particular individual aspect (disease, which exact subtype).

**Genetic Shift**  Random change of larger sequence regions or even entire genes that can influence function, e.g. catalytic domain or functional site. In contrast, gene drift is a smaller random change in allele frequency within the gene pool of a population.

**Global** *Alignment*  Two sequences are compared with each other by comparing which amino acid residues are modified and which are conserved. One can either compare over the whole length or only a part of the sequence (see local *alignment*). For global *alignment,* there is an exact method, Needleman and Wunsch search, which is slow but accurate, and various heuristic methods (inexact, but few errors and much faster). For example, for phylogenetic methods, CLUSTAL search works fast.

**Gödel, Gödel's Incompleteness Theorem**  see second Gödel's incompleteness theorem.

**Hardware**   All the physical aspects of a computer (chassis, transistors, input and output devices such as printers, memory disks, etc.).

**Hidden Markov model**   Stochastic model that calculates certain transition probabilities from an observable system state on the basis of a Markov chain and allows statements about the *hidden* states.

**Homology Protein Structure Prediction**   Method for predicting protein structure, using known protein structures as templates.

**Hypothesis-Free**   On the positive side, this is research that looks at the results without bias and develops appropriate conclusions. On the other hand, it usually means expensive omics experiments. For these, it is not a good idea to just produce a lot of data without an initial scientific hypothesis and hope that the statistician will then find something "significant". The following paper explains this problem very well: Ioannidis JPA (2005) Why most published research findings are false. PLoS Med 2(8): e124.

*In Silico* **Analysis**   with the computer, computer-assisted analysis. Literally: "in silicon" (Latin).

**Internet Protocol (IP for Short)**   Used to exchange data packets from computer to computer, so that this is done as quickly as possible, making use of all free space and also coping with a major computer failure.

**IP**   Abbreviation for the Internet Protocol (see there).    Knowledge-based

**Bioinformatics work, such as the creation of a network, based on literature and expert knowledge.**

**Last Universal Common Ancestor (LUCA) or LCA**   Last common ancestor of all life, which can be inferred bioinformatically via protein sequences and protein families (lived about 2.5 billion years ago, 1000–1500 basic protein genes, the COGs [clusters of orthologous genes; bacterial sequence families to genes with the same function] give an approximate impression of these oldest genes).

**Local Alignment**   Two sequences are compared by contrasting which amino acid residues are modified and which are conserved (see also global *alignment*). In local *alignment*, one looks for only one piece, especially very well, to catch a local piece that is particularly similar, especially the domain that has the highest sequence similarity, i.e. a characteristic domain of the protein. But then please remember that after that you should also examine the other pieces in the protein and assign the function. For local *alignment* there is an exact method, the Smith and Waterman search. This is slow, but accurate. Various heuristic methods (not quite exact, but much faster with only a few errors) use this method, where after a (FASTA; fast *alignment*) or a double (BLAST) fast index search then an exact Smith and Waterman alignment is done.

**Long Branch Attraction** Form of systematic error (tree-building error) in which distantly related taxa are incorrectly considered to be closely related or closely related to be unrelated, resulting from comparing sequences of different lengths or when a single sequence is quite long and the taxa have different numbers of mutations.

**Markov Chain** This is the name given to a random process (Markov chain; also Markov process, after Andrei Andreyevich Markov; other spellings: Markov chain, Markoff chain, Markof chain). A good example is the random results of dice rolls. By knowing only a limited past history (e.g. the last three throws), it will only be possible to make predictions about future developments that are as good as those made by knowing the entire past history of all previous throws: Each new throw has a random result, and the probability of a given number of dice is always one-sixth.

**Markov Process** see Markov chain.

**Mathematical Modelling** Mathematical modelling describes the representation of experimental data with mathematical equations. Here, there are the Boolean/discrete, quantitative and semi-quantitative methods. In principle, these methods consider the nodes (proteins) of a network according to their activation state, i.e. either activated (*on;* maximally activated = 1) or inhibited (*off;* maximally inhibited = 0). According to the initial state (how strongly is the node activated/deactivated), the further temporal course, i.e. how does the state of the node change over time, is calculated for each individual node of the network. In this way, the behavior or the network interconnection can be examined in more detail, whereby corresponding network effects, i.e. the respective effect of a node, also become clear. Boolean modeling always considers the *on/off* (1/0) state of a system, i.e., the node is either activated (*on;* 1) or inhibited (*off;* 0). Quantitative modeling is useful for kinetic data, such as Michaelis–Menten kinetics (example software: PottersWheel). Here, the system state of a network is considered using exact concentrations and mathematical differential equations, but this requires information about the kinetics. Semiquantitative modeling combines both methods, which enables one to consider the system state in the interval between 0 and 1, which can also be done without knowledge about the kinetics (example software: SQUAD and Jimena).

**Maximum Likelihood Method** Phylogenetic method in which the most probable pathway is calculated for all mutations (every single mutation is taken into account) (very computationally intensive and time-consuming, but particularly accurate).

**Medical Informatics** In common parlance, this is computer support in the clinic. In particular, this includes computers in intensive care monitoring and anaesthesia, the electronic infrastructure for patient documentation (doctor's letters, findings, electronic medical records) and expert systems (such as databases on antidotes for poisoning or infections) as well as educational software (e.g. for anaesthesia, anatomy). In contrast, the modelling of diseases would be directly attributed to bioinformatics.

**MEDLINE (Medical Literature Analysis and Retrieval System Online)** A major bio-informatics database operated by the National Center for Biotechnology Information (NCBI) that collects and provides medical information and literature.

**Metabolomics** Analysis of the metabolome, the totality of all metabolic products (metabolite).

**Modelling** Field of application of bioinformatics that deals with the computer-aided mathematical description of experimental data and system properties/effects.

**Modular** Recurring elements (units) in biological systems, e.g. amino acids for proteins. These are parts of molecular networks and form filaments that always form new patterns and properties (combinatorics).

**Nanocellulose Chip** New technology that replaces silicon computer chips with nanocel-lulose chips. By using light-controlled proteins, transistors are replaced. They have the potential to function much better, in particular to be more environmentally friendly, more durable (thousands of years), faster (petahertz frequency) and to have better storage prop-erties (exabyte memory). DNA serves as the storage medium, polymerases to synthesize and read in sequences and exonucleases to degrade and read out, the nanocellulose is the matrix for the enzymes and the DNA. Polymerases, exonucleases and other molecules are controlled by light-controlled protein domains.

**Natural Computing (also called "***Analog Computing***")** Branch of synthetic biology that describes computing with molecules or even with entire living organisms.

**Neighbor Joining** Phylogenetic procedure in which the family tree is based on neighbor similarity and the respective ancestors are calculated for direct neighbors.

**NetworkAnalyzer** Cytoscape plugin, which allows an analysis of the network topology, e.g. with respect to network interconnection (average number of interaction neighbors) or robustness *(network centrality)*.

**Neural Networks** see *neural computing*.

**Neuronal Computing** Field of application/software in bioinformatics in which a pro-gram recognizes certain patterns and properties of information processing in known data using artificial neuronal networks (neurons and their interconnections) and learns to pre-dict these accordingly for unknown data sets.

**Node Computers** The information that is passed from computer to computer on the Internet is bundled at central points. These central computers are then called Internet nodes.

**NP Problems (Non-deterministic Polynomial Complexity)** Mathematical problems that are very computationally expensive and whose possibilities lead combinatorially to an exponential growth of possibilities, e.g., the traveling salesman's problem of driving to numerous cities along the most optimal route possible. The same is true for the prediction of protein structure (whether *ab initio* in three dimensions or as a homology model), the calculation of stable system states for *pathways* (for example, in the cancer cell), and metabolic modeling. Many exciting biological problems are NP problems. Typically, in an NP problem, I don't know exactly when I will find the solution, no matter what computer algorithm I use. However, if I am shown the solution, I can confirm it in polynomial time (i.e., rather quickly).

**Omics**  Branch of biology that deals with the analysis of large amounts of biological data. Examples are proteomics, metabolomics, genomics, RNAomics, interactomics, which deal with big data about proteins, metabolites, genomes, RNA and interactions.

**Ordered Systems**  Description of systems by simple mathematical equations, where the behaviour is predictable and can be described exactly for the entire period of the flight or train journey, e.g. flight of a rocket.

**Pandemic**  An epidemic spreading worldwide (epidemic = contagious disease, infectious disease). Currently (2020) a pandemic is caused by the virus Sars-CoV-2 (severe acute respiratory syndrome coronavirus 2). This is a coronavirus (cause respiratory disease; they have a ring, a "corona" of appendages around the spherical body). Pedigree analyses show high relatedness to the SARS virus (2002/2003 pandemic). Another relative is the MERS-CoV (Middle East respiratory syndrome coronavirus). A pandemic with many millions of deaths was the "Spanish flu" after World War 1. Important for a pandemic are factors that ensure a steady spread of the disease across many national borders and thus worldwide (the factor R0, the infection rate per infected person always remains at least slightly above 1, so that there are always more people infected). This can be well modelled bioinformatically, as can the effect of control measures, mutation rates, mortality, changes in the infection rate.

**Parsimony**  Phylogenetic procedure in which the mostly not directly observable ancestors are calculated in such a way that all observed present-day sequences can be generated with as few mutations of these ancestor sequences as possible.

**Polymerase**  enzyme that produces a new nucleic acid. This is usually done according to a *template*. There are RNA-producing RNA polymerases and DNA-producing DNA polymerases.

**Polymerase Chain Reaction (PCR)**  Method of molecular biology that serves to double the genetic information (DNA) by means of a chain reaction (constant doubling of the DNA strands). In order to achieve this specifically for a certain DNA sequence, one needs the start of the desired sequence and determines a complementary initial sequence (start primer) and at the end of the sequence on the opposite strand a reverse primer that is again complementary to it. With the help of the two primers, one specifies where a new strand is to be synthesized for the polymerase that is also required. After about 1 min of DNA synthesis, the freshly synthesized molecules are separated again by heating each 1000 base pairs, and the two primers are then allowed to reattach to the new strands, thus obtaining more and more identical DNA strands as long as there are enough primer molecules and polymerase for the PCR.

**Polynomial Complexity** Problems that are not very computationally intensive and whose mathematical description is done with a polynomial (computation time depends on the length) (see also P-problems).

**Positive/Negative Feedback Loops** Feedback loops in networks that reinforce everything and stabilize the system (positive *feedback loops*) or dampen excessive regulation (negative feedback *loops*).

**P-problems** Problems that do grow polynomial as the sequence or number of units increases for which I make my prediction, but it happens exactly and clearly. So there is a safe solution strategy with clear computational timing. Examples are database searches or RNA folding. Both grow quadratically with sequence length for many related algorithms, where for databases it is the product of search sequence and total database length (see also polynomial complexity).

**Precision** (specificity) Correct-negative rate, i.e. the proportion of correctly negatively grouped hits out of the total of correctly negative hits, e.g. predicted as non-interacting interaction partners that also show no interaction experimentally, or people grouped as healthy who are actually healthy.

**Primary Database** Databases that contain only the basic data and information, such as protein sequences.

**Primary Metabolites** See secondary metabolism.

**Primary Metabolism** see Secondary metabolism.

**Principal Component Analysis** (PCA).

**Principal Component Analysis (PCA)** Method of multivariate statistics. It aims to transform high-dimensional data into a new coordinate system (usually 2D) in order to reduce complexity in the data and extract relevant variables (principal components).

**Programs** Tools (software) used to examine and analyze data sets or data from experiments. Programs first consist of a declaration part, which defines the variables and data fields used, followed by the calculation part. Typically, the computational part consists of a read/input part, a main loop, and an output part. The main loop reads the input data and performs the calculations, often accesses subroutines itself, writes results to the output part, and monitors the sequence of the program through logical queries until the program finally processes everything and stops. In practice, stopping or halting a program is not always easy to predict.

**Programming Language** Instructions (syntax) to a computer to perform a calculation (algorithm or program), such as sequence analysis. Popular programming languages are, for example, Perl, Java, Python or C + +.

**Promoter** gene readout start sequence.

**Protein** see Proteins.

**Protein Domain** A self-contained folding unit in a protein, is about 100–150 amino acids in size, has a specific molecular function, e.g. catalytic function, cofactor binding (e.g. for cofactors such as NADH, FAD), interaction domain (between two domains in the protein or a whole protein) and regulatory function (e.g. DNA binding or transmitting a signal).

**Proteins** Proteins. Most important building material of the cell (*proteos,* the first): all enzymes (accelerate biochemical reactions in the cell) and important structural proteins (collagen, e.g. hair; albumin in the blood, etc.). Proteins are made up of amino acids. The 20 most important ("proteinogenic amino acids") are assembled with the help of the genetic code according to the building instructions of the mRNA in the ribosome. Proteins are therefore products of translation. Afterwards, proteins can be modified further (post-translational modifications, e.g. sugar residues or lipid residues are retained). Proteins are large molecules (macromolecules) with specific functions, for example as enzymes or transcription factors. Their three-dimensional structure and amino acid sequence are decisive for the function.

**Protein Kinases** Enzymes that transfer phosphate residues (phosphotransferases) and have an activating effect.

**Protein Phosphatases** Enzymes that remove phosphate residues and have an inhibitory effect.

**Protein Structure** Structure of a protein that is responsible for its function. A distinction is made between primary structure (amino acid sequence), secondary structure ($\alpha$-helix, $\beta$-sheet), tertiary structure (single protein chain from several secondary structures) and quaternary structure (several tertiary structures, i.e. several protein chains, often important for cooperative structural adaptation).

**Proteomics** Analysis of the proteome, the totality of all proteins.  PSSM (Position-Specific Scoring Matrix; Often also Called Position-Specific Weight Matrix, PSWM)

Alternative to the consensus sequence, which specifically allows the prediction of these motif patterns in other or new unknown sequences, such as transcription factor binding sites (see also consensus sequence).

**p-Value** Statistical parameter indicating how likely it is to get the hit by chance.

**Quaternary Structure** see protein structure.

**Ramachandran Plot**  Calculates the phi and psi torsion angles in the protein, providing a graphical overview of the distribution of α-helices and β-sheets.

**Random Systems**  Description of systems where the behavior is unpredictable for a short time, such as dice roll (you can't predict the next roll, but the result space can be predicted, can only be one to six).

**Recall(Sensitivity)**  Indicates how many of the hits are also stored in the database as real entries (correct-positive rate, i.e. proportion of correctly grouped hits out of the total of correct hits, e.g. potentially predicted interaction partners that are experimentally validated or persons grouped as ill who are actually ill).

**Regression (Regression Analysis)**  Statistical procedure to find correlations and relationships between a dependent (explained variable, regressand) and independent (explanatory variable, regressor) variable(s). The most common are linear regression, logistic regression or Cox regression (survival time analysis).

**Regulatory RNA Elements**  Motifs of RNA (characterized by specific sequence, structure and folding energy) that perform important regulatory functions and regulate transcription and translation, e.g. *iron-responsive elements* (IRE, regulate *iron* metabolism in humans and animals depending on the iron content of the cell) and *riboswitches* (regulate gene expression in prokaryotes).

**RNA (Ribonucleic Acid)**  Biochemically, a mixture of nucleotides that are all linked by a ribose sugar and a phosphate "backbone" to form a long molecule, the RNA single strand. Product of transcription and serves as an information carrier (mRNA) for the synthesis of proteins. RNA can simultaneously store information, but also form secondary structures and, when folded appropriately, accelerate reactions like an enzyme, forming an RNA enzyme, called a ribozyme for short. Therefore, even before the genetic code, a few hundred million years after the origin of life, there was the RNA world, in which RNA organisms with ribozymes and RNA genomes were important forms of life.

**RNA World**  see RNA.

**Sars-CoV**  see Pandemic.

**Sars-CoV-2**  see Pandemic.

**Sars Virus**  see Pandemic.

**SBML (Systems Biology Markup Language)**  Storage format of bioinformatics for systems biology. For example, this explains networks (e.g. metabolic or regulatory) well, making them machine-readable (as XML), such as for CellDesigner and SQUAD.

**SCOP (Structural Classification of Proteins)**  Classification of protein structure, based on structure and sequence and involving direct expert analysis by protein structure experts (in particular Alexey Murzin). There was initially the classical database (https://scop.mrc-lmb.cam.ac.uk/legacy/). A comprehensive reclassification is SCOP2 (https://scop.mrc-lmb.cam.ac.uk). For practical use (protein structure prediction and classification), SCOPe (Structural Classification of Proteins – extended; https://scop.berkeley.edu) at the University of Berkeley (Universität Berkeley) is recommended, because the old classification is simply extended and, for example, the ASTRAL structure databases are also used.

**Second Gödel's Incompleteness Theorem**  shows that sufficiently strong noncontradictory systems cannot prove their own noncontradiction (a computer thus remains in the undecidable). The outstanding mathematician Kurt Gödel (1906–1978) deserves credit for proving, by means of mathematics, the existence of fixed limits for formal systems.

**Secondary Databases**  Databases that integrate data and information from primary databases and use them for further analyses, such as protein sequences for predicting protein structures or domains.

**Secondary Metabolism**  Primary metabolites are central to metabolism and are found in many or nearly all (central metabolites) organisms. In particular, primary metabolism includes central carbohydrate metabolism (glycolysis, pentose phosphate pathway, and citric acid cycle), lipid metabolism (synthesis and beta-oxidation), and amino acid synthesis and degradation, as well as nucleotide production, degradation, and recycling *(salvage)*. Secondary metabolites are additional metabolites that only occur in specific organisms and then have specific effects (pharmacological, neurotransmitters, ecological, signalling, etc.).

**Secondary Structure**  In proteins, two important secondary structures, helices and *beta strands*, form from the sequence (also called primary sequence or primary structure) via hydrogen bonds. The latter can also assemble into beta-sheets. Here we can distinguish parallel and antiparallel ones, in the case of helices the frequent alpha helices (every 3.6 amino acids one turn; i to i + 4 hydrogen bond, discovered by Pauling) and narrower ones ($3_{10}$-helix, every 3 amino acids one turn; i to i + 3 bridge) and wider ones (pi-helix, every 4 amino acids one turn, i to i + 5 bridge). The secondary structure can be subdivided much more finely. Loops, the third type of secondary structure, are also more finely divided into bends, disordered regions, and typical loops. RNA also forms secondary structures, especially loops, stems, and pseudoknots (loop contact; true, stable knots would block RNA and do not occur in biology).

**Sequence Comparison**  Two sequences are compared by contrasting which amino acid residues are altered and which are conserved. You can either compare over the whole length (see global *alignment*), which is especially good for phylogenetic analyses, or only a piece (see local *alignment*), which is especially good to catch a local piece that is particularly similar, especially the domain that has the highest sequence similarity, i.e. a characteristic domain of the protein. But then please remember that after that you should also examine the other pieces in the protein and assign the function.

**Server**  see database.

**Shannon Entropy**  Measure of the information content of a message (unit: bit).

**Signal Cascade**  Signal path.

**Signalling Pthway**  Signalling pathway.

**Signalling Pathway**  Biological network that transmits a signal in which, for example, kinases and phosphatases (or several enzymes) interact (are interconnected) and regulate each other or each other in turn (e.g. activate and typically amplify cellular signals) and are responsible for certain functions and processes in the cell, e.g. cell growth and cell differentiation.

**Simulation**  see Modelling.

**Single Nucleotide Polymorphisms (SNPs)**  Changes in the DNA sequence at exactly one nucleotide, which can lead to neutral, negative and positive health effects. Genes, proteins and sequences that are involved, for example, in the structure of the nervous system, such as at synapses, are thus assigned to their function with the aid of genome, sequence and domain analysis. A detailed overview of SNPs is provided, for example, by the database "Online Mendelian Inheritance in Man" (OMIM).

**Software**  Commands (instructions) in the computer that are arranged in a meaningful way and perform a specific task, e.g. programs, databases, simulations, models (see also programs).

**Synthetic Biology**  A branch of biology that deals with the technical use of biological processes.

**Systems Biology Modelling**  see mathematical modelling.

**Tertiary Structure**  see protein structure.

**Tipping Point**  A new system state (attractor) is sought when the tipping point is passed. Because the old system state has been left far enough, the system will then enter a new state because the new attractor will then stabilize and reinforce itself again if the system is disturbed or changed enough.

**Transcription**  Part of gene expression that leads to the formation of RNA from DNA using a polymerase after the gene start sequence (promoter).

**Transcription Factor Binding Sites**  DNA motifs in the promoter to which a transcription factor specifically binds.

**Transcription Factors**  Bind in the promoter to specific DNA binding sites (DNA motifs, transcription factor binding sites) and regulate the transcription of a gene.

**Transcriptomics**  Analysis of the transcriptome, the totality of all transcripts.

**Translation**  Part of gene expression that leads to the formation of proteins using an mRNA and the genetic code in the ribosome.

**Tree Formation Error**  see *long branch attraction.*

**Turing-Computable**  All calculations that a Turing machine can perform are therefore also called Turing-computable. The famous mathematician Alan Turing had considered and proved, how all possible mathematical calculations, especially of algebra, can be represented by five basic operations and a very long calculating tape. Of course, this also clearly defines the limits for formal systems and computability, such as for computers. In particular, aesthetic, ethical or moral judgments, but also self-reflection and, in mathematics, all numbers, quantities and concepts that cannot be described by an algorithm (calculation rule for computers) are not Turing-computable.

**XML (Extensible Markup Language)**  Machine-readable language for Internet pages, used in bioinformatics to represent data as text files.

# Tutorial: An Overview of Important Databases and Programs

# 19

**Abstract**

The tutorials are designed to walk you through important analysis steps and associated software that appear in the book, so that you learn to do the right thing as you practice. In doing so, we have tried to provide some tips as well. As always, practice is necessary! It's pretty easy to learn the software once by clicking, but learning to use it is only learned by repetition and by learning about the underlying algorithms and parameters. The tutorials and tips here are merely an aid to being able to do this a little more easily, but are of course no substitute for a course. Nevertheless, you can get to know the programs much better by practicing and using the tutorials alongside the book than by just reading the book: Because as everywhere, the same applies to bioinformatics: practice, practice, practice and look closely.

## 19.1 Genomic Data: From Sequence to Structure and Function

**Where Can I Find Genomic Data and Related Information?**
Genome informatics is a "perennial" in bioinformatics. Classical genome databases such as Ensembl and UCSC provide an overview of annotation and genomic position. A well-structured database is NCBI, which is a collection of various databases. It is a helpful entry site where one can find information on publications, genes, proteins, sequences, genomic positions, etc. For example, if one selects the gene database and searches for a gene (e.g. enter HIV-1 gag pol in the search window; Fig. 19.1), NCBI gives an overview of the genomic context, but also provides further information, such as interactions (Fig. 19.2). The NCBI also offers an initial introduction to individual gene variants via the OMIM database (https://www.ncbi.nlm.nih.gov/omim). Here you can look up individual mutations and genetic diseases or dispositions. To do this, simply select the OMIM database

© Springer-Verlag GmbH Germany, part of Springer Nature 2023
T. Dandekar, M. Kunz, *Bioinformatics*,
https://doi.org/10.1007/978-3-662-65036-3_19

**Fig. 19.1**   Searching the gene database with the HIV-1 gag pol gene

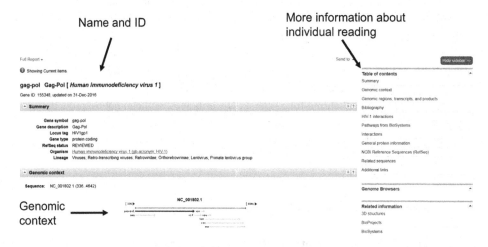

**Fig. 19.2**   Overview of the genomic context of the HIV-1 gag pol gene

and enter a search term (e.g. human immunodeficiency virus 1; Fig. 19.3). OMIM then displays information on pathogenesis and clinical data, for example, but also on the genes involved (e.g. IL-10) (Fig. 19.4). Interestingly, many of the statements also apply to homologous proteins in organisms that are not too distantly related (mammals, vertebrates). Genome variability is studied more systematically in the 1000 Genomes Project (https://www.internationalgenome.org; including SNPs, insertions, deletions, *copy number variation*), but also in the ENCODE Project (https://www.encodeproject.org/).

A collection of datasets (e.g. *microarray* and next-generation *sequencing,* summary of experiment, download) can be found, for example, in the Gene Expression Omnibus

Select OMIM in
NCBI database

Search Keyword: Human
immunodeficiency virus 1

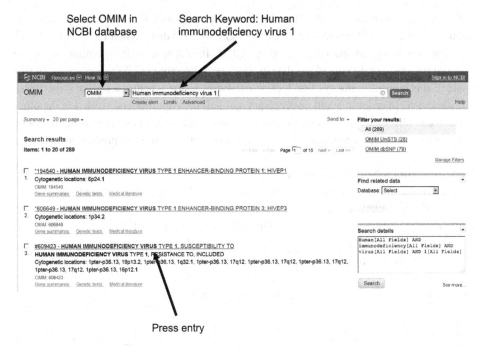

Press entry

**Fig. 19.3**  Searching the OMIM database with HIV-1

Further clinical
information

All information about
individual reading

IL10 gene has an
influence

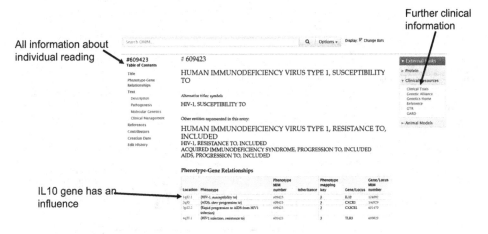

**Fig. 19.4**  Overview of HIV-1 in the OMIM database

(GEO), cBioPortal, TCGA and GENEVESTIGATOR databases. For example, the GEO database contains a large number of experimental datasets from numerous publications. In addition, it offers the option of storing even unpublished datasets and using them for internal work, but without making them publicly available to anyone. GEO is accessible via NCBI and allows direct searches with keywords (e.g. gene name or terms; Fig. 19.5). In our example, searching with IL-10 as the gene name yields 2097 records, e.g., an experiment in the mouse with a total of five records (GSE56444) that studied the effect of IL-10 receptor deficiency on macrophages (Fig. 19.5). In addition to information about the experiment, one can download the data or analyze it directly via GEO (Fig. 19.6). In this context, it is helpful to always first look at databases to see whether a similar experiment already exists, in order to avoid unnecessary and lengthy experimental work. On the other hand, it can also be advantageous for bioinformatic work to use already existing datasets, for example to set up or validate a bioinformatic model or to support predictions.

### How Do I Find a Nucleotide Sequence?

You can find a sequence in NCBI and any genome database (e.g. UCSC and Ensembl). To do this, simply select the nucleotide database in NCBI and search for a term in the search window (e.g. IL-10 *human;* Fig. 19.7). On the results page, press FASTA to go directly to the desired sequence (see Fig. 19.7; if you press GenBank and Graphics, you can also display additional information on the genomic position).

**Fig. 19.5** Searching the GEO database with IL-10

**Fig. 19.6** Overview of the results in the GEO database on IL-10

| Select nucleotide | Search Keyword: | Press FASTA | Promoter region of the IL-10 gene |
| database in NCBI | IL-10 human | | |

**Fig. 19.7** Overview of how to find a nucleotide sequence in the NCBI database

**How Can I Evaluate Omics Data?**

In the book, we learned about some omics techniques, such as genomics, transcriptomics, or proteomics, and how they are related. It is important to know that systems analysis is not so easy to formalize. However, it is easy to recognize important ingredients about biological systems if you have enough biological knowledge. In practice, however, one is mostly occupied with collecting and evaluating omics data (e.g. own experiments or from databases, such as GEO). In most cases, the statistical software R is used, which allows analysis and graphical representation and is also widely used, e.g. with the Bioconductor tool for high-throughput data analysis (see Sect. 19.6). There are numerous online tutorials and already prescribed scripts, it is best to simply go to https://www.r-project.org/ and https://www.bioconductor.org/ for information. In addition, several genome analysis pipelines exist, e.g. GensearchNGS, in which we collaborate (Wolf B, Kuonen P, Dandekar T et al (2015) DNAseq workflow in a diagnostic context and an example of a user friendly implementation. Biomed Res Int 2015:403–497. https://doi.org/10.1155/2015/403497). For proteome and transcriptome, our two papers Stojanović SD, Fuchs M, Fiedler J et al. (2020) Comprehensive bioinformatics identifies key microRNA players in ATG7-deficient lung fibroblasts. Int J Mol Sci 21(11):4126. https://doi.org/10.3390/ijms21114126) and Fuchs M, Kreutzer FP, Kapsner LA et al (2020) Integrative bioinformatic analyses of global transcriptome data decipher novel molecular insights into cardiac anti-fibrotic therapies. Int J Mol Sci 21(13):4727. https://doi.org/10.3390/ijms21134727) provide a good overview. For this it is best to look at the publication, there you will find instructions and you can practice yourself.

If you want to look a little more into machine learning, you can check out our analysis pipeline for diagnostic and prognostic signatures (Vey J, Kapsner LA, Fuchs M et al (2019) A toolbox for functional analysis and the systematic identification of diagnostic and prognostic gene expression signatures combining meta-analysis and machine learning. Cancers [Basel], 11(10). pii: E1606. https://doi.org/10.3390/cancers11101606). A nice application example is also shown in the paper Schweitzer S, Kunz M, Kurlbaum M et al (2019) Plasma steroid metabolome profiling for the diagnosis of adrenocortical carcinoma. Eur J Endocrinol 180(2):117–125. https://doi.org/10.1530/EJE-18-0782).

In addition, there is further special software that evaluates mass spectroscopy data, for example lipids with the software Lipid-Pro (Ahmed Z, Mayr M, Zeeshan S et al (2015) Lipid-Pro: a computational lipid identification solution for untargeted lipidomics on data-independent acquisition tandem mass spectrometry platforms. Bioinformatics. 2015;31(7):1150–1153. https://doi.org/10.1093/bioinformatics/btu796. PubMed PMID: 25,433,698). The software and a tutorial can be found at https://www.neurogenetics. biozentrum.uni-wuerzburg.de/services/lipidpro/. On the other hand, isotopolog data can be analyzed using Isotopo (Ahmed Z, Saman Z, Huber C et al (2014) 'Isotopo' a Database Application for Facile Analysis and Management of Mass Isotopomer Data, Database: The Journal of Biological Databases and Curation, Oxford University Press). For the software and a tutorial go to https://www.tr34.uni-wuerzburg.de/software_developments/iso-topo/. These are just three examples from my own work. There is a great deal of other software and development work available for analysing omics data from many research groups around the world (this is where individual searches help, according to the biological problem).

**How Do I Perform a Sequence Analysis?**

The best way to do this is to use BLAST (Basic Local Alignment Search Tool). As we have already learned, BLAST is a heuristic search algorithm that allows nucleotide and protein sequences to be compared with a large database in terms of their local similarity (*two-hit method*). Thus, homologous genes can be identified and the individual positions compared in order to annotate unknown sequences, but also to find corresponding differences in other organisms (e.g. for the development of an animal model). Important parameters are the *E-value* and *p-value*. The *E-value* (expected value) indicates how likely it is that the match with a similar or better score will be found again in the database (depending on the size of the database), whereas the *p-value* indicates how random the match found is. If you want to find a similar sequence in the database, the hit should always have the lowest possible *E-value* and *p-value* (at least <0.05) and a high identity. BLAST can perform a number of searches, such as blastn for a nucleotide sequence and blastp for a protein sequence (see Figs. 19.8 and 19.10). But it can do much more: blastx translates a nucleotide sequence into a protein sequence and then searches against the protein database, tblastn searches with a protein sequence against a translated nucleotide database, and tblastx searches with a translated nucleotide sequence against a translated nucleotide database. What is important to practice at this point? Going through the practice tutorial, learning the two stages of function mapping, first via domain mapping by searching the Conserved Domain Server at least at NCBI on the website (a few thousand domains in a database, that's very fast), and only then searching the large database with millions of sequence entries and billions of nucleotides. There are also very good tutorials available on the NIH site to practice these different types of BLAST. You should practice several searches with BLAST (even try a meaningless sequence; see Sect. 12.1). For comparison, you can also try other programs or servers (e.g. the BLAST at EMBL or in Switzerland). We will illustrate the sequence analysis with two examples (see Figs. 19.8, 19.9, 19.10 and 19.11). Since our

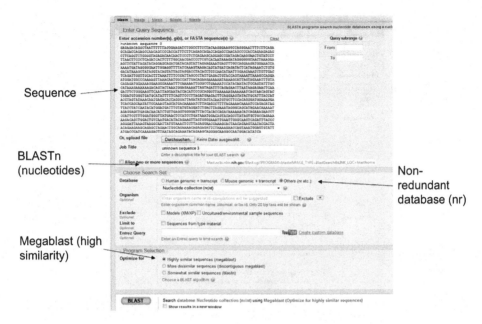

Sequence

BLASTn
(nucleotides)

Non-
redundant
database (nr)

Megablast (high
similarity)

**Fig. 19.8**  Search with a nucleotide sequence in blastn

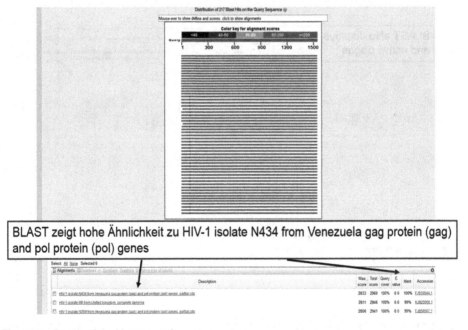

BLAST zeigt hohe Ähnlichkeit zu HIV-1 isolate N434 from Venezuela gag protein (gag)
and pol protein (pol) genes

**Fig. 19.9**  Result of the blastn search

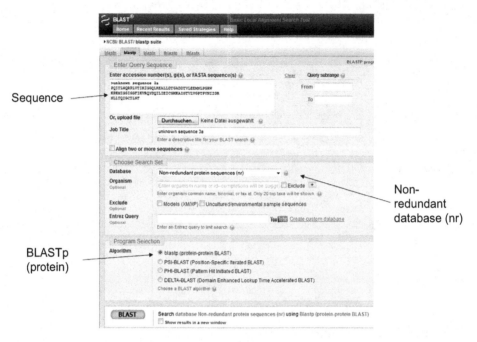

**Fig. 19.10** Searching with a protein sequence in blastp

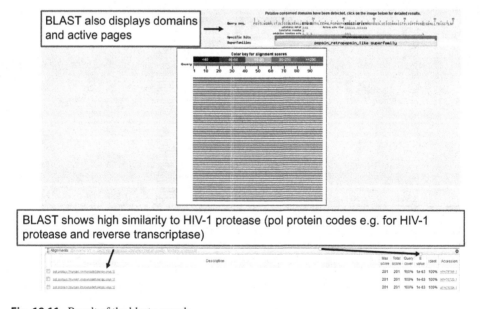

**Fig. 19.11** Result of the blastp search

first unknown sequence is a nucleotide sequence, we choose a blastn search (Fig. 19.8). In addition, we choose megablast so as to achieve a high similarity of our sequence with the entries deposited in the database. To avoid duplicate result hits, we also select the non-redundant database (each entry exists only once). As a result of the BLAST search, we get an overview of all hits found (here 217) including further information e.g. on the identity or the *E-value*. In our example, BLAST finds a high similarity (*E-value:* 0, identity: 100%) to the HIV-1 isolate N434 from Venezuela (Fig. 19.9).

In our second sequence example, we choose a blastp search and also the non-redundant database (Fig. 19.10). Our result indicates a high similarity (*E-value:* close to 0, identity: 100%) to the HIV-1 protease (Fig. 19.11). Furthermore, BLAST also identifies conserved domains (e.g. catalytic domain at position 25–27, DTG).

**What Else Should Be Considered in a Sequence Analysis?**
Importantly, now that you have found a hit, you should verify your result and back it up with additional software searches, such as whether all the motifs or amino acid residues are there that I need for the protein function (see also the PROSITE and AnDom examples). It is also important whether BLAST allows searching back, i.e. whether there is a hit in the database again when I re-enter the sequence I hit with my query. Otherwise, it is important that all the clues to my sequence (all the tests, all the data, all the other program results) match each other and thus confirm the search. For example, a transcription factor (according to the BLAST search, because it was most similar to it) should then also have at least one DNA-binding domain in the domain composition, but also a protein localization signal in the protein sequence, for example.

**How Do I Perform a Simple Genome Annotation?**
In addition to sequence analysis, for example to find out which gene is present, it is necessary to annotate the promoter region and to examine it for transcription factor binding sites (TFBS). The promoter region can be detected with the software Berkeley Drosophila Genome Project (https://www.fruitfly.org/seq_tools/promoter.html; for prokaryotes and eukaryotes) and PRODORIC (Prokaryotic Database of Gene Regulation; https://www.prodoric.de/vfp/), the corresponding DNA motifs, e.g. TATA box, can be recognized and a promoter can be identified. Transcription factors bind to specific DNA motifs (DNA binding sites) in the promoter, called TFBS, and thus regulate transcription. If I know the consensus sequence of TFBS, I can easily bioinformatically screen an unknown promoter sequence for possible binding sites, which I can use for further experimental investigations. For this purpose, besides programs that list experimentally validated TFBS (such as MotifMap), there are also numerous programs that predict TFBS, e.g. ALGGEN PROMO, PRODORIC, TESS, TRANSFAC, JASPAR or Genomatix. We will briefly show this for the two *tools* MotifMap and ALGGEN PROMO. MotifMap offers three ways to find TFBSs, e.g. via a direct gene search (here with IL-10 in humans [Genome hg19]), displaying the corresponding TFBSs (e.g. NFAT2 with additional information such as position, motif and region) (Fig. 19.12).

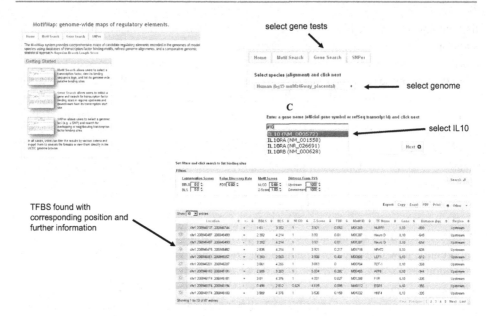

**Fig. 19.12** Searching for TFBS with MotifMap

ALGGEN PROMO, on the other hand, is a direct prediction tool for a corresponding sequence, for which TFBS matrices from TRANSFAC are used. After selecting SearchSite in ALGGEN PROMO, you can insert your sequence into the search window. An important parameter to consider is the *"dissimilarity rate"*, which indicates how high the tolerated deviation of the predicted TFBS to the used matrix may be (here 15%; feel free to set other values and see how the number of hits changes). As a result, all predicted TFBS including position and *dissimilarity* (here, e.g., for NF-AT2 five TFBS; the corresponding consensus sequence and matrix can also be seen) are displayed in ALGGEN PROMO (Fig. 19.13).

As can be seen, these prediction programs are quite easy to use and provide a relatively quick first insight into possible TFBS, such as unknown sequences, but usually show a high abundance of predicted binding sites. In this context, it is important to know the exact parameters of the individual programs in order to obtain meaningful results for further experimental investigation. If one is careless and chooses, for example, a too high *"dissimilarity rate"*, I may get hits that are biologically none at all. Consequently, for further investigations, the position with the lowest *dissimilarity rate* should always be selected for the desired TFBS, i.e. the one with a high match to the search template (here for NF-AT2 e.g. position 632–640 with a *dissimilarity rate of <5%*). In any case, it is necessary to validate bioinformatically predicted TFBSs experimentally. Only then can I be sure that the transcription factor found actually has an effect on gene expression, otherwise only the DNA nucleotides of the prediction match (which is why I got a hit), but this has no biological relevance.

Finally, another option is to label the genome sequence, examine it with BLAST, and thereby immediately identify the proteins it contains. For example, Psi-BLAST allows me

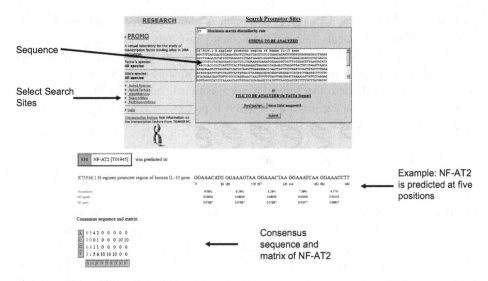

**Fig. 19.13**   Search for TFBS with ALGGEN PROMO

to annotate an unknown protein on the second BLAST run by including the unknown sequences in my search on the first run, i.e. doing a position-specific iterative matrix search (which is what Psi is in Psi-BLAST). But if you want to do this well, you should make many additional predictions, including looking at the structure, the localization, the domains, and only then you get quite good results and insights into the function (see also examples PROSITE and AnDom). The work of Gaudermann P, Vogl I, Zientz E et al (2006) Analysis of and function predictions for previously conserved hypothetical or putative proteins in Blochmannia floridanus. BMC Microbiol. 2006;6:1). If one wants to be more precise, like the ENCODE consortium, and find all regulatory elements in a genome (and not just the proteins or genes), then it is advisable to map out conserved regions via closely related genomes and also to use active motif search programs such as motif-based sequence analysis tools (MEME) (for this, read the paper https://www.sdsc.edu/~tbailey/papers/meme.ml.pdf and refer to the web site https://meme-suite.org/doc/meme.html).

Very handy to identify repetitive elements (recurring units) is the general software RepeatMasker (https://www.repeatmasker.org). We have also developed our own server, L1base, which finds LINE elements, i.e. large, repetitive, *selfish* DNA sequences (https://line1.bioapps.biozentrum.uni-wuerzburg.de/; here you are redirected to the Charité page, https://l1base.charite.de, which shows the current further development of the server and a documentation). Another possibility is to search for *repeats* in protein sequences, where the *tool* REPRO (based on local *alignment*, Smith-Waterman, and subsequent iterative *clustering;* https://www.ibi.vu.nl/programs/reprowww/) is very useful. Again, the documentation on the website is recommended. Genome annotation then quickly becomes a science in itself. For the human genome, relevant sites are already recommended in the book chapter, but also mentioned here. The ENCODE entry page already mentioned also provides a wealth of information. It is important to remember that these are only functional annotations of elements. Some of the elements have only weak or no selection pressure. For a comparison between vertebrates including humans, the UCSC genome browser is recommended (https://genome.ucsc.edu), which meanwhile compares a whole zoo of different genomes with each other (https://genome-euro.ucsc.edu/cgi-bin/hgGateway), but also includes information e.g. from the ENCODE project, such as methylation data, or predictions by RepeatMasker, such as LINE.

**How Can I Create a Phylogenetic Family Tree?**
Phylogenetic trees provide an overview of functional and evolutionary relationships. A number of software options have been described in the book for this purpose. It is important that even a simple program like CLUSTAL (https://www.ebi.ac.uk/Tools/msa/clustalo/ [newest version: CLUSTAL omega]; https://www.genome.jp/tools/clustalw/ [somewhat older version, aligns pairwise sequences over their whole length quite fast and draws a phylogenetic tree]) with experience brings better results (with CLUSTAL it is important to take sequences of approximately the same length; in addition, depending on the presumed evolutionary distance, one can correct with matrices here). The more complex softwares are correspondingly more complex to use. An example for accurate phylogenetic tree analysis is the PHYLogeny Inference Package (PHYLIP; https://evolution.genetics.washington.edu/phylip.html), which allows the construction of phylogenetic trees from sequences based on various methods, such as parsimony, likelihood, and bootstrapping (see the website for detailed documentation). Another option is the software MUSCLE (Multiple Sequence Comparison by Log-Expectation; https://www.drive5.com/muscle/), which, in addition to multiple *alignment,* computes a phylogenetic tree based, for example, on the methods UPGMA (Unweighted Pair Group Method with Arithmetic Mean; fast method if there are many sequences) or Neighbor joining (better approximation to the true tree, but slow if there are too many sequences). The results from MUSCLE can also be saved in a format compatible with PHYLIP (Newick) and used there. Detailed documentation on MUSCLE can be found on MUSCLE (https://www.drive5.com/muscle/manual/) or on the EBI website (https://www.ebi.ac.uk/Tools/msa/muscle/help/).

## 19.2   RNA: Sequence, Structure Analysis and Control of Gene Expression

**How Do I Find and Analyze an RNA Sequence and Structure?**
During transcription, an RNA is produced that has a secondary structure. One important database is Rfam. It is easy to look up and use and gives an overview of different RNA families including sequence and structure. There are different functional RNA classes, such as miRNAs and lncRNAs, which have an impact on gene expression. Important databases include miRBase (https://www.mirbase.org/) and LNCipedia (https://www.lncipedia.org/), which provide specific information on sequence, structure and functional

interaction partners. Numerous RNA interaction partners can also be predicted bioinformatically, for example with miRanda or TargetScan. It is important to note that these prediction algorithms are based on different parameters, such as prediction by *seed region* and/or folding energy, and often find too many potential interaction partners and have a high false positive rate. For a basic introduction and further information, we recommend our two review articles (https://www.ncbi.nlm.nih.gov/pubmed/25486579; https://www.ncbi.nlm.nih.gov/pubmed/28035947), which introduce numerous databases and programs for the functional analysis of miRNAs and lncRNAs and discuss problems in the analysis. For practice, one can take a look at our analysis pipeline (https://academic.oup.com/bib/article/21/4/1391/5553031).

If one has the RNA sequence, one can also have the secondary structures predicted bioinformatically (e.g., structure determination by energy minimization). RNA secondary structure folding is a complex process: In addition to a complementary sequence, the folding energy must also be considered. Not every fold is also thermodynamically optimal (should always have a low folding energy), especially since there are also many secondary structures (e.g. *stem-, hairpin-* and *interior-loop;* RNA is therefore not a linear structure). RNAfold and mFold are softwares that can do this and give you a reliable and easy RNA fold. Let's show an example in RNAfold. After calling the server, one can enter the sequence in the search window, using the preset parameters minimum free energy of folding and partition function to get a single secondary structure (Fig. 19.14). As a result, one

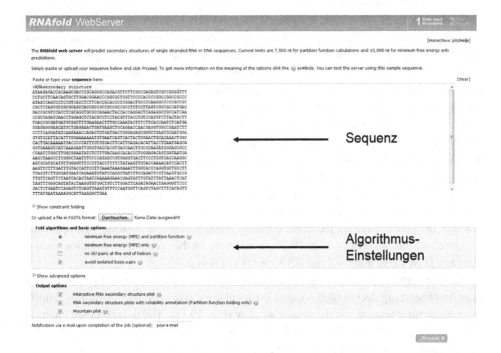

**Fig. 19.14**  Secondary structure folding with RNAfold

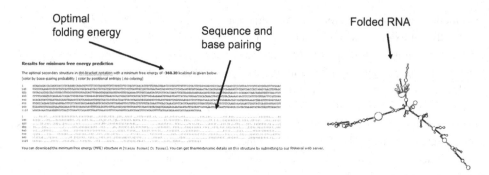

**Fig. 19.15** Result of secondary structure folding with RNAfold

then gets the optimal RNA folding including colored marking of the matching pairing as well as the corresponding base pairing and energy displayed (Fig. 19.15). However, in order to assess these correctly, it is important to have several folding variants displayed and to compare the conservation of the five best energy folds of the RNA in order to better assess the stability of the individual regions. It is therefore advisable to change the parameters for this purpose or to use mFold as well. If, in addition, a protein can presumably bind, the single-strand regions and protein-binding motifs are important (shown, for example, by RNAAnalyzer or RegRNA). Overall, therefore, more difficult, rather something for experienced users and users.

In addition, there are certain RNA motifs, e.g. regulatory RNA elements, such as *iron-responsive elements* (IRE) and *riboswitches,* which occur in humans, but also in bacteria. They take on regulatory functions and, for example, control translation depending on the iron content in order to regulate iron metabolism in humans and animals, e.g. IRE. Bioinformatically, it is of course also possible to find RNA motifs. It is best to use several criteria, such as sequence, structure and folding energy. Only if all parameters are correct, i.e. lie within a certain limit, should the program give a corresponding hit, thus achieving a higher accuracy. The RNAAnalyzer (https://rnaanalyzer.bioapps.biozentrum. uni-wuerzburg.de/) gives a good overview of which regulatory elements are hidden in the RNA, but also whether it is a catalytic RNA, for example (also translates RNA into protein and allows structural analyses via AnDom). To do this, simply call up the page and enter a sequence in the search window (Fig. 19.16). The RNAAnalyzer then examines this for possible IREs and indicates the corresponding position (here at position 71), sequence, structure and folding energy (Fig. 19.17; in addition, further general information such as exon and UTR range and catalytic RNA motifs are also given).

Another helpful software is the Riboswitch Finder (https://riboswitch.bioapps.biozen-trum.uni-wuerzburg.de/), which focuses on regulatory *riboswitches.* To do this, simply enter the desired sequence in the search window (Fig. 19.18). The program then indicates the *riboswitch* found with the corresponding position (here at position 1288), sequence, structure and folding energy (Fig. 19.19).

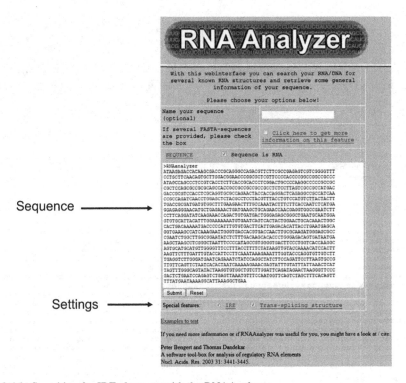

**Fig. 19.16** Searching for IRE elements with the RNAAnalyzer

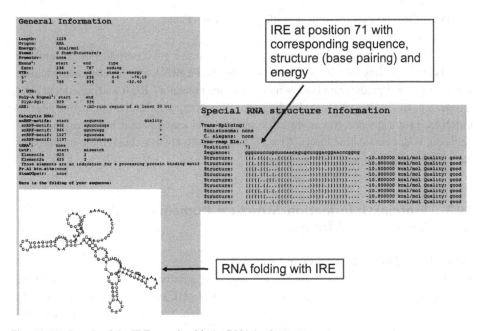

**Fig. 19.17** Result of the IRE search with the RNAAnalyzer

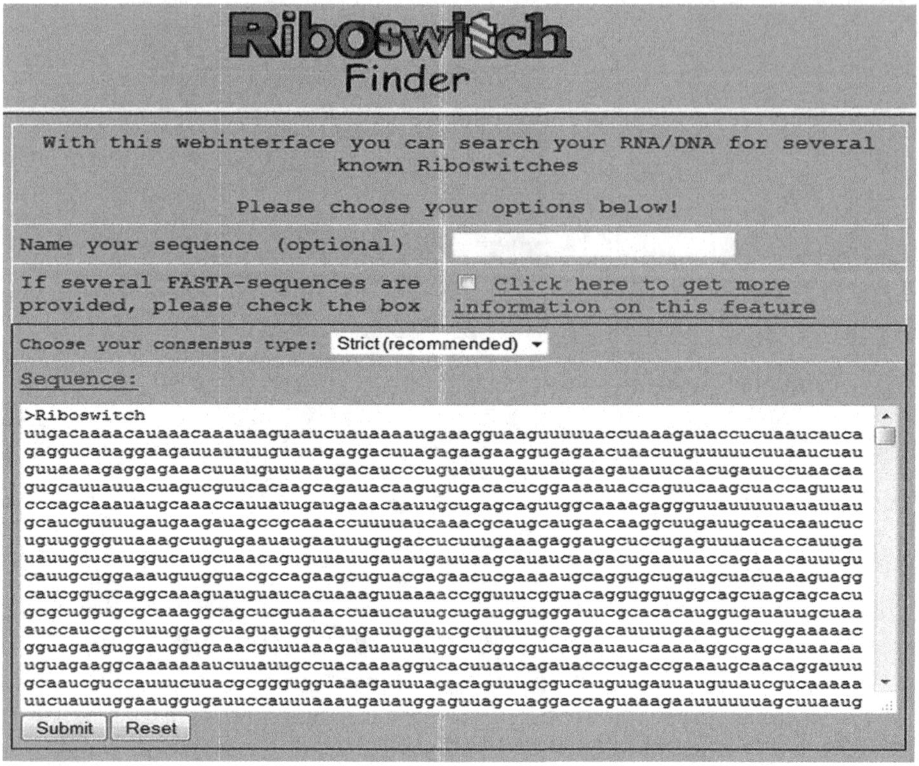

**Fig. 19.18** Searching for riboswitches with the Riboswitch Finder

These are two examples from our own work and form a good introduction. In addition, there is also a variety of other software, such as the RegRNA server (https://regrna.mbc. nctu.edu.tw/html/prediction.html), which identifies regulatory RNA and functional RNA motifs, e.g. splice site, ribosome binding site, *riboswitches* and RNA interaction sites for miRNAs, in a sequence (for more information see documentation https://regrna2.mbc. nctu.edu.tw/documentation.html). As with all analyses, always check the output against each other using independent programs.

## 19.3    Proteins: Information, Structure, Domains, Localization, Secretion and Transport

**Where Can I Find Information on Proteins?**
Information, for example on sequence and structure, can be found in the PDB database, where protein structures (experimentally determined structures) are deposited. This is a good entry page where you can get a lot of information. You can look at individual proteins, but you can also directly download the three-dimensional structure as a PDB file or

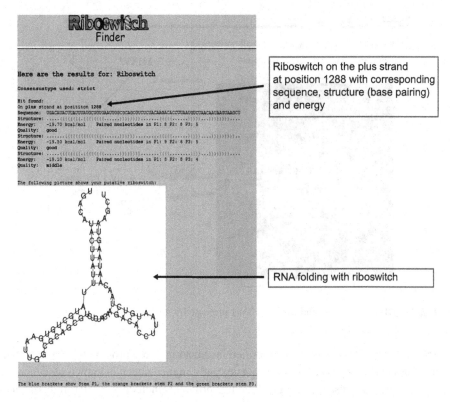

**Fig. 19.19**   Result of the riboswitch search with the Riboswitch-Finder

FASTA with all the important information. In our tutorial we show the inhibition of the HIV-1 protease by ritonavir (Fig. 19.20). To do this, one can simply search for a keyword or directly with the PBD ID in the database (here 1HXW; Fig. 19.20). On the results page, one can, for example, view the structure of the complex (shown here), but also display further information, such as the sequence and annotation (see tab in the figure above) (the *download* can be accessed to the right of the PDB ID).

Another useful database is UniProt, which, in addition to sequence and structure, provides information on domains, interaction partners and function. Here, too, it is recommended to consult different databases for the relevant question and to compare the information. In addition, there are more specific databases and software that collect information on, for example, localisation, secretion and transport, e.g. the SPdb (Signal Peptide database) database. SPdb lists signal peptides and associated DNA and protein sequences from archaea, prokaryotes and eukaryotes based on Swiss-Prot and EMBL (a detailed tutorial can be found on the website https://proline.bic.nus.edu.sg/spdb/help.html). For example, if I know the localisation, I can make statements about the function and then in turn carry out further analyses, e.g. find all the enzymes and interaction partners involved for a signalling pathway, e.g. using the KEGG database. Bioinformatically, I use sequence and programs with neural networks or hidden Markov models. Using a training dataset of

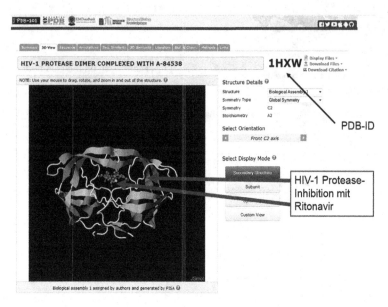

**Fig. 19.20**   PDB entry of the inhibition of HIV-1 protease by ritonavir

proteins with known, experimentally verified localization, these programs learn to predict a particular localization based on the sequence and can then match unknown sequences or sequences of new interest accordingly. Example programs include SignalP (prediction and localization of signal peptides in prokaryotes and eukaryotes), ChloroP (prediction of chloroplast transit proteins), TargetP (prediction of subcellular localization of proteins in eukaryotes; based on ChloroP and SignalP), and TMHMM (prediction of transmembrane domains). For details on each program and underlying parameters, it is best to refer to the respective websites. In this tutorial we want to show an example with the TargetP server and choose a signal peptide (human ER protein 44; ERP44; UniProt ID Q9BS26) and again the HIV-1 protease (proteases localized in lysosomes or cytosol). On the page, we paste both sequences as FASTA format into the corresponding search window and select all default parameters (not shown here). On the results page, we then get an overview of the function and localization of the protein, distinguishing between mitochondrial *targeting, chloroplast* transit and signal peptide and another function *(other)* (Fig. 19.21), the localization accordingly between mitochondrion, chloroplast and ER (signal peptide). For our example, it shows that ERP44 is a signal peptide (score column: SP close to 1) and is localized in the ER (secretory pathway; S in column Loc), whereas HIV-1 protease has none of the defined functions (high score in column *other*), i.e., it does not contain mitochondrial *targeting, chloroplast transit,* and signal peptide (Fig. 19.21). To look if ERP44 is a signal peptide, one can still check this afterwards with SignalP (not shown here). Thus, it can be seen that TargetP groups the two proteins correctly, so that it can assume that this would give meaningful predictions in the case of an unknown sequence. It is therefore

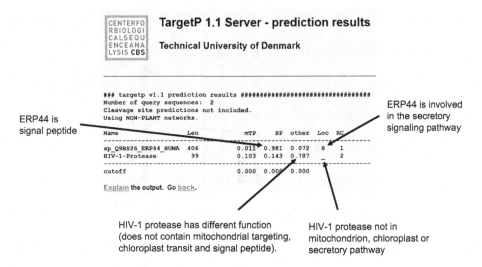

**Fig. 19.21**  Signal peptide prediction with TargetP

helpful in this context to consult proteins with known function and localization for the analysis, in order to check on one's own and make sure that the software also provides trustworthy predictions.

### Which Software Can I Use to Display/Visualize a Protein Structure?

You can visualize and animate the protein structure with PyMOL or RasMol (enter file or PDB-ID). For this purpose, please work through the individual tutorials of the programs yourself. Useful especially for unknown structures is a Ramachandran plot, which shows the distribution of phi and psi torsion angles. An example software is the FOS server. Here one can enter a PDB ID (here 1hho, human oxyhemoglobin; multiple IDs are also allowed; Fig. 19.22, left). Since there is a different distribution of the phi and psi torsion angles of the proteins in the helix and leaflet structure (for more information on protein geometry and Ramachandran plot, see the website https://www.cryst.bbk.ac.uk/PPS95/course/3_geometry/index.html), the software is able to calculate this defined arrangement. As a result, the distribution of proteins is displayed in a plot (Fig. 19.22, right). In our Fig. 19.22, a cloud of dots can be seen in the lower left square, indicating primarily right-handed alpha helices (if the cloud of dots were in the upper left, they would be primarily beta helices).

### How Do I Find and Analyse Protein Structures/Domains (e.g. Tyrosine Kinases)?

Proteins have a very specific structure (e.g. secondary structure of helix and leaflet) and consist of domains, independent folding units that are responsible for very specific functions. To get direct information about protein structures, you can use the databases SCOP (Structural Classification of Proteins) and CATH (Class Architecture Topology Homology)

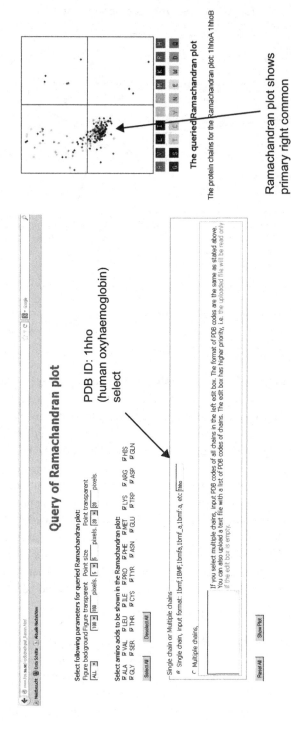

**Fig. 19.22** Ramachandran plot of 1hho (human oxyhaemoglobin)

(SCOP is shown later with AnDom). In these databases you can search for a protein and look at the corresponding structure and class, but also find, for example, related proteins of these families and associated homology models in other organisms. Have a look in the database yourself. The online tutorial (supplement) of our DrumPID database (https://database.oxfordjournals.org/content/2016/baw041.full) is also helpful. Both of these databases are knowledge-based and classify known proteins. To perform domain analysis, for example, the PROSITE or AnDom databases are helpful. AnDom is a domain annotation (domain analysis) software that uses SCOP classification. The ExPASy server (PROSITE) is the website from the Swiss Bioinformatics Institute (Schweizer Bioinformatik-Institut). As explained in the book, this is an expert system for protein sequence analysis. For example, I can use various software options here to check whether my protein sequence is really the enzyme I suspect, if that would be the BLAST result, for example. To do this, you can use PROSITE to see if all the important catalytic amino acid residues are there, or *"peptide properties"* to check if the amino acid composition to the protein is correct, e.g. if the protein has enough hydrophobic amino acids to fit into the membrane at all. This can then be checked with AnDom. Let us illustrate this with an example (Figs. 19.23 and 19.24). As a basis, we take the result of the blastp search, where we found a high similarity to the HIV-1 protease (e.g. catalytic domain at position 25–27, DTG) (see Fig. 19.11). To verify this, we analyze the sequence using PROSITE (Fig. 19.23). To do this, simply enter the protein sequence in the search window and start the search. As a result, a protease domain at position 20–89 and an active site of catalytic protease activation at position 25 can be detected (Fig. 19.23), which corresponds to the BLAST result.

To check this and additionally obtain the structure classification, one can additionally examine the sequence with AnDom (Fig. 19.24). To do this, also insert the sequence into the

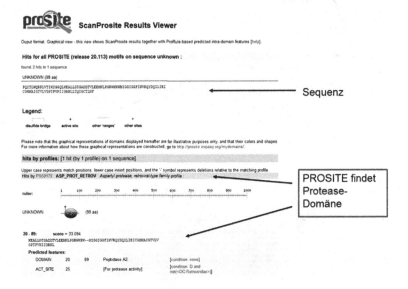

**Fig. 19.23**   PROSITE search with the HIV-1 protease

search window and select the preset BLAST settings. AnDom then computes the available domains and associated SCOP classification based on a Reversed Position Specific BLAST (RPS-BLAST), a very fast sequence comparison against a domain database (alternatively, AnDom allows searching with the IMPALA algorithm, but this is 10 to 100 times slower). In our example, we again obtain an HIV-1 protease and the SCOP ID b.50.1.1 (Fig. 19.24).

In the further course, one can follow this SCOP ID and find out about the protein family (here retropepsin), its structural composition and associated protein domains, but also view and download the corresponding structure in more detail (Fig. 19.25). Alternatively, one can also search directly in the SCOP database for the HIV-1 protease or with the corresponding ID (not shown here).

Other useful databases are SMART (Simple Modular Architecture Research Tool; for annotation, search by name or sequence), Pfam or ProDom (Protein-Domain; based on consensus or multiple *alignment* of a protein or DNA sequence), please inform yourself on the corresponding pages.

In addition, domains and functional motifs (based on SMART and Pfam) of eukaryotic proteins can be identified with the ELM server (Eucaryotic Linear Motif), which can also predict signal peptides and provide information on secondary structure. Alternatively, one can also perform a multiple *alignment* of several sequences oneself in order to find domains or conserved regions and specific differences. For this purpose, the already mentioned software MUSCLE is a useful tool (see tutorial section phylogenetic tree). It can be used online on the EBI website (https://www.ebi.ac.uk/Tools/msa/muscle/), alternatively it can be installed and used locally (https://www.drive5.com/muscle/).

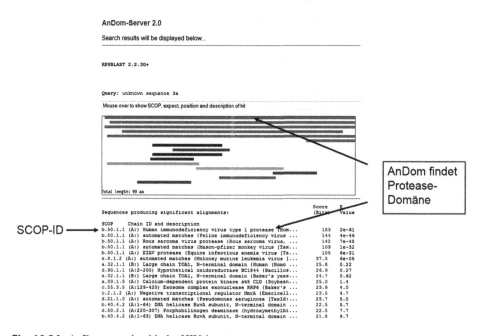

**Fig. 19.24**  AnDom search with the HIV-1 protease

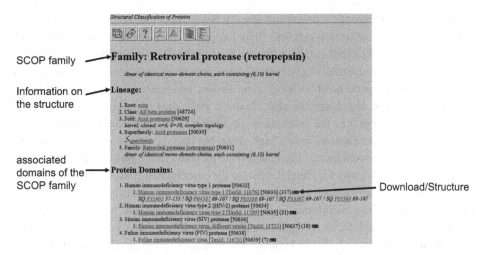

**Fig. 19.25**  Overview of the HIV-1 protease in the SCOP database

## 19.4    Cellular Communication, Signalling Cascades, Metabolism, Shannon Entropy

Communication in prokaryotes usually takes place via two-component systems, which enables direct control (sensor activates *responder,* which then immediately initiates transcription – thus responding quickly to an external stimulus). The situation is much more complex in eukaryotes: Here it is mostly indirect and connected with intracellular communication, e.g. via glucocorticoids and *second messengers.* Often there is also combinatorial regulation via complex signalling cascades. To understand this, it is advantageous to take a closer look at the RNA, DNA and protein networks.

**Topological and Dynamic Modelling of Regulatory Networks**

Protein–protein interactions (PPI) play an important role in the organism. One example is signaling cascades, in which different proteins interact with each other (e.g. activate one after the other) and typically regulate or amplify cellular signals. In addition to pairwise interactions (number of possible interactions $[n^2 - n]/2$), there are of course also complexes (number of possible complexes is $2^n$) between proteins, so that a large number of possible PPIs exist, which makes it difficult to detect all interaction partners experimentally or to predict them bioinformatically. In some cases, there are also tissue-specific interactions.

**How Do I Find and Analyse Protein Interactions and Networks?**

The context of molecules is a current topic in bioinformatics, which is the so-called interactomics. Examples of PPI can be found in the STRING (experimental and predicted PPI), PlateletWeb, KEGG, iHOP and HPRD databases (see individual tutorials on the pages).

Finally, entire interactome contexts are already being modeled. A nice example is the *E-Cell Project* (https://www.e-cell.org). It is important to know the basic properties of each database. The STRING database (Search Tool for the Retrieval of Interacting Genes/Proteins) contains numerous experimentally determined or bioinformatically predicted (based on existing gene neighborhood, gene fusion, *co-occurrence* and *co-expression*) protein interactions. However, one also obtains a great deal of information about their functions. STRING scores the interactions according to its own scoring system (0–1). Good, trustworthy interactions have a high score, but one should always look where and how the interactions were found. The PlateletWeb database (our developed database) contains interactions primarily for platelet, but also for other human cells, such as phosphorylations. The listed interactions in PlateletWeb are based on experimental datasets, e.g. proteomics, so are not bioinformatically predicted. We will show an example for the platelet-derived growth factor receptor beta (PDGFRB) (Fig. 19.26). For this, a total of 66 interaction partners were found in human cells, of which 46 are platelet-specific interaction partners, e.g. interaction with ARAF, based on proteomic data (Fig. 19.26). Another important database is KEGG (Kyoto Encyclopedia of Genes and Genomes). This contains numerous network maps for important signaling pathways, e.g., Wnt, MAPK, Ras-Raf-Mek-Erk signaling pathways, apoptosis, or the cell cycle, but also provides a great deal of additional information, e.g., orthologous genes, on metabolism, enzymes, diseases, and drugs. The databases iHOP (information hyperlinked over proteins) and HPRD (Human Protein Reference Database) contain experimentally determined protein interactions and numerous related information. With all these databases, it is very easy to find individual interaction partners or entire signalling pathways (e.g. for a disease) with which one can then put together a network. This can then be studied in more detail, for example to

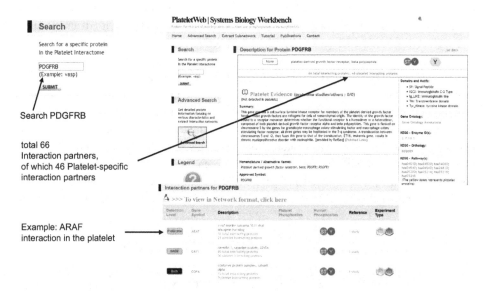

**Fig. 19.26**  PlateletWeb database overview

understand the network topology or intercellular communication: What goes wrong in the network so that a tumor develops? What happens after an infection with a pathogen? What is the effect of a drug? What is a potential *drug target?* But you can also easily find orthologous genes to identify commonalities or differences. This can be helpful, for example if I want to do experiments in the mouse, it is beneficial to know what the commonalities are in the interactions and networks. The network can also be used for dynamic modeling (*in silico* simulation) to better understand the behavior of the network after, say, infection or drug administration. With all that said, one should be aware that one needs to be careful even with the topology (machine-readable drawing software: CellDesigner, Cytoscape): If I miss an important network node (a regulatory protein, e.g. a kinase), all predictions will be wrong – at the edge of the network, in side cascades, I am much more likely to make a mistake, the network is robust here. In the same way, the semiquantitative models (e.g. with Jimena, SQUAD) are only approximate descriptions. If you want to describe a system more precisely, for example with differential equations, you need a lot of and time-resolved data. But even with the semiquantitative models, a number of cycles together with experimental verification (usually in collaborations) are necessary to ensure that I have sufficiently considered all the connections and components in my network that are important for the biological process.

**How Do I Perform Network Analysis and Modeling of Regulatory Networks?**
Best done by (a) network reconstruction, (b) network analysis, and (c) temporal analysis of a network (biological system).

(a) A network reconstruction can be performed using a transcriptome-interactome mapping. For this, significantly expressed genes from an experimental dataset (own experiments or e.g. *microarrays* from the GEO database; please always check here what exactly was investigated and how the experiment was performed [which cell and array type etc.]) are mapped to the interactome (e.g. PPI from STRING database). This can be done using, for example, a mySQL database, which is free database creation software. To do this, simply upload the *microarray dataset* and the interactome as a table and reconstruct the network using the command Select * From 'TableMA', 'TableI' Where ('TableMA'.id = 'TableI'.A OR 'TableMA'.id = 'TableI'.C) (MA = *microarray;* I = *interactome;* OR, since protein ID can occur in column A or C). Then save the result as .sif *(simple interaction file)* to be able to read it into Cytoscape later. For step a), it is best to refer to our two papers (https://www.ncbi.nlm.nih.gov/pubmed/24558299; https://www.ncbi.nlm.nih.gov/pubmed/28265997), which provide a methodological overview and a tutorial.

(b) To visualize and functionally analyze the created network, you can use CellDesigner and Cytoscape. CellDesigner is a classical software for the creation of gene regulatory and biochemical networks, but also includes numerous *tools* and *packages,* e.g. for dynamic simulations and analyses. Cytoscape also allows the visualisation of a network, but also data integration (e.g. *microarray*) and also has numerous *plugins* for additional analyses, e.g. network and functional analysis (Cytoscape allows different formats to be loaded, for example .sif). The two *plugins* BiNGO (GO annotation) and ClueGO (GO annotation and other databases such as KEGG) are particularly helpful for detecting the associated functions and signalling pathways for all proteins. From this, you can in turn create a subnetwork of all proteins for a particular process or signaling pathway and examine it in detail. To examine the network topology, the *plugin* NetworkAnalyzer is suitable. Important topological network parameters are, for example, the average number of interaction partners, network centrality (provides information on robustness) and heterogeneity (provides information on organization/distribution), which you should examine in any case. With the help of the analyses in Cytoscape, you can get a general overview of the network topology, but you are also able to detect important hubs (potential targets), which you can further investigate by means of mathematical modeling c) and then validate experimentally. For step b), please also refer to our two papers (https://www.ncbi.nlm.nih.gov/pubmed/24558299; https://www.ncbi.nlm.nih.gov/pubmed/28265997). Another option is provided by our paper miRNAs (https://pubmed.ncbi.nlm.nih.gov/30421407/). There is also a very helpful online tutorial on Cytoscape where you can learn about features, *plugins*, etc. In addition, other tools for functional analysis and visualization of omics data exist, such as g:Profiler, GSEA and EnrichmentMap (a nice overview is shown in the paper [https://www.nature.com/articles/s41596-018-0103-9]).

(c) The mathematical modelling of regulatory networks is a widespread field of application in bioinformatics. This enables us to analyze the behavior of a network over time in order to validate experimental data or to simulate experiments *in silico* in advance. For the mathematical modeling of regulatory networks, there are the Boolean, quantitative and semiquantitative methods. In principle, these methods consider the nodes (proteins) of a network according to their state between 0 and 1, i.e. either activated (*On;* maximally activated = 1) or inhibited (*Off;* maximally inhibited = 0). According to the initial state (how much is the node activated/deactivated), the further time course, i.e. how does the state of the node change over time, is calculated for each individual node of the network. In this way, the behavior or the network interconnection can be examined in more detail, whereby corresponding network effects, i.e. the respective effect of a node, also become clear. Boolean modeling always considers the *On/Off* -(0/1-) state of a system, i.e. the node is either activated (*On;* 1) or inhibited (*Off;* 0). Quantitative modeling is useful for kinetic data, such as Michaelis–Menten kinetics. Here one can look at the system state of a network in the interval between 0 and 1, but this requires information about the kinetics. A combination of both methods is semiquantitative modeling, whereby one is able to consider the system state in the interval between 0 and 1 as well, but this can be done without knowledge about the kinetics. An example software for semiquantitative modeling is SQUAD, where the system state of a network is first represented using a discrete system (Boolean system), identifying all *steady-state states,* which is then transformed into a dynamic system (differential equation, exponential function). SQUAD identifies all *steady states* using the reduced-order binary decision diagram (ROBDD) algorithm. A *steady state* is a network state to which the network returns, i.e. a stable state that is reached again even after changes or disturbances and does not change. Especially helpful is the perturbation function in SQUAD, with which one can write one's own protocol and define exactly which activation a certain node has at a certain point in time, in order to map or predict the simulation e.g. according to the experimental data and the mutation background (administration of a drug, *knockouts,* activation of receptors). For step c), there is a good tutorial for SQUAD and an example network (T-helper cell network) that you can practice with to get started. In addition, you can practice a bioinformatics *in silico simulation* on your own by watching our online tutorial (https://www.ncbi. nlm.nih.gov/pubmed/27077967). Here you will be shown all the necessary steps and can "recreate" it yourself (scripts for simulation can be found there as well). An alternative is our own software Jimena, which also has a nice online tutorial (https://www. bioinfo.biozentrum.uni-wuerzburg.de/computing/jimena_c/).

**How Do I Perform Metabolic Modeling of Metabolic Pathways/Fluxes?**
It should be noted that one needs as input file for the elementary mode analysis a list of all enzymes (reversible or irreversible should be decided according to the physiological conditions) and a list of all enzyme substrates. Then the given algorithms can calculate all modes effortlessly. But unfortunately, an enzyme can have more substrates than known in the KEGG database (https://www.genome.jp/kegg/). So, in addition, one has to consider biochemical knowledge, literature and databases like the BRENDA database (https:// www.brenda-enzymes.de), which collects very many substrates for an enzyme, along with information about Michaelis–Menten constant and biochemistry. Finally, metabolic enzymes without substrate or under special conditions (e.g. without iron) can suddenly acquire new regulatory functions.

It is interesting to note that dynamic modelling using gene expression data is only an approximation of the true fluxes, but in practice such gene expression data are much more likely to be available than the laborious determination of metabolite concentrations. Dynamic modelling can then also look at true concentrations and kinetics for metabolites, for example using the software PLAS (Power Law Analysis Software – modelled with power functions; https://enzymology.fc.ul.pt/software/plas/). In addition, for the calculation of metabolic pathways/fluxes (elementary mode analysis and flux mode calculation) there are our developed programs Metatool (calculation of all possible metabolic pathways; the Metatool input files have to be edited exactly, otherwise the simple program crashes. It is recommended to start with a simple example, see online tutorial, and then adapt the example file step by step) and YANAsquare (calculation possible for certain situations, e.g. exponential growth with glucose as nutrient source or without oxygen: which *pathways* are then active and how strongly, see exercise tasks for elementary mode analysis). As a first introduction and good basis for metabolic analysis, the online tutorials for Metatool   (https://www.bioinfo.biozentrum.uni-wuerzburg.de/computing/metatool_4_5/;

https://pinguin.biologie.uni-jena.de/bioinformatik/networks/) and YANAsquare (https://www.bioinfo.biozentrum.uni-wuerzburg.de/computing/yanasquare/) are recommended at this point. Of course, one can look at enzymes in more detail, for example with the help of metabolic control theory. A good introduction is the book by David Fell and Keith Snell (1997), which shows how to calculate the strength with which an enzyme controls a metabolic flux, regulatory coefficients and the like: *Understanding the Control of Metabolism.* Also helpful is the book by Reinhart Heinrich and Stefan Schuster (1996) *The Regulation of Cellular Systems.* More recent results can be found in numerous individual publications (just browse the Internet yourself).

### How Can I Better Understand Signal Cascades by Measuring the Encoded Information?

Here we have learned about Shannon entropy. An encoding is done with bits, and there are different levels of encoding. The paper by Heinrich et al. (2002) nicely translates the signal-to-noise problem into a biological application example, kinase signal cascade (Heinrich R, Neel BG, Rapoport TA (2002) Mathematical models of protein kinase signal transduction. Mol Cell 9(5):957–970).

The decoding of protein or nucleotide sequences using the genetic code is fast and reliable, but the other codes are much more difficult to decipher. For example, the three-dimensional structure is difficult to predict from the protein sequence (something for specialists; the accuracy for the best methods [e.g. Zhang lab, QUARK server: https://zhanglab.ccmb.med.umich.edu/QUARK/ as well as David Baker lab, Robetta: https://robetta.bakerlab.org], if the structure is not too complex and unknown, is about 4–6 angstroms). Therefore, we focus more on 3-D predictions by homology modeling. True 3-D predictions for RNA (more degrees of freedom) are even more difficult. Sugar code decoding is just beginning (https://www.ncbi.nlm.nih.gov/books/NBK1965/; NIH Bookshelf Glycomics; Chauhan JS, Bhat AH, Raghava GP, Rao A (2012) GlycoPP: a webserver for prediction of N- and O-glycosites in prokaryotic protein sequences. PLoS One 7(7):e40155. https://doi.org/10.1371/journal.pone.0040155). And the lipid code is even less understood.

### Are There Also Problems for the Computer, and When Does It Become Difficult for the Computer?

This is an exciting topic for computer scientists. In practice one should be careful to think that there are simple general solutions how fast a computer will solve a given task. The Wikipedia page (https://en.wikipedia.org/wiki/P_versus_NP_problem) about this is already very instructive. But Gerhard J. Woeginger's page (https://www.win.tue.nl/~gwoegi/P-versus-NP.htm) only opens the eyes how difficult, exciting and versatile this seemingly simple topic is, especially in the formulation: *If the solution to a problem is easy to check for correctness, is the problem itself easy to solve?* If so, all NP-problems are convertible into P-problems; but probably this is not the case, or at least it has been stubbornly open as a question for decades.

## 19.5    Life Always Invents New Levels of Language

**Synthetic Biology: Where Can I Find Suitable Databases and Literature?**
Concrete insights into the field of work could be mentioned here:

(a) The GoSynthetic database (compares natural and engineered processes): Liang C, Krüger B, Dandekar T (2013) GoSynthetic database tool to analyse natural and engineered molecular processes. Database (Oxford) 2013:bat043. https://doi.org/10.1093/database/bat043. PubMed PMID: 23,813,641; https://gosyn.bioapps.biozentrum.uni-wuerzburg.de.

(b) Our PCT application and description of nanocellulose computer chip: Dandekar T (2015) Invention "Intelligent nanocellulose film for improved smart cards" 04/27/2015 File number DE 102015 005307.8 received. Dandekar T (2016) Modified bacterial nanocellulose and its uses in chip cards and medicine PCT U30719WO, published 3rd Nov 2016.

(c) However, it is also very exciting to work through the other references (Church, Grass, Goldman) on the DNA topic or the current developments in the field of nanocellulose: Dumanlı AG (2016) Nanocellulose and its composites for biomedical applications. Curr Med Chem. PubMed PMID: 27,758,719; Abitbol T, Rivkin A, Cao Y et al (2016) Nanocellulose, a tiny fiber with huge applications. Curr Opin Biotechnol 39: 76–88. https://doi.org/10.1016/j.copbio.2016.01.002. Review. PubMed PMID: 26,930,621.

(d) It is even better to put this in perspective and comparison with similar new developments. This makes it even better clear that it is a general development that we will soon have a new molecular technology between molecular biology, *computational* biology, electronics and nanotechnology, which will start a new industrial revolution after the computer, from which we can greatly benefit. Important technologies in this respect are in particular:

**Optogenetics**

Mühlhäuser WW, Fischer A, Weber W et al (2016) Optogenetics – bringing light into the darkness of mammalian signal transduction. Biochim Biophys Acta 1864(2):280–292. https://doi.org/10.1016/j.bbamcr.2016.11.009. [Epub ahead of print] Review. PubMed PMID: 27,845,208.

**3D Printer**

Biomaterials: 'Bones' made with 3D printer. Nature. 2016 Oct 6; 538(7623): 8. https://doi.org/10.1038/538008a. PubMed PMID: 27,708,302.

Coulais C, Teomy E, de Reus K et al (2016) Combinatorial design of textured mechanical metamaterials. Nature 535(7613):529–532. https://doi.org/10.1038/nature18960

Wehner M, Truby RL, Fitzgerald DJ et al (2016) An integrated design and fabrication strategy for entirely soft, autonomous robots. Nature 536(7617): 451–455. https://doi.org/10.1038/nature19100

**Molecular Imprinting**

Cutiongco MF, Goh SH, Aid-Launais R et al (2016) Planar and tubular patterning of micro and nano-topographies on poly(vinyl alcohol) hydrogel for improved endothelial cell responses. Biomaterials 84:184–195. https://doi.org/10.1016/j.biomaterials.2016.01.036

**Molecular machines** (the 2016 Nobel Prize in Chemistry, after all!):

https://www.nobelprize.org/nobel_prizes/chemistry/laureates/2016/popular-chemistryprize2016.pdf

Capecelatro AN (2007) From Auld Reekie to the City of Angels, and all the Meccano in between: A Glimpse into the Life and Mind of Sir Fraser Stoddart. The UCLA USJ, 20, 1–7.

Feringa BL (2011) Ben L. Feringa. Angew. Chem. Int. ed., 50, 1470–1472.

Peplow M (2015) The Tiniest Lego: a tale of nanoscale motors, rotors, switches and pumps. Nature, 525, 18–21.

Stoddart JF (2009) The master of chemical topology. Chem Soc Rev., 38, 1521–1529.

Weber L, Feringa BL (2009) We must be able to show how science is beneficial to society. Chimia 63(6):352–356.

Current insights into this new subject area are developing at a rapid pace, incidentally also due to optogenetics:

Howe MW, Dombeck DA (2016) Rapid signalling in distinct dopaminergic axons during locomotion and reward. Nature 535(7613):505–510. PubMed PMID: 27,398,617; PubMed Central PMCID: PMC4970879.

Li N, Daie K, Svoboda K et al. (2016) Robust neuronal dynamics in premotor cortex during motor planning. Nature 532 (7600):459–464. https://doi.org/10.1038/nature17643. PubMed PMID: 27,074,502; PubMed Central PMCID: PMC5081260.

Tovote P, Esposito MS, Botta P et al (2016) Midbrain circuits for defensive behavior. Nature 534(7606): 206–212. https://doi.org/10.1038/nature17996. PubMed PMID: 27,279,213.

Bioinformatics models and approaches are then built on this data, which can then explain entirely new levels and answer questions, such as how our brain and consciousness function, or model global problems and master global digitalisation. Starting with simple bioinformatics applications and calculations, these examples show a global context. Another way to globally integrate bioinformatics and your own results is the WikiMedia Foundation, especially Wikipedia (everyone can and should co-edit if they can contribute knowledge to a term). But it also makes sense to get familiar with the WikiMedia infrastructure, e.g. Wikidata (https://www.wikidata.org/wiki/Wikidata:Main_Page) and Query Service (https://query.wikidata.org).

## 19.6   Introduction to Programming (Meta Tutorial)

Our book does not focus on programming. This is because we are initially concerned with the fascination of the topics, all of which can be dealt with using bioinformatics, and because we have had more and more bioinformatics software on the net since about 1995. So the difficulty is rather to keep track of the different possibilities of analysis and to use the right software.

Nevertheless, it is quite typical for bioinformaticians, after they have become sufficiently familiar with their field, to program new software themselves, which then searches for exactly the motifs that interest them, or a database with exactly the data that they are investigating in detail, or a model, for example of a signal cascade. For this reason, we have compiled introductory materials here for readers fascinated by programming.

The areas in which writing your own programs for bioinformatics can happen are already clear from the book:

- **Collect and store data** (i.e. build your own database),
- **Examine/analyze data** (i.e., write programs in the strictest sense),
- **Understand** (or model) **data.**

Every program works the same way. After a header part, where the variables are declared, the actual program starts: There is a read-in part (e.g. all sequences), a main loop (processing part for the calculations/tests), which can call further loops, and the output part (displays the calculation/results).

### Collect and Store Data
A common task in bioinformatics would be to set up a web server yourself (https://perl-webserver.sourceforge.net; https://sourceforge.net/projects/perlwebserver/files/perlwebserver/). It is equally important to set up a database yourself (https://perlmaven.com/simple-database-access-using-perl-dbi-and-sql). Depending on the needs and requirements, knowledge of common programming languages such as SQL, HTML and Java is necessary. However, there are helpful tutorials and ready-made scripts for this, which can

be used individually in each case. Another standard task is to run a web server that offers e.g. a software like BLAST, which in turn can be used by other users for their analyses. To do this, you can simply install the BLAST server yourself by downloading it from the NCBI website: https://blast.ncbi.nlm.nih.gov/Blast.cgi?PAGE_TYPE=BlastDocs&DOC_TYPE=Download. However, it must be said that this requires some prior knowledge.

**Examine/Analyze Data**
Data can be analyzed particularly well with R, Perl and Python. In particular, Perl allows you to formulate so-called *"regular expressions"*, which test whether a certain expression occurs in a text, file, or database (e.g., with an A at the beginning or exactly two "t"s at the end of a line). https://regexr.com explains in more detail how to describe these searches correctly.

A tutorial for Perl is available here (https://perlmaven.com/perl-tutorial), which covers typical introductory topics for programming with Perl such as installation, debugging and command line, and explains terms such as *scalars (strings, operators)*, *files* and *arrays*. Also useful is the book Perl in 21 days (by Patrick Ditchen; https://www.google.com/url? sa=t&rct=j&q=&esrc=s&source=web&cd=&cad=rja&uact=8&ved=2ahUKEwjI2PbduI HrAhUDx4sKHZmCCwUQFjACegQIARAB&url=http%3A%2F%2Fstarkill.synology. me%2Fuwe%2FPerl21%2F26393_21t_perl.pdf&usg=AOvVaw0FnAy8RM8Pv3CWeM HN2r0-), which gives a step by step introduction to programming with Perl.

In addition, there are other programming languages such as C++ and Julia, but also web-based programming environments such as Jupyter Notebook, which support various languages such as Julia, Python and R. For this you are welcome to inform yourself on the internet depending on your interest.

**Understand (or Model) Data**
We have explained this in detail in the book and tutorial. This can be done especially well with the *tools* Cytoscape, CellDesigner and SQUAD or Jimena – and then more elaborately with self-written programs.

This is already a brief overview of programming. But if you want to know more and practice, you will now find an overview of programming languages and what they do, including tutorials.

## Programming Languages and Tutorials

### Perl
Perl (Practical Extraction and Report Language): This programming language is very popular in bioinformatics because it is very good for processing long lists and compiling them into new lists.

You can learn this relatively easily with textbooks, such as the **Perl Cookbook** by Tom Christiansen and Nathan Torkington (2003). This book is just very well written and provides a very good introduction to the PERL programming language.

Or the book **Beginning Perl for Bioinformatics** by James Tisdall (2001).

There are of course countless tutorials on the net, e.g.

- https://www.perl.org/learn.html
- https://www.tutorialspoint.com/perl/
- https://wiki.selfhtml.org/wiki/Perl
- https://www-cgi.cs.cmu.edu/cgi-bin/perl-man

To be able to program faster in PERL yourself, there are also the BioPerl modules:

- https://en.wikipedia.org/wiki/BioPerl
- https://bioperl.org (is the entry page).

Here are three aspects of such recipes listed as article examples:

Angly FE, Fields CJ, Tyson GW (2014) The Bio-Community Perl toolkit for microbial ecology. Bioinformatics 30(13):1926–1927. https://doi.org/10.1093/bioinformatics/btu130

Vos RA, Caravas J, Hartmann K et al (2011) BIO: phylo-phyloinformatic analysis using perl. BMC Bioinformatics 12:63. https://doi.org/10.1186/1471-2105-12-63

Stajich JE, Block D, Boulez K et al (2002) The Bioperl toolkit: Perl modules for the life sciences. Genome Res 12(10):1611–1618.

**Java**

This programming language by James Gosling (1991) runs on every major operating system ("platform", Windows, Mac and LINUX) and is so popular because you can write it once and then run it (especially over the Internet) on any platform. It is an object-oriented, modern programming language, so "objects" as complex concepts are central to it. Java's syntax is similar to C or C++, but Java's comfortable, high level language does not make it as easy to refer to single bit instructions (the machine language) as it is with C or C+ +.

Here, too, there is Biojava, i.e. ready-made program modules for bioinformatics:

https://biojava.org

And in addition a number of recipes and program modules (routines):

https://biojava.org/wiki/Main_Page/

As soon as more complex calculations are the focus (instead of lists, data, web servers, databases or sequence properties), more languages such as C or C++ (computer languages that are also used intensively and developed further today, newer is e.g. *c#,* pronounced: *c sharp,* and similar more) and **Fortran** (Formula Translation) are used in bioinformatics, old but constantly modernised. For example, Fortran 2003 is object-oriented, and Fortran 2008 even allows *"concurrent programming",* i.e. parallel, instead of serial programming.

## MATLAB

https://de.mathworks.com/products/matlab.html

Allows complex computations to be efficiently expressed in this language in a mathematical way.

MATLAB is matrix-based. Linear algebra in MATLAB looks like linear algebra in a textbook. This makes the code for these calculations easier to write, read, and analyze, and easy to manage. Numerical analyses are also easy to write. Another advantage is that computations are distributed across multiple processors ("cores"), making them much faster. This makes parallelization easy. More information can be found here https://de.mathworks.com/.

## Programming Language R

If, on the other hand, the calculations are of a more statistical nature, i.e. deal directly with the analysis of large amounts of data, R is often used in bioinformatics:

https://www.r-project.org

R is also very easy to learn by following the link, installing R right there and learning it too. R is freely available and very useful for statistical analysis and graphical representation of biological data (results and graphs can also be used for scientific publications). It is command line based and can be used on different platforms and operating systems (e.g. Windows, Linux). In short, R is a really nice and easy to learn programming language, best try it yourself (there are also numerous online codes to use). Moreover, it is interconnected with other programming languages and platforms, such as Bioconductor, for even more specialized data analysis.

A good example of high-throughput data analysis is Bioconductor (https://www.bioconductor.org), which now has 1881 (as of August 4, 2020) software packages (https://www.bioconductor.org/packages/devel/BiocViews.html).

In the following we want to show five short examples in R. The scripts are kept very simple and introductory, so that you can solve the problem quickly. However, the problem and the scripts are usually much more difficult and extensive.

---

**Example 19.1**

Concentrations of ozone (per 100 million particles, pphm) were measured in three gardens on ten summer days and summarized in Table 19.1.

---

We can now use a t-test to determine whether the mean ozone content in Gardens A and B is significantly different.

To answer this question, we need to formulate a test hypothesis. The corresponding null hypothesis ($H_0$) would be: The ozone level is not significantly different, and the corresponding alternative hypothesis (H1) would be: The ozone level is significantly different. We can reject the null hypothesis at a significance level of 5% ($p\text{-value} < 0.05$). We can calculate the test statistic in R. To do this, we would read in the data as follows:

```
> GardenA = c(3, 4, 4, 3, 2, 3, 1, 3, 5, 2)
> GardenB = c(5, 5, 6, 7, 4, 4, 3, 5, 6, 5)
> gardenC = c(3, 3, 2, 1, 10, 4, 3, 11, 3, 10)
```

For the t-test, we can use the R command t.test:

```
> t.test(gardenA, gardenB, var.equal=T)
```

**Table 19.1**  Ozone content in gardens (per 100 million particles, pphm)

| Garden A | Garden B | Garden C |
|----------|----------|----------|
| 3 | 5 | 3 |
| 4 | 5 | 3 |
| 4 | 6 | 2 |
| 3 | 7 | 1 |
| 2 | 4 | 10 |
| 3 | 4 | 4 |
| 1 | 3 | 3 |
| 3 | 5 | 11 |
| 5 | 6 | 3 |
| 2 | 5 | 10 |

As a result we get:

```
Two Sample t-test
data: GardenA and GardenB
t = -3.873, df = 18, p-value = 0.001115
alternative hypothesis: true difference in means is not equal to
095 percent confidence interval:
-3.0849115 -0.9150885
sample estimates:
mean of x mean of y
3      5
```

Thus, we obtain a *p-value* of 0.001115. This means that we can reject the null hypothesis. Accordingly, the mean ozone concentration in garden B is significantly higher than in garden A.

**Example 19.2**
The trial of a new therapy yielded the following data (Table 19.2).

We can now use a t-test to determine whether the new therapy shows a significant improvement. Analogous to Example 19.1, we would first formulate the test hypothesis (*p-value* < 0.05). The null hypothesis ($H_0$) would be: The new therapy does not affect or prolong the average duration of illness. The corresponding alternative hypothesis H1: The new therapy shortens the average disease duration (one-sided test).

In R, we would use the following script:

```
> groupA = c(7, 8, 11, 10, 9, 11, 13)
> groupB = c(9, 7, 9, 11, 6, 11, 11, 8)
> t.test(groupA,groupB,var.equal=T)
```

**Table 19.2** Effect of a new therapy (group A = conventional therapy; group B = new therapy; duration of illness in days)

| Group A | Group B |
|---------|---------|
| 7       | 9       |
| 8       | 7       |
| 11      | 9       |
| 11      | 11      |
| 10      | 6       |
| 9       | 11      |
| 11      | 11      |
| 13      | 8       |

And would get as a result:

```
Two Sample t-test
data: GroupA and GroupB
t = 1.0377, df = 14, p-value = 0.317
alternative hypothesis: true difference in means is not equal to
095 percent confidence interval:
-1.066768 3.066768
sample estimates:
mean of x mean of y
10        9
```

A *p-value* of 0.317 was found, which means that we cannot reject the null hypothesis. Thus, the new therapy does not show any significant improvement with regard to the average duration of illness.

**Example 19.3**
In an investigation it should be determined whether there is a correlation between the airbag and car type (Table 19.3; see also library [MASS], car.data in R).

We can now use a chi-square test (test for independence) to determine whether the two variables are independent, that is, whether the number of airbags is independent of the type of car.

Analogous to the two previous examples, we must also formulate the test hypothesis here ($p\text{-}value < 0.05$). The null hypothesis ($H_0$) would be: Both variables are independent. The alternative hypothesis H1: The number of airbags depends on the type of car.

In R we would use the following script (*clipboard* loads data from cache, just copy the table to do this):

```
> table<-read.table(clipboard)
> chisq.test(table)
```

And would get as a result:

```
Pearson's Chi-squared test
data: table
```

**Table 19.3** Number of airbags in different car types

|                  | Compact | Large | Medium | Small | Sports | Van |
|------------------|---------|-------|--------|-------|--------|-----|
| Driver/passenger | 2       | 4     | 7      | 0     | 3      | 0   |
| Driver           | 9       | 7     | 11     | 5     | 8      | 3   |
| None             | 5       | 0     | 4      | 16    | 3      | 6   |

```
X-squared = 33.0009, df = 10, p-value = 0.0002723
Warning message:
In chisq.test(table) : Chi-square approximation may be incorrect
```

The warning message in this case is due to the insufficient number of samples, can be neglected here (exercise example). The *p-value* is 0.0,002,723, so we can reject the null hypothesis, i.e. the number of airbags depends on the type of car.

**Example 19.4**

The state of biological systems can be described with mathematical formulas. The formula $f(x) = -\cos(x) - 0.1x^2$ describes (in a very simplified way) the equilibrium of the erythrocyte production in the body. Here, the x-axis shows the amount of red cells in the body, and the y-axis represents the energy the body invests to get back into balance. Small disturbances are easily compensated by the system, large disturbances affect the vital functions and can no longer be compensated.

We want to draw the graph in R and look at it more closely (e.g. attractors, stable state of the system [local minimum], tolerated disturbances that the system can still compensate for [local maximum]).

The R script would look like this:

```
> x<-seq(-5, 5, by=0.1)
> plot(x, -cos(x)-0.1*x^2, type="l")
```

The plot is shown in Fig. 19.27. The x-axis shows the blood loss or decay, the y-axis the blood production. The zero point reflects the maximum production when still healthy. An attractor would be the healthy state: If the disruption is not too severe, the system falls back to the minimum (x-axis at 0) (no erythrocytes are produced). If erythrocytes perish again, new ones are produced. Another attractor would be the sick state: as soon as the disturbance exceeds the two maxima (x-axis at ±2.596), catastrophe (sick state) occurs. To calculate the exact local minima and maxima, you have to set the first derivative to zero (not shown here).

**Example 19.5**

A cyclist is injected with a dose of erythropoietin (Epo). At the start time of the measurement t0, n0 molecules of Epo dock onto each hematopoietic cell. As Epo detaches from the receptors over time and is broken down by the body, only $n0 \cdot e^{-t}$ molecules are still docking at time t. Each molecule of Epo docked to the cell activates alpha-STAT transcription factors via a signaling cascade per unit time t by phosphorylating them. Phosphatases are permanently active in the cell, which remove the phosphate residue from the STAT transcription factors and thus deactivate them. The phosphatases deactivate beta-% of the active transcription factors per time unit t – and this already from time t0.

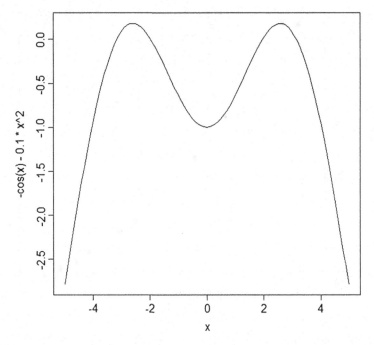

**Fig. 19.27** Graphical representation of erythrocyte production in the body (x-axis: blood loss or decay, y-axis: blood production)

We want to know how many activated molecules STAT are present at time t = 3. To do this, we can calculate the solution in R and draw the graph of STAT concentration (in this example, n0 = 10, alpha = 5 and beta = 10%).

```
> n0 <-10 # number of Epo molecules at time t0.
> a <-5    # alpha
> b <-0.1 # beta
> stat <- 0
> stat[1] <-a*n0*(1-b) # Actually time t0, but R doesn't
like 0 as index.
> for (t in 2:20) {stat[t] <-(stat[t-1]+a*n0*exp(-
t+1))*(1-b)} # A loop that always increments t by 1 and
performs the calculation.
```

Looking at stat now, we always have to subtract 1 from the index to get the correct value. After all, we had already stored t0 as t1.

```
> stat
[1]  45.00000  57.05457  57.43921  53.93570  49.36634  44.73291
40.37116 36.37508
```

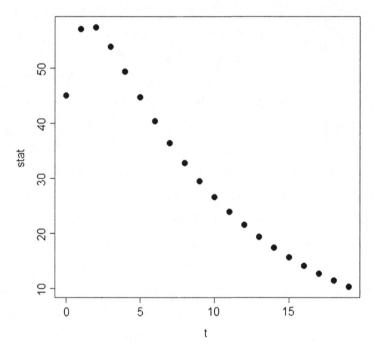

**Fig. 19.28** Graphical representation of STAT concentration (x-axis: time, y-axis: STAT concentration)

```
[9]   32.75267  29.48296  26.53670  23.88378  21.49568  19.34622  17.41163
15.67048
[17]  14.10344  12.69310  11.42379  10.28141
```

You can see it better in the graphic:

```
> t <- 0:19
> plot(t,stat, cex=1.5, pch=16)
```

The plot is shown in Fig. 19.28. The STAT concentration (y-axis) per time (x-axis) is shown. It can be seen that after a short sharp rise, the concentration slowly levels off again. In our example, 53.93,570 molecules of STAT are present at time t = 3.

# Solutions to the Exercises

# 20

**Abstract**

In this part, we give suggested solutions and additional explanations to the exercises.

## 20.1 Sequence Analysis: Deciphering the Language of Life

**Questions 1.1 and 1.2**

Bioinformatics, or *computational biology,* attempts to solve biological problems with the computer. The aim is to secure information and knowledge about organs and diseases in databases and make them accessible to everyone, but also to identify and understand the molecular causes associated with a disease and develop suitable models based on this. This means that the aim is to understand biological function on the basis of information about DNA, RNA and proteins through programs and software. This is done, for example, by sequence analyses in order to obtain information about a pathogen, but also by genome comparisons in order to obtain differences between the organisms involved (e.g. humans and parasites). This in turn enables the creation and comparison of metabolic networks and, finally, the calculation of drugs for important proteins in the parasite that optimally block the parasitic protein but are tolerated by humans. Bioinformatics is thus able to better answer basic medical questions based on theoretical knowledge, such as why people age and die.

Three main areas can be distinguished: (i) Databases and servers integrate and collect biological data. (ii) Programs and software to study and analyse datasets or experiments. (iii) Bioinformatics models for modelling and simulation. This can then be used to understand biological functioning, such as modelling the interaction of a drug with its *target,* or

© Springer-Verlag GmbH Germany, part of Springer Nature 2023
T. Dandekar, M. Kunz, *Bioinformatics,*
https://doi.org/10.1007/978-3-662-65036-3_20

simulating metabolism to understand how the metabolic signalling network works. Databases would include PubMed, Gene Expression Omnibus (GEO) and GENEVESTIGATOR.

---

### Example 1.3

1. Question: Answer B
2. Question: Answer A
3. Question: Answer D

---

### Reply Comment

If you didn't find the right answer, here is the corresponding protein sequence: https://www. ncbi.nlm.nih.gov/protein/AAX29205.1. To do this, it is best to select Protein next to the search bar in PubMed and type HIV into the search bar, after which you should find the entry "TAR, partial [synthetic construct], Accession: AAX29205.1". Here you will find all the information about the answers.

---

### Question 1.4

The BLAST (Basic Local Alignment Search Tool) algorithm allows protein and nucleotide sequences to be compared with a large database in terms of their local similarity. In this process, a sequence is compared for its similarity with reference sequences in a database and can provide information on which virus a patient has contracted. BLAST uses a heuristic search and the *two-hit method*: A short word list (so-called *lookup table*) is first compared with the short word lists of the database (indexed database). If at least one matching short word is found in an entry, the algorithm immediately checks whether there is another short word hit in the vicinity (fixed distance), and only then calculates the *alignment*. In all other cases, the algorithm *blasts ahead* to the next database entry.

With a BLAST search, one is thus able to identify homologous genes and compare the individual positions in order to be able to identify unknown sequences, but also to find corresponding differences in other organisms (e.g. for the development of an animal model). However, sequence analysis can be taken much further bioinformatically. For example, the patient's virus can be compared with other patient isolates, related viruses (HIV-1, HIV-2, etc.) and other sequences. In the clinic, by the way, HI viruses are now even routinely sequenced according to resistance mutations, so that it is possible to recognise in good time how the virus population changes under antiretroviral therapy, in order to change and optimise the therapy accordingly. For further information, please use the link to BLAST (https://blast.ncbi.nlm.nih.gov/Blast.cgi).

---

### Question 1.5

So, in your own program, you would first read in the sequence (input part), then use an algorithm ("*two-hit method*") to calculate the similarity to the entries in the database (processing or calculation part; of course, you have to have created the reference database first), and finally there is a nice output list (list of hits and statistical parameters).

---

### Question 1.6

Answer A, C, D

### Example 1.7
Answer B.

The BLAST algorithm can perform a number of searches, e.g. blastn for a nucleotide sequence and blastp for a protein sequence. But it can do much more, e.g. blastx translates a nucleotide sequence into a protein sequence and then searches against the protein database, tblastn searches with a protein sequence against a translated nucleotide database, and tblastx searches with a translated nucleotide sequence against a translated nucleotide database.

### Example 1.8
Answer A, D.

The sequence comparison with BLAST first tells what the function of the sequence is (which piece of which virus is here as a sequence). In the example, the blastp search should have found the pol protein and protease of HIV-1. Another important output is the *E-value* (expected value). This indicates that my output *alignment* will be found again in the database with a similar or better score, so it depends on the size of the database (unlike the *p-value*). If you are looking for the highest possible similarity, the selected BLAST hit should have the smallest possible *e-value* and a high identity.

If the blastp search did not find the pol protein and the protease of HIV-1, then try it best like this: Since it is a protein sequence, please select a blastp search and copy the unknown sequence into the search window, then simply start the BLAST search (please see if the non-redundant protein sequence database is set as default). As an example, four hits are shown as a result (Fig. 20.1).

---

### Question 1.9 and Example 1.10

A *dotplot* allows you to compare two sequences on a graph (x–/y-axis) to find similar areas (represented as a dot). In both cases (by hand and software), your *dotplot* should find similar areas between the two exercise sequences.

---

## 20.2   Magic RNA

### Example 2.1
1. Answer C, E

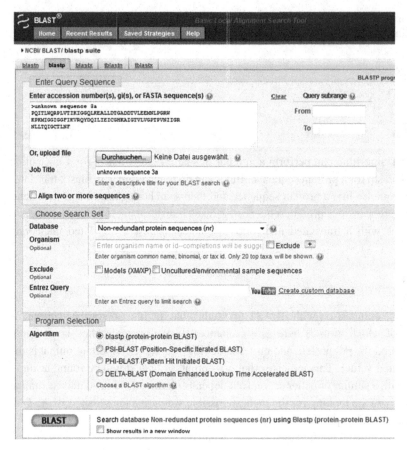

**Fig. 20.1**  blastp search with HIV-1

The secondary structures exert important functions of RNA in the regulation of transcription, such as catalytic activity of ribosomes (ribozymes). RNA secondary structure folding is a complex process; in addition to a complementary sequence, the folding energy must also be considered. Whereby the following always applies: A pairs with U (two hydrogen bonds), G pairs with C (three hydrogen bonds). But there are also other rules, for example G pairs with U, only one hydrogen bond. In addition, the folding energy must also be considered. Not every folding is thermodynamically optimal (it should always have a low folding energy, because the lower the free energy, the more stable the structure), especially since there are several secondary structure forms (e.g. *stem-, hairpin- and interior-loop*). Secondary structures can be predicted bioinformatically, but this is not easy. There are various algorithms for this purpose, which are all based on dynamic programming methods, but nevertheless work differently. For example, the Nussinov algorithm first calculates the maximum number of base pairs and then uses this information to calculate the secondary structure with the maximum base pairing. However, since RNA structures do not always have the maximum possible base pairing, this method does not always give useful results. A more optimal and faster solution for structure determination is provided by algorithms based on energy minimization. The Zuker algorithm calculates the optimal secondary structure with the minimum free energy, based on a thermodynamic model, e.g. mFold server. On the other hand, the Sankoff algorithm simultaneously folds and aligns two sequences using an energy model to minimize the free energy, e.g. LocARNA program. A useful online web server for secondary structure prediction is ViennaRNA Webservices (https://rna.tbi.univie.ac.at/). There are many more *tools* for RNA analysis here. For additional information, see the book section or Kunz et al. (2015).

In the exercise example, RNAfold (also in ViennaRNA Webservices, also based on energy minimization) should find a possible secondary structure fold with a *minimum free energy of* − 360.20 kcal/mol.

If you did not get a secondary structure for the sequence example, your result should look like (Fig. 20.2; to search, simply copy the example sequence into the search window and use the default parameters).

2. Here it is important to see that the change in energy released is not automatically equal to the sequence length, e.g. it is not double. For example, the sequence ATGCTACGCGATGCATCGAGCGCAT has an energy of −3.5 kcal/mol and twice the

**Fig. 20.2**  RNA folding with RNAfold

sequence     length     of     −21.5     kcal/mol,     whereas     the     sequence
GCATGACGTAGCAGCCGTACGATAT has an energy of −2.10 kcal/mol and twice
the length of −12.40 kcal/mol.

Regulatory RNA elements are found in humans, but also in other organisms such as bacteria. Examples of regulatory RNA elements are *iron-responsive elements* (IRE) and *riboswitches*. They perform regulatory functions and control transcription and translation. IRE regulate iron metabolism in humans and animals, depending on the iron content of the cell. *Riboswitches* regulate gene expression in prokaryotes. Metabolites specifically bind a *riboswitch,* which leads to a conformational change of the *riboswitch* and thus switches genes on or off. However, there are also other RNA elements in prokaryotes, e.g. the 6 S-RNA (general STOP signal) and the ppGpp *(messenger).* An important database for RNA families is Rfam, which lists quite a few different families (best take a look).

Bioinformatically, it is of course also possible to find regulatory RNA motifs. Here, it is best to combine several criteria, such as sequence, structure and folding energy, in order to achieve a higher degree of accuracy. An IRE can be recognised by these three criteria, among others:

- Matching consensus sequence "CAGUGN" and a C alone, without G as a partner in the opposite strand *("bulged"),*
- structure (loop stem structure, *stem-loop*) of two stems on top of each other, in between is the unpaired C and
- energy (when this structural part is considered as a whole, −2.1 to −6.7 kcal/mol).

Only when all parameters are fulfilled, i.e. when all criteria for an RNA molecule are met, should the bioinformatic motif search produce a corresponding hit.

Regulatory RNA elements can be identified using programs such as RNAAnalyzer, Riboswitch Finder or RegRNA. Of course, a bioinformatically predicted hit should be checked experimentally. This is the only way to be sure that the element found actually performs a biological function.

Answer C, D, E (please also look at the previous answers)
**Example 2.6**

1. Question: You should find three positions (CAGTGC, CAGTGA, CAGTGC)
2. Question: Answer A, C, D

The RNAAnalyzer finds an IRE at position 71 in the exercise example (you should also find it by hand). However, it also finds a catalytic RNA.

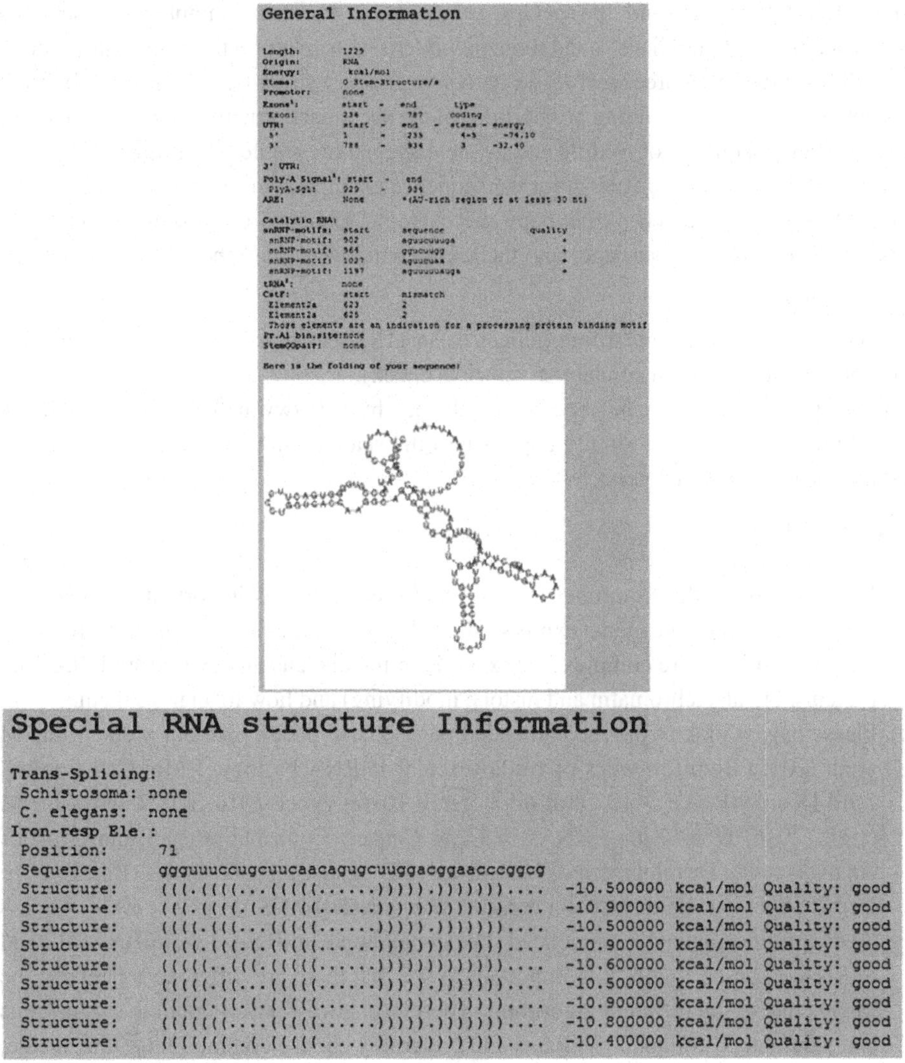

**Fig. 20.3**  IRE example

If you did not get an IRE for the sequence example, your result should look like (Fig. 20.3); for search please just copy the example sequence into the search window and use the default parameters:

**Example 2.7**

Answer C, D.

*Riboswitches* regulate gene expression in prokaryotes. Metabolites specifically bind a *riboswitch,* which leads to a conformational change of the *riboswitch* and thus switches genes on or off. However, there are other RNA elements in prokaryotes, e.g. the 6 S-RNA (general STOP signal) and the ppGpp *(messenger)*. For the example sequence, the riboswitch finder should have found three possible *riboswitches* on the plus strand at position 1288, which have three *stem-loops* in their secondary structure. These three different hits come from different folding possibilities of the secondary structure, where in this case two show a good quality of folding energy, i.e. have a more stable structure.

If you did not find a *riboswitch* for the sequence example, your result should look like (Fig. 20.4). For the search please copy the example sequence (corresponding example from the Riboswitch Finder page) into the search window and use the preset parameters.

**Example 2.8**

You should find the typical four helices of an ITS2 secondary structure (you can also use RNAfold to fold the secondary structure to check).

(Example sequence can be found here: https://www.ncbi.nlm.nih.gov/nuccor e/260206998?report=fasta; simply copy it into the search window of the ITS2 database and use the default parameters).

**Example 2.9**

(a) For this, look at the recommended sites and literature. It is important to know how both classes regulate gene expression (miRNAs in the nucleus mRNA binding; lncRNAs much more complex, for example in the nucleus and cytoplasm RNAs and proteins, but also chromatin and histone modifying) and how to analyze them.

(b) Please have a look at the recommended pages. Then please also our work: Kunz M et al. (2015) Bioinformatics of cardiovascular miRNA biology. J Mol Cell Cardiol. 2015 Dec; 89(Pt A): 3–10. https://doi.org/10.1016/j.yjmcc.2014.11.027 and Kunz M et al. (2016) Non-Coding RNAs in Lung Cancer: Contribution of Bioinformatics Analysis to the Development of Non-Invasive Diagnostic Tools. Genes (Basel). 2016 Dec 26; 8(1). pii: E8. https://doi.org/10.3390/genes8010008.

(c) To do this, look at the recommended sites, for example, miRNA-132, miRNA-212 and miRNA-7 can be found. It is important to see that there are differences between the *targets* due to the different algorithms. Therefore, always know about the algorithms and parameters, compare programs and preferably choose common hits (if available, use experimentally validated hits).

## 20.3   Genomes – Molecular Maps of Living Organisms

**Question 3.1**

For this you should know: 3.2 billion base pairs, about 23,700 genes, 2–3% of the genome for protein reading frames, most of it is "ballast" (*selfish DNA,* LINE and SINE). Best to read the book chapter again.

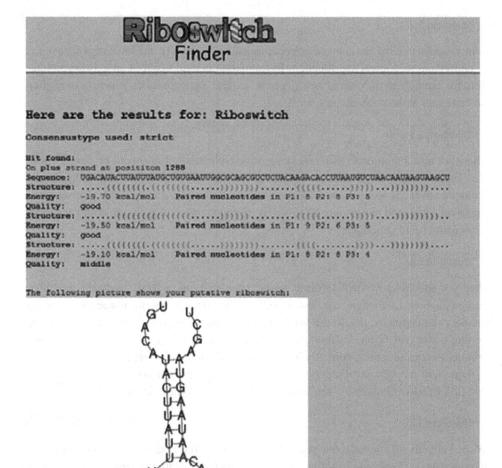

**Riboswitch** Finder

**Here are the results for: Riboswitch**

Consensustype used: strict

Hit found:
On plus strand at posititon 1288
Sequence:   UGACAUACUUAUUUAUGCUGUGAAUUGGCGCAGCGUCUCUACAAGACACCUUAAUGUCUAACAAUAAGUAAGCU
Structure:  .....(((((((((.(((((((((.....))))))).......(((((.....)))))...))))))))))....
Energy:     -19.70 kcal/mol    Paired nucleotides in P1: 8 P2: 8 P3: 5
Quality:    good
Structure:  .......(((((((((((((((.....)))))).........(((((.....)))))....))))))))))..
Energy:     -19.50 kcal/mol    Paired nucleotides in P1: 9 P2: 6 P3: 5
Quality:    good
Structure:  .....(((((((((.(((((((((.....))))))).......((((.......)))...))))))))))....
Energy:     -19.10 kcal/mol    Paired nucleotides in P1: 8 P2: 8 P3: 4
Quality:    middle


The following picture shows your putative riboswitch:

The blue brackets show Stem P1, the orange brackets stem P2 and the green brackets stem P3.

**Fig. 20.4**  Riboswitch example

## Question 3.2

Annotation is to label the genome or genome sequence (usually unknown organisms), i.e. to understand its content and function. Reannotation is used to check an existing annotation, for example in the case of new genes or sequencing techniques. It is best to read up on this again in the book chapter.

## Question 3.3

There are now a number of different sequencing techniques. You should be familiar with the classical sequencing technique according to Sanger (chain termination synthesis) and the more advanced methods, so-called *next generation sequencing,* such as pyrosequencing (Roche 454) and *sequencing by synthesis (*Illumina) (second generation) and nanopore sequencing (third generation).

## Question 3.4

Here you should know DNA labeling, sequencing (shotgun method), *mapping,* assembly and annotation. Challenges are, among others, the computer performance, but also the memory requirements, due to the flood of sequences, by sequencing techniques that are getting better and faster. Furthermore, there are also problems, especially with *repeat regions,* to represent them correctly in their length and number of repeats.

Important pioneers are Frederick Sanger (chain termination synthesis), Craig Venter and Erik Lander (first *"draft"* sequence of the human genome in 2001).

## Question 3.5

We had already mentioned a few points here in Task 3.4 (computer performance, memory requirements). The secure handling and use of data (confidential treatment, avoidance of data misuse, data protection) etc. are certainly also important.

## Question 3.6

Here, for example, there are the online libraries NCBI and EBI, but also the genome browsers UCSC and Ensembl.

## Questions 3.7 and 3.8

The cell nucleus, also called the nucleus, contains the entire genetic material of an organism, i.e. the DNA. The DNA consists of genes that code for specific proteins, such as enzymes, hormones or transcription factors, which fulfil important tasks in the organism. In order to form a protein, i.e. the active form of a gene, the information of a gene must first be read. This process is also known as gene expression, whereby a distinction is made between the two processes of transcription (formation of mRNA on the basis of a DNA sequence) and translation. The analysis of DNA sequences is important, for example to investigate the promoter region for transcription factor binding sites (TFBS). Transcription factors (TFs) recognize and bind to specific DNA motifs (DNA binding sites) in the promoter, called TFBSs, and thus regulate transcription. If I know the consensus sequence of the TFBS (template), i.e. the DNA nucleotides to which the TF binds, I can also easily bioinformatically investigate an unknown sequence for possible binding sites, which I can then use for further experimental investigations. Appropriate software is already available for this purpose. Apart from programs that list experimentally validated TFBS (such as MotifMap), there are also numerous programs that predict TFBS, e.g. ALGGEN PROMO, PRODORIC (Prokaryotic Database of Gene Regulation), TESS (Transcription Element Search System) or Genomatix. It is useful to always use several programs to compare results and find common TFBS. As these programs disappear so often from the open accessible internet as they can be commercially used and sold, we recently published AIModules, which offers TFBS analysis including conserved TFBS modules in different promotor regions (Aydinli et al., 2022; https://aimodules.heinzelab.de/#/)

A computer program for promoter analyses would first "learn" the TFBS, this is done using stochastic models, e.g. PSSMs or HMMs. In a further step, the program would then read in a promoter sequence (read-in part) and then search for similarities with the consensus sequence found (internal calculation part, e.g. with a BLAST), which are then in turn output as hits (output part).

Possible challenges and sources of error are, for example, that several DNA sequences are necessary to create the template, i.e. the more binding sites the training data set contains, the more accurately the template can also be trained. Statistical parameters should also be considered. TFs also often bind to DNA combinatorially at a certain distance from each other, and there are also other elements that influence transcription, such as *enhancers*. All these factors and challenges should be taken into account by a program to enable accurate prediction. In any case, it is advisable to validate bioinformatically predicted TFBS experimentally. Only then can I be sure that the TF actually has an effect on transcription. Otherwise, only the DNA nucleotides of the prediction match (that's why I got a hit; false positive hits), but this has no biological relevance.

### Example 3.9

C, D (please also look at the previous answers).
ALGGEN PROMO should find numerous TFBS for the example sequence, including NF-AT2 [T01945].

If something did not work for you, then try it best like this. In ALGGEN PROMO, select the option *"SearchSites"* (under *Step 2*) and copy the sequence into the search window, then start the search (please make sure that the default *"Maximum matrix dissimilarity rate"* is set to 15; this specifies the maximum deviation from the actual DNA nucleotide sequence [template] of the TFBS that is allowed, you can also change this parameter yourself and observe what happens). As *output* you will see all TFBS found, their position and score (under Data [txt] you can also display a list of the TFBS found and the corresponding TF).

### Example 3.10

Hidden Markov models are stochastic probability models that predict hidden system states (e.g. exon, intron) from a sequence (observations, e.g. ATCCCTG...) using a Markov chain (Bayesian network; supervised machine learning). We learned about several application examples in the book, such as for genome annotation, protein domain prediction, and network regulation. For more details and information, see the article Sean R Eddy (2004; What is a hidden Markov model? Nature Biotechnology volume 22, pages 1315–1316. https://doi.org/10.1038/nbt1004-1315).

## 20.4    Modeling Metabolism and Finding New Antibiotics

### Questions 4.1 to 4.5

One algorithm to calculate metabolic fluxes is elementary mode analysis. It calculates enzyme chains that keep all internal metabolites in balance. That is, the enzymes consume as much of an internal metabolite as other enzymes involved in that metabolic pathway produce. External metabolites are source (e.g. glucose) and sink metabolites (e.g. pyruvate as the end product of glycolysis), these cannot and do not need to be kept in balance. Before starting the calculation, one makes a list of all enzymes and reactions, the stoichiometric matrix, which compiles the number of molecules each reaction consumes or produces. To assemble the metabolic enzymes and reactions correctly, one performs the metabolic reconstruction. One looks over which enzymes should be present in the genome based on the sequence analysis or completes this with further sequence analysis. Then you can compile a list of all reactions and enzymes that are known for the metabolic pathway (or metabolic network) you want to reconstruct in that organism. If I am careless and overlook enzymes that are encoded in the genome, it may happen that individual reactions are not connected to the metabolic network at all or that I assume wrong reactions that cannot happen in the genome at all (best to always use and compare several databases). Enzymes and reactions can be obtained e.g. from the KEGG database (https://www.genome.jp/kegg/pathway.html; with EC numbers for all enzymes) and the ExPASy Biochemical Pathways database (https://web.expasy.org/pathways). Enzymes found only in bacteria but not in humans are potentially interesting antibiotic targets. Example software for metabolic modeling is Metatool and YANAsquare/YANAvergence (faculty-owned software). However, there are also other programs, e.g. CellNetAnalyzer (https://www2.mpi-magdeburg.mpg.de/projects/cna/cna.html).

**Examples 4.6 and 4.7**

A detailed description including a tutorial can be found at https://www.bioinfo.biozentrum.uni-wuerzburg.de/computing/metatool_4_5/ or https://pinguin.biologie.uni-jena.de/bioinformatik/networks/metatool/metatool.html.

## 20.5   **Systems Biology Helps to Discover the Causes of Disease**

**Task 5.1: Answers B-E are correct**

Protein–protein interactions (PPI) play an important role in the organism. One example is signaling cascades, in which different proteins interact with each other (e.g. activate one after the other) and regulate or typically amplify cellular signals. In addition to pairwise interactions (number of possible interactions $[n^2 - n]/2$), there are of course also complexes (number of possible complexes is $2^n$) between proteins, so there are a large number of possible PPIs, making it difficult to detect all interaction partners experimentally. In some cases, there are also tissue-specific interactions. Here, it can be quite useful to focus on only a few interesting interactions, such as tissue- or disease-specific, where interaction databases are helpful. The STRING database is an interaction database and contains numerous experimentally determined and bioinformatically predicted protein interactions. Thus, individual interaction partners can be found very well, e.g. to perform network analyses or to better evaluate *microarray experiments*, but also to obtain orthologous genes (search for *Cluster of Orthologous Groups*, COG, included). The KEGG database contains network maps for important signalling pathways, e.g. Wnt, MAPK, Ras-Raf-Mek-Erk signalling pathway, apoptosis or the cell cycle, but also provides a lot of additional information, e.g. on metabolism, enzymes and diseases. Thus, one can find entire signaling pathways (e.g. for a disease) and thus, for example, identify metabolic pathways involved or find *drug targets*. The PlateletWeb database (own chair database) contains protein interactions primarily for the platelet, but also for other human cells, such as phosphorylations. The listed interactions in PlateletWeb are based on experimental data sets, e.g. proteomics data. The databases iHOP (information hyperlinked over proteins) and HPRD (Human Protein Reference Database) also contain experimentally determined protein interactions and numerous related information. However, there are many other databases, such as IntAct (https://www.ebi.ac.uk/intact), MINT (https://mint.bio.uniroma2.it/mint), and BioGRID (https://www.thebiogrid.org). In all these databases you can find numerous interactions with which you can then build a network. However, it is always advisable to use several databases and compare them with each other in order to find common and trustworthy interactions.

A protein–protein interaction network can be constructed using the following steps: i) network reconstruction and ii) network analysis. i) Network reconstruction can be done using different databases, e.g. protein–protein interactions from the STRING database. This network can be saved as .sif *(simple interaction file)* to be read into Cytoscape. ii) Network analysis can be done e.g. with the software Cytoscape (please have a look at www.cytoscape.org for a short description of Cytoscape). For this purpose Cytoscape has numerous *plugins* to choose from, e.g. BiNGO (biological process analysis), AllegroMCODE (analysis of functional modules and complexes) and NetworkAnalyzer (topology analysis). This allows the identification of important biological functions or functional network proteins (potential therapeutic *targets*), which can then be further characterized and experimentally validated.

The network created can then be examined in more detail later on, e.g. to understand intercellular networks and communication (what goes wrong in the network so that a tumour develops; what is a potential *drug target*). It can also be used for dynamic modelling (*in silico* simulation) to better understand the behaviour of the network, e.g. what happens after an infection or what effect does a drug have?

**Example 5.6**

Correct is B and D. To find all human interaction partners for BRCA1, you should enter BRCA1 as the search term and search, then select human as the organism.

**Question 5.7**

To do this, you need to reconstruct a network (e.g. STRING and KEGG databases), then load the network (e.g. as a .sif file) into Cytoscape and examine it with the BiNGO *plugin* (alternatively also ClueGO *plugin*) (see also previous answers).

**Example 5.8**

Answer A.

Correct is A. The network should look like this (Fig. 20.5):

**Example 5.9**

Answer B, C.

BiNGO identifies overrepresented biological functions (with *p-value* and corresponding genes), so-called *Gene Ontology* (GO), in a network (https://www.ncbi.nlm.nih.gov/pubmed/15972284). In the GO groups, genes are grouped according to their species-specific known function into the categories of biological processes, cellular component and molecular function (https://www.geneontology.org/). One can thus find all processes involved for the network, which allows one to detect, for example, functions and proteins involved specifically for a process, such as the cell cycle. From this, one can then in turn create a subnetwork of all proteins for this process and investigate it in detail. In this case, the BiNGO analysis shows a large number of biological processes (well over 100), including BRCA1 involvement in the *cell cycle checkpoint* (GO-ID 75).

**Question 5.10**

A *Gene Ontology* is a species-specific functional grouping (biological process, cellular component and molecular function) of genes *(term)*. Allows a functional annotation (see also question 5.9).

**Example 5.11**

Answer A, C, E.

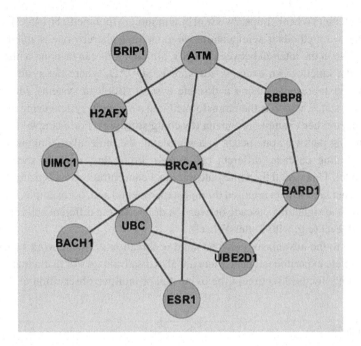

**Fig. 20.5** Network

Within the scope of a network analysis, the network topology should also be examined. This provides an overview of the network behavior, e.g. the interconnection and networking of the nodes. This can help to detect important functional network nodes, so-called hub proteins or hubs. Hubs are highly interconnected nodes in a network, which are suitable as potential therapeutic *drug targets*, for example.

---

**Questions 5.12 to 5.15**

For the mathematical modelling of regulatory networks, there are the Boolean/discrete, quantitative and semi-quantitative methods. In principle, these methods consider the nodes *(proteins)* of a network according to their activation state, i.e. either activated (On; maximally activated = 1) or inhibited (Off; maximally inhibited = 0). According to the initial state (how much is the node activated/deactivated), the further temporal course, i.e. how does the state of the node change over time, is calculated for each individual node of the network. In this way, the behavior or the network interconnection can be examined in more detail, whereby corresponding network effects, i.e. the respective effect of a node, also become clear. Boolean modeling always considers the on/off (1/0) state of a system, i.e., the node is either activated (On; 1) or inhibited (Off; 0). Quantitative modeling is useful for kinetic data, such as Michaelis–Menten kinetics. Here, the system state of a network is considered using exact concentrations and mathematical differential equations, but this requires information about the kinetics. An example software for quantitative modeling is PottersWheel (https://www.ncbi.nlm.nih.gov/pubmed/18614583). A combination of the two methods is semiquantitative modeling, whereby one is able to consider the system state in the interval between 0 and 1, although this can be done without knowledge about the kinetics. An example software is SQUAD, where the system state of a network is first represented using a discrete system (Boolean system), identifying all "*steady state*" states, which is then transformed into a dynamic system using an exponential function. Another example is Jimena (teaching software). For example, to model the cAMP signaling pathway, one needs to assemble all the molecular components (cAMP, receptor, signaling cascade, different cell types). From this, one can then develop a dynamic model. This would then integrate the exact concentration levels using differential equations (exact kinetic data required through experiments) and, for example, model either the activity of the signalling cascade or even the drug effect in different cells as a function of the cAMP level (e.g. with PottersWheel).

In addition to the advantages, such as rapid observation of the network behavior even without complete experimental data, there are also disadvantages of mathematical modeling, such as only focused section of the living cell or intuitive observation of the network behavior.

### Question 5.16

Answer A, B, D (see also previous answers).

### Question 5.17

A *steady state* describes the network state to which the network returns, i.e. a stable state that is reached again even after changes/disturbances or does not change (see also previous answers).

## 20.6   Extremely Fast Sequence Comparisons Identify all the Molecules that Are Present in the Cell

### Question 6.1

When I want to look something up in a book, I can either flip through the book from front to back quickly (it's easy to miss something). Or I look in the index, for example in this book under "super-fast sequence comparisons". That means I can find the right page right away via the index. The acceleration of BLAST works the same way. Only those indices are examined more closely (with exact *alignment*) that are promising. Of course, this only works if the index is there. Creating a keyword directory for a database is called "indexing", because an index (a keyword directory) is created. This has to happen with each new version of a database (i.e. a list of sequences) before the BLAST search can then go over the database so quickly.

### Question 6.2

With the BLAST search there is a double acceleration, because a second good index hit must be there before the exact *alignment* is started.

Here again as a reminder the tutorial on how to find a sequence:

https://blast.ncbi.nlm.nih.gov/blastcgihelp.shtml

Now for comparison, here is a FASTA server that only works with one *hit:*

https://fasta.bioch.virginia.edu/fasta_www2/

For an unknown sequence, it can make sense to try both options, since both servers produce different results depending on the sequence. However, the BLAST server is faster.

Finally, the hits found can also be used in an *alignment* for the overall search:

https://www.ncbi.nlm.nih.gov/books/NBK2590/

https://blast.ncbi.nlm.nih.gov/Blast.cgi?CMD=Web&PAGE=Proteins&PROGRAM=blastp&RUN_PSIBLAST=on

### Question 6.3

(a) A further acceleration of the sequence comparison is e.g. the BLAT search:https://genome.ucsc.edu/FAQ/FAQblat.html

(b) The tutorial also explains the advantages, namely even faster than the BLAST search, index search goes over a whole genome. Disadvantage: Less "depth", so distant similarities are not detected as reliably.

### Question 6.4

An analogy: Nothing goes faster than the speed of light. That's why you have to be prepared for long waiting times (years!) when travelling to the stars. Therefore, the fastest way is not to go at all, but to *think!*

All our sequence comparisons try to find out which protein is present, i.e. what its annotation (bioinformatic functional description) or function is. We have just learned about examples: BLAST, Psi-Blast, FASTA, other BLAST variants. All these searches are heuristic, i.e. fast, but not quite exact. There are also exact searches. This is global sequence comparison using Needleman and Wunsch algorithm and local sequence comparison using Smith-Waterman algorithm. Further possibilities are searches in domain databases like SMART, ProDom, protein family databases like Pfam, finally also specialized searches like BLOCKS similarity search – but (see above): The fastest way is to use the correct annotation. Where is the best place to find it? Investigate this right away in task 6.5.

### Question 6.5

• Annotation in GenBank is a very good standard annotation (detailed description of the properties of the gene or protein or RNA molecule). Here, however, the annotation is filed by the author after checking and proofreading by NCBI. In this respect, there are differences in the depth or detail of the annotation. This is particularly evident in the case of *draft* genome sequences, where often little is known about the function and properties of the individual genes, or even when only an automatic annotation is given, for example only by sequence comparison to a domain database.

- The UCSC Genome Browser allows a detailed view of the human genome along with details on the properties of a gene. Particularly detailed are mRNA, exons, gene length information, etc. In addition, the UCSC Genome Browser systematically provides comparisons with other genomes (mammals, vertebrates).
- Swiss-Prot/UniProt: Here proteins are described particularly precisely and accurately. Originally (Swiss-Prot), all proteins were annotated and examined by hand by experts, but even now all annotations are checked in detail here.

The question of which annotation is most suitable cannot be answered in a general way, because this always depends on the biological question. Let us briefly illustrate this with our example of glutathione reductase. If we compare the annotation of the three different databases, we can see differences in the database ID for glutathione reductase, for example in GenBank and in the UCSD Genome Browser X54507 or in UniProt Q03504, despite the uniform name designations. Thus, when using different databases, it is advisable to always check if you actually have the same gene/protein between them. However, when looking closer at the databases, it also becomes clear that UCSC and GenBank focus on genomic position, whereas UniProt focuses more on biological function and interaction context. So if you are only interested in the sequence, e.g. protein, you will find this in GenBank and UniProt, whereas if you want to find out about the genomic region of the gene, e.g. *antisense* or neighbouring genes, you should rather use UCSC and GenBank (graphically visible). If you are looking for functional domains or interaction partners, you will find more information in the UniProt database. So you can see that databases are partly structured differently and have different foci, so it is up to you which database is best suited. But there is one thing you should always keep in mind: Comparing several databases is advisable in any case, because this way you can be sure to have found the right information.

### Question 6.6

Here you can think for yourself. It is important that the database/server should contain trustworthy data (that it is also traceable where the data comes from), the user interface should be easy to use, understandable and clear, but also up to date. It is also advantageous to avoid overloaded pages and rather focus on one topic area, but provide further links (but make sure that the links are always up to date) for individual analyses (a nice example is our DrumPID database, which focuses on *drugtarget interactions*).

### Question 6.7

It is important to check at regular intervals whether the data is still up to date or to enter new data, but also to see whether the methodology is still up to date or whether there are better procedures. Furthermore, you should make sure that cross-links to other websites or websites necessary for the operation of the database are up to date and functioning.

Ideally, the project leader or first author should check the database at regular intervals to ensure that the data is still up-to-date. It also makes sense to link the databases to ongoing projects and to keep developing them so that they remain up to date. Useful is also information about the last *update* (when and ideally also specifically what was done) of the website (include when programming), so that the *user* has an overview.

## 20.7 How to Better Understand Signal Cascades and Measure the Encoded Information

### Question 7.1

Well, I need three bits to do that, because LLL or 111 is the representation of the number seven in dual numbers.

### Question 7.2

The representation starts with the 512 bit (2 to the ninth power), then there are 488 left, which fits the 256 (2 to the eighth power), then there are 232 left, with that I can fill the 2 to the seventh power (128 bit), then there are 104 left, which corresponds to the 2 to the sixth power (64), then there are 40 left, which fits the 32 (2 to the fifth power), then there are 8 left (2 to the third power), all remaining digits are zeros: 1,111,101,000.

### Question 7.3

An example is https://www.binaryhexconverter.com/decimal-to-binary-converter, which then gives as answer for the decimal number 1000 (one thousand): 0000001111101000.

### Question 7.4

A letter has on average about 4.7 bits, i.e. one needs (depending on the coding scheme more bits) at least 5 bits to be able to represent a total of 32 different characters (26 letters, then there are ö, ä, ü, ß, comma and dot). The word "word" needs four times as much, i.e. at least 20 bits, to be encoded.

### Question 7.5

What is important is the secure encoding of information with the aid of solidly stored bits. The bits thus measure the indispensably necessary amount of information for the encoding. For example, each nucleotide has 2 bits because there are four nucleotides. One can also see how in RNA molecules this information content per bit is increased when biologically necessary, especially by nucleotide modifications, especially methylations (e.g. pseudouridine in tRNA) and other modifications. Here, this bit increase is important to increase the safety of protein synthesis, i.e. to increase the reading accuracy of the tRNA.

### Question 7.6

Important examples include: Signaling cascade amplifies signals, despite noise (technical term: noise) in the cell. Nice examples are the Ras-Raf-Mek-Erk cascade (intracellular cascade; high signal amplification; shutdown phosphatase; disease, by B-Raf mutation about cascade constantly on, leads to melanoma) or blood coagulation (extracellular cascade; also very good signal amplification, extrinsic and intrinsic stimulus uptake, the various clotting factors amplify the signal, finally this amplification via thrombin then produces fibrin polymers; here it is interesting that the opposite action, the dissolution of a blood clot, is again a cascade, via the plasminogen system).

### Question 7.7

The Ras-Raf-Mek-Erk cascade is shown schematically in Fig. 20.6. The signal is amplified by a factor of ten in each case, so I get:

1 molecule Ras

10 molecules of Raf

100 molecules of Mek

1000 molecules of Erk

A 1000-fold stronger signal in the cell than at the beginning. Note: Exact data and kinetic modeling for this cascade can be found in Robubi et al. (2005).

### Question 7.8

Growth signals are passed on via this cascade to further locations in the cell, in particular to transcription factors, which then switch on genes in the cell nucleus that then lead to cell growth. It is important that the signal is switched off again. This generally happens through phosphatases. A nice example is the Ras-Raf-Mek-Erk signalling pathway. Ras is a kinase that, when activated, regulates all other downstream components of the signalling pathway, for example Raf and Mek, and thus influences proliferation, i.e. cell growth. However, if a mutation prevents the cellular Raf from being switched off, for example, the growth signal remains on all the time. A biological example is melanoma. Here, a B-Raf mutation is present, and then the cellular phosphatases can no longer switch off the cascade and set it to zero.

### Question 7.9

Since the Ras signalling pathway is often deregulated in tumours and leads to unchecked tumour growth, it has developed as an interesting approach in research. In this context, the Ras signalling pathway can also be described mathematically, for example by means of a differential equation, in order to model the temporal behaviour of the entire signalling cascade on the computer. Our chair has already investigated the signal pathway in a paper from 2006 and used the following differential equation (Fig. 20.7; from Robubi et al. 2005):

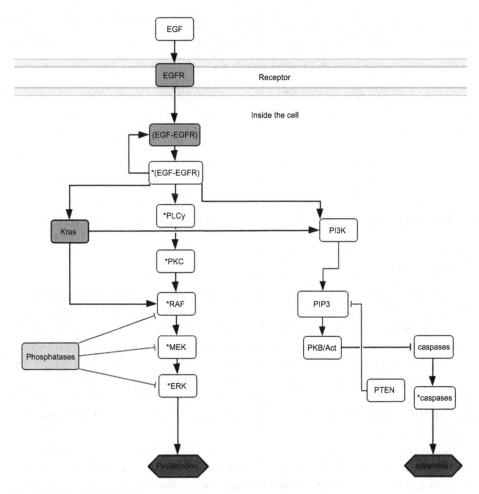

**Fig. 20.6** Ras-Raf-Mek-Erk cascade

The figure shows the simplified wiring of the cascade (top) from Ras to Erk and its activation over time (input → signal path → output) as well as its mathematical description (bottom). Here, the concentration of the activated kinase (X) depends on the time (t). In addition, α describes the phosphorylation, β the corresponding dephosphorylation and C the total concentration of the kinase. This formula can be used to describe the whole cascade in a simplified way and then model it in the computer, for example to better understand tumor growth (e.g., how does overexpression of the kinase affect proliferation or to derive new therapeutic approaches). For more information, see also Robubi et al. (2005) B-Raf and C-Raf signaling investigated in a simplified model of the mitogenic kinase cascade. Biol Chem. 386(11): 1165–1171.

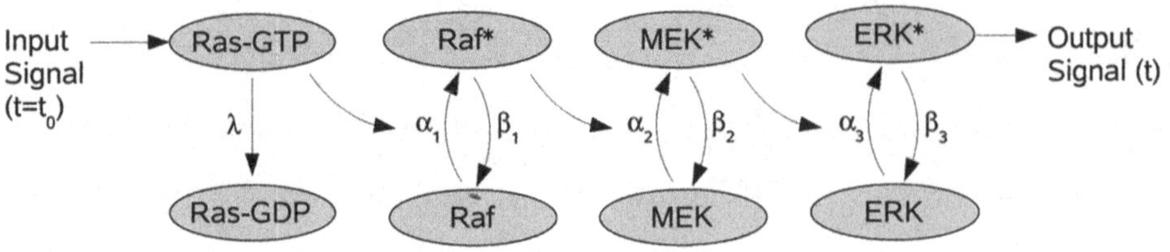

$$\frac{dX_i}{dt} = \alpha_i X_{i-1}\left(1 - \frac{X_i}{C_i}\right) - \beta_i X_i.$$

**Fig. 20.7** Mathematical modelling of the Ras-Raf-Mek-Erk cascade. (From Robubi et al. 2005)

---

### Question 7.10

As mentioned in the textbook section, there is a nice article on this by Heinrich et al. (2002). It describes the properties of phosphatases and kinases, in particular signal amplitude, signal frequency, signal duration (phosphatase must switch off fast enough for both) and of signal amplitude and signal height (kinase must amplify strongly enough) in signal cascades. Interestingly, this allows one to develop a mathematical theory that defines the regulation of the signaling cascade as a function of a finite number of key parameters. These models can then be used for linear kinase-phosphatase cascades, but also take into account *feedback* interactions, *crosstalk* to other *pathways*, the cytoskeletal framework and G proteins.

This mathematical description then shows even more clearly that phosphatases are more important in their effects on signal rate and duration, whereas signal amplitude is primarily controlled by kinases. Simple pathway models show good signal amplification (tasks here directly before) only at the expense of speed.

However, more complicated, realistic pathway models can also achieve high amplification and signal rate. For this, a stable, switched-off state of the cascade is important. Moreover, different agonists can trigger either transient or continuous signals in the same signaling pathway. The accumulated knowledge of such a model can also be used for the design of signaling cascades.

---

### Question 7.11

The signal must first be strong enough for this, i.e. the metabolic flux through this metabolic pathway must be high enough. This is regulated by pacemaker enzymes that control the metabolic flux particularly strongly, i.e. have a particularly high metabolic control coefficient (see also the contributions by David Fell (2005), metabolic control analysis;

https://link.springer.com/chapter/10.1007%2Fb137745 or even more recently). Typically, such enzymes are located at the beginning or end of a metabolic pathway. Equally important is the exactness of the coding (see above, e.g. Task 7.1.). If the enzymes involved have a broader specificity or the enzyme is poorly positioned (for example, at a metabolic branch), several *pathways* are thus altered at once, but this is sometimes biologically intended.

---

**Question 7.12**

---

Several enzymes are radically switched *("moonlighting")*. As long as there is sufficient substrate, they function as metabolic enzymes. If there is too little substrate, however, these enzymes become regulatory active. A nice example is aconitase, which normally produces isocitrate from citrate as the first step in the citric acid cycle. In the case of iron deficiency, the iron-sulfur cluster in the active site is missing, and the enzyme then functions instead as *iron-responsive element-binding protein 1* and binds to RNA, namely *iron-responsive elements.*

A good link to this are *moonlighting databases,* for example: https://www.moonlightingproteins.org (even the cover image fits there ☺). Also nice is: https://www.uniprot.org/database/DB-0189.

---

**Question 7.13**

---

(a) On the one hand, this can be used for energy production, for example glycolysis and gluconeogenesis occur simultaneously *("futile cycles")*. Happens in the brown fat of newborns (and other young mammals). This leads to a much more sensitive response to metabolic changes when both *pathways* run simultaneously (again, like the initial example of glycolysis and gluconeogenesis). Therefore, it is even possible to determine the *futile cycles* with the help of software such as Metatool or YANA (see Chap. 4), and the enzymes involved there are then quite often the enzymes that play a special role in regulation.

(b) For example, flow 100 in BOTH directions. Net result is then zero, nothing is moved. But if I now have 10% enzyme change, without the *futile cycle* I would only have 10% change in one direction. So now that I have sacrificed some metabolic energy through the futile *cycle,* I get a much higher sensitivity: execution changes from 100 to 110%. But the reverse direction changes from 100 to 90%. So now the net difference is twice as much, 20% regulation. Of course, this also goes further "down" for each real situation, so e.g. glycolysis is just at 110% and gluconeogenesis at 90%, the net result for glycolysis is then 20%. If I now have another 10% change in regulation, glycolysis changes to 120% and gluconeogenesis only has 80%, but with that even a total of 40% difference and increase to glycolysis.

## 20.8   When Does the Computer Stop Calculating?

Here some algorithms are compared in terms of their computation time, it results:

(a)  RNAfold with small RNA and large RNA (quadratic increase with sequence)
(b)  BLAST search (grows linearly with search sequence and database)

Short peptide example, long protein example. Search in the NRDB database, and only in the human sequences (use species option).

The *E-value* moves favorably downward, toward smaller values, for a smaller database. Why? Well, the larger the database, the higher the probability of random hits. So the expected value *(E-value)* for a random, non-biological, non-relevant hit gets higher. So the better I can narrow down where I expect my hit to be (e.g. a species-specific database), the more significant and meaningful my result will be.

(a) Protein folding

This is an NP-hard problem, which means that the computation time increases many times with each additional amino acid. It is thus not at all clear how long the computer will take (non-polynomial complex problem), but at least if you get a solution you can determine in polynomial time how good it is. Nevertheless, protein structures can be predicted for many practical purposes, e.g. by comparison with known structures, e.g. with SWISS-MODEL (but already here the answer comes only by e-mail, it takes time), or somewhat more precisely, but more computationally expensive, with MODELLER or actually *"ab initio"*, i.e. from the sequence, by folding, calculated by the Zhang lab (with QUARK etc.).

A nice answer is given by this Youtube video, but unfortunately it is in English: https://www.youtube.com/watch?v=SC5CX8drAtU.

Here are compared:

Greedy strategy: locally optimal choice at each stage; at each stage visit an unvisited city nearest to the current city. This heuristic need not find a best solution, but terminates in a reasonable number of steps; finding an optimal solution typically requires unreasonably many steps. In mathematical optimization, greedy algorithms solve combinatorial problems having the properties of matroids (a structure that captures and generalizes the notion of linear independence in vector spaces).

*Local search strategy:* Local search is a generic term for a number of metaheuristic search methods in combinatorial optimization. The methods are used in many variations to solve complicated optimization problems approximately (e.g., the traveling salesman problem). The basic principle is to find a better solution starting from a given initial solution by locally changing the current solution to find a better solution from the neighborhood under consideration.

*Simulated annealing strategy*: In each case, to solve the *traveling salesman* problem as well as possible.

### Question 8.3

- Monte Carlo
- *simulated annealing* (since pointing to the protein folders, e.g. at SWISS-MODEL there is a *refinement,* which should be in the direction)
- evolutionary strategies
- genetic algorithm (is implemented in YANASquare, mention that then)
- *optimizer* (*steepest descent,* also mention the procedure for YANAvergence, the Broyden-Fletcher ...).

### Question 8.4

A difficult computational problem is a bioinformatics problem in which many possibilities lead combinatorially to an exponential growth of possibilities, e.g., the traveling salesman's problem of traveling to numerous cities along the most optimal route possible. These exponentially complex problems with very, very long computation time for systematic trial and error (longer than the universe exists, etc.) are contrasted with easier problems where the computation time grows only polynomially (P-problems), e.g. quadratically or cubically with the length of the query, such as the sequence length. However, almost all interesting bioinformatics problems are combinatorial (e.g. protein folding or possible protein complexes). It has also been shown that they are all analogous to the traveling salesman problem, i.e., they require non-polynomial computation time, are NP-hard.

## 20.9   Complex Systems Behave Fundamentally in a Similar Way

### Question 9.1

The behavior of ordered system is predictable and exactly describable for the whole period. Random systems are unpredictable in the short term, but the outcome space can be predicted (such as a dice, can only be one to six). In addition, there are chaotic systems that can only be described exactly over short periods of time, but remain within fixed limits (attractor) over the long term.

### Question 9.2

Here we have learned about numerous systeming ingredients in the book: Modular units (nucleic and amino acids) have interactions and in turn form complexes and networks (e.g. *feedback* or *feedforward loops*), from which filaments, organelles, tissues and ultimately cells, an organism and entire ecosystems develop (new patterns and properties always emerge, emergence).

### Question 9.3

In the book, we learned about numerous methods such as genomics, transcriptomics, proteomics and metagenomics (please refer to Sect. 9.2).

### Question 9.4

New properties and effects that result from components coming together but are not attributable to the individual components (a system is much more than the sum of its parts). An example is the circulatory system (supplies body with nutrients and oxygen and has pulse and blood pressure, results from interaction of many individual blood and heart muscle cells).

### Question 9.5

The best way to do this is to look at Fig. 9.4 (links) and link the two networks.

### Question 9.6

The EPO production with the help of quadratic function (see the task on R in the tutorial).

### Question 9.7

The simplest way is to look at water and its flow behaviour: If it stands still, flow is dead (so also at living systems). If pressure is not too strong (e.g. look at Main at Würzburg, when flowing within its riverbed with normal amount of water), flow is nice and steady ("healthy state" at living systems). If the pressure is even stronger (e.g. at the weir under the old Main bridge), then the flow becomes swirled ("turbulent") and uneven, chaotic (a sign of stress in chaotic systems). There are numerous educational films about systems biology, especially in English, e.g. Systems biology explained (Weizmann institute); https://www.youtube.com/watch?v=HCFoZDlV4FY.

### Question 9.8

System state 1: Heart at rest, all is well.
System state 2: Heart in sympathetic activation, heart beats faster, but normal load, for example during sports (healthy).

System state 3: Heart has too much work, third Erk phosphorylation is activated (a *tipping point,* when then more and more heart cells are switched this way, hypertrophy, is currently irreversible).

System state 4: Cardiac hypertrophy, now the heart has too little oxygen, therefore simultaneous activation of both activation pathways. Myocardial infarction, collapse: not shown here, but of course the late consequence of untreated heart failure.

More details can be found in the paper (including details on the semiquantitative simulation of the different system states): Brietz A et al. (2016) Analyzing ERK 1/2 signalling and targets. *Mol Biosyst.*

## 20.10 Understand Evolution Better Applying the Computer

### Question 10.1

Evolution is the change in characteristics of living organisms over time. Important mechanisms are, for example, mutations, selection, *gene drift* and separation.

### Questions 10.2 and 10.3

One color always prevails in the end. We have a Darwinian evolutionary approach here. The probability of being hit is directly proportional to the number of individuals. Random fluctuations, however, lead to the random extinction of individual colors until eventually only one color remains. This "game" vividly reproduces neutral evolution (all colors have exactly the same chance of winning at the start, and as colors become fewer, their die-off rate becomes proportionally lower). So just pure fluctuation, and yet one color eventually prevails. This simulates genetic *drift* very nicely.

Of course, it is also very easy to simulate selection for the "fitter" by modifying the rules of the game, e.g. that one color (red) simply gets two offspring for each hit and you always randomly roll two individuals for this case. Then red always wins. How fast this happens depends on randomness. So the result here is predictable, but the sequence of the individual steps is not.

True evolution is always a mixture of both, lots of *drift* involved, as perfectly illustrated in Stephen Jay Gould's *"A wonderful life"*.

### Question 10.4

Now the probability that a tandem of two colors asserts itself is proportional to the product of both colors. Thus, the more individuals there are for a tandem, the quadratically better the rates. This is why a "once and for all" selection occurs. Quite quickly a tandem of two colors asserts itself, and no other tandem can grow so high, because no population can compete with the super-exponential reproduction rate.

This simulation model nicely illustrates how over-exponential growth prior to the first, delimited cells led to selection from a population of mutually catalyzing molecules. In particular, it explains very well why only one genetic code (with minimal dialects) remained.

Additional task for those interested: write R code to recreate the three games (not difficult, but takes some time).

---

### Question 10.5

- *Parsimony*: The phylogenetic tree is calculated in such a way that the observed diversity from the (not observed, but only calculated) precursor sequences is correctly reproduced with as little parsing as possible.
- ML, *Maximum likelihood* the phylogenetic tree is calculated as it probably has been (single probabilities for each nucleotide exchange are considered). Calculation point out, ideally take the same FASTA multisequence file.

---

### Question 10.6

Take the NCBI download and also the *taxonomy option of* BLAST. First use a keyword search to find the HI virus together with the complete polymerase sequence, e.g. *HIV1 human;*

https://www.ncbi.nlm.nih.gov/protein/?term=HIV1+and+human+and+polymerase+co mplete. Is so already feasible. But if you, for example, simply take HIV and *protein* and *human* as search *terms,* then you can search yourself to death with so many hits. They then find for man:

```
>gi|1906384|gb|AAB50259.1|  pol  polyprotein   (NH2-terminus
uncertain) [Human immunodeficiency virus 1]
M S L P G R W K P K M I G G I G G F I K V R Q Y D Q I L I E I C G H K A
I G T V L V G P T P V N I I G R N L L T Q I G C T L N F P I S P I E T V P V K L K P G M D G P K V K Q W
P L T E E K I K A L V E I C T E M E K E G K I S K I G P E N P Y N T P V F A
I K K K D S T K W R K L V D F R E L N K R T Q D F W E V Q L G I P H P A G
L K K K K S V T V L D V G D A Y F S V P L D E D F R K Y T A F T I P S I N N E T
P G I R Y Q Y N V L P Q G W K G S P A I F Q S S M T K I L E P F R K Q N P D I V I Y Q
Y M D D L Y V G S D L E I G Q H R T K I E E L R Q H L L R W G L T T P D K K H Q K
E P P F L W M G Y E L H P D K W T V Q P I V L P E K D S W T V N D I Q K L V
G K L N W A S Q I Y P G I K V R Q L C K L L R G T K A L T E V I P L T E E A E L E L A
E N R E I L K E P V H G V Y Y D P S K D L I A E I Q K Q G Q G Q W T Y Q I Y Q E P F
K N L K T G K Y A R M R G A H T N D V K Q L T E A V Q K I T T E S I V I W G K T
P K F K L P I Q K E T W E T W W T E Y W Q A T W I P E W E F V N T P P L V K L
W Y Q L E K E P I V G A E T F Y V D G A A N R E T K L G K A G Y V T N R G R Q
K V V T L T D T T N Q K T E L Q A I Y L A L Q D S G L E V N I V T D S Q Y A L G I
I Q A Q P D Q S E S E L V N Q I I E Q L I K K E K V Y L A W V P A H K G I G G N E
Q V D K L V S A G I R K V L F L D G I D K A Q D E H E K Y H S N W R A M A S D F
N L P P V V A K E I V A S C D K C Q L K G E A M H G Q V D C S P G I W Q L D C T
H L E G K V I L V A V H V A S G Y I E A E V I P A E T G Q E T A Y F L L K L A G R W
P V K T I H T D N G S N F T G A T V R A A C W W A G I K Q E F G I P Y N P Q S Q G
V V E S M N K E L K K I I G Q V R D Q A E H L K T A V Q M A V F I H N F K R K G G
I G G Y S A G E R I V D I I A T D I Q T K E L Q K Q I T K I Q N F R V Y Y R D S R N P L
W K G P A K L L W K G E G A V V I Q D N S D I K V V P R R K A K I I R
DYGKQMAGDDCVASRQDED
```

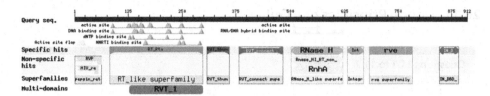

**Fig. 20.8** Domain analysis

Now use BLAST:

https://blast.ncbi.nlm.nih.gov/Blast.cgi

Pay attention to the protein BLAST:

https://blast.ncbi.nlm.nih.gov/Blast.cgi?PROGRAM=blastp&PAGE_TYPE=BlastSearch&LINK_LOC=blasthome

Paste sequence into question form.

After the BLAST search has been performed, the analysis of the domains can be seen in the top result section (Fig. 20.8):

The phylogenetic tree should now show that the domains are well conserved. For this, you can look at the *alignments* in detail (lower part of the BLAST result). But very helpful is the phylogenetic tree report (click on *"Taxonomy Report"*). In particular, you will find all species listed and the number of found, related species sorted by organism groups (here, of course, because searched with HIV, mainly HIV polymerase sequences).

### Question 10.7

Proceed analogously as in 10.6, but here the species richness is much greater, nice family tree.

### Question 10.8

CLUSTAL has the following link: https://www.ebi.ac.uk/Tools/msa/clustalo/. MUSCLE can be found here https://www.ebi.ac.uk/Tools/msa/muscle/. For orientation, Parsimony and ML are referred to here (see Sect. 10.5).

### Question 10.9

With a multiple *alignment,* you can compare multiple sequences and identify similar or dissimilar regions.

### Question 10.10

For this, all you have to do is look closely at the SMART domain analysis website and seek out the *seed alignment.* In particular, also look at the conserved and less conserved residues.

## 20.11　Design Principles of a Cell

---

Transfer RNA (tRNA) mediate the translation of the correct amino acids from the RNA code, this happens at the ribosomes. Biophysical laws determine the structure (e.g. hydrogen bonds, hydrophobic interaction), but also other effects such as *crowding*. However, these are so complex that the exact process of the formation of the three-dimensional protein structure has not yet been completely deciphered (e.g. via *"molten globule"*state). However, since many protein sequences and protein domains are known, much information about function and structure can be obtained from databases. For example, much information and resolved three-dimensional structural coordinates together with annotations for the protein can be found in the PDB (https://www.rcsb.org/pdb/home/home.do) and UniProt (https://www.uniprot.org/) databases. In addition, there are also classification databases, for example according to sequence and structural similarity such as SCOP (structural classification of proteins; https://scop.mrc-lmb.cam.ac.uk/scop/, from 2010 continued with SCOP extended; https://scop.berkeley.edu) and CATH (*classification by class, architecture, topology and homology;* https://www.cathdb.info/), or according to protein families and function the databases PROSITE (https://prosite.expasy.org/) and Pfam (https://pfam.xfam.org/). Thus, it is possible to obtain predictions of protein structure and function through experiments and bioinformatic modelling (e.g. differential equations and simulations). In this context, there are different approaches to predict protein structure from a sequence, e.g. *ab-initio* and comparative predictions (e.g. homology modeling, *threading*). Ab-initio predictions are based on the biophysical properties of proteins, whereas homology modeling uses known protein structures. There are many useful softwares to visualize (e.g., hydrogen bonds or hydrophobic regions) and analyze (e.g., *docking* and modeling) protein structures, such as PyMOL (https://www.pymol.org/), RasMol (https://www.openrasmol.org/), and Swiss-PdbViewer (https://spdbv.vital-it.ch/). A protein structure analysis can be performed bioinformatically, e.g. with AnDom (contains three-dimensional structural domains based on SCOP classification), SWISS-MODEL (https://swissmodel.expasy.org/), I-TASSER (Iterative Threading ASSEmbly Refinement; https://zhanglab.ccmb.med.umich.edu/I-TASSER/) or with a Ramachandran plot, which provides information about possible structures, domains and function. A Ramachandran plot (e.g., RAMPAGE software; https://mordred.bioc.cam.ac.uk/~rapper/rampage.php) calculates the phi and psi torsion angles in the protein, thus providing a graphical overview of the distribution of alpha helices and beta leaflets.

---

I can find a possible function for a protein if I look in the sequence for possible sequence motifs and protein domains, i.e. independent folding units. This shows me, for example, whether an active site, a regulatory domain or interaction domains are present in my protein, thus giving me information about the possible function of the protein. Example databases/programs include PROSITE, AnDom, SMART (https://smart.embl-heidelberg. de/), and the ELM server (*eukaryotic linear motifs;* https://elm.eu.org/index.html). It is always best to use several programs and compare the results, because this is the only way to be sure that you have found a trustworthy match. Recurring conserved regions in multiple sequences can be found using multiple *alignments.* These allow to compare (align) several sequences with each other. There are various programs for this, such as MUSCLE (Multiple Sequence Comparison by Log-Expectation; https://www.ebi.ac.uk/Tools/msa/ muscle), MAFFT (Multiple Alignment using Fast Fourier Transform; https://www.ebi. ac.uk/Tools/msa/mafft/) and Clustal Omega (https://www.ebi.ac.uk/Tools/msa/clustalo/). Multiple sequence alignments can be used to find conserved regions, possible domains, or specific differences between different sequences. Another method is phylogenetic trees, which can be created with PHYLIP (Phylogeny Inference Package; https://evolution. genetics.washington.edu/phylip.html), for example. In addition to a multiple sequence alignment, one can also find the evolutionary relationship between the sequences.

### Question 11.12

Answer A, C, D.
In the chosen example for the "TAR protein", both programs should have found a *double stranded RNA-binding domain* (dsRBD), suggesting that binding occurs via double stranded RNA molecules.

   If something did not work for you, then try it best like this (Fig. 20.9). The corresponding protein sequence can be found below the genebank number, then click on FASTA, which will automatically redirect you to the FASTA sequence (see also https://www.ncbi. nlm.nih.gov/protein/60653021?report=fasta). Then copy this sequence and paste it into the search windows at PROSITE and AnDom. The output of both pages can be found in Fig. 20.9 below.

### Question 11.13

For this: https://www.rcsb.org/pdb/explore/explore.do?structureId=1HSG. Then: https://thegrantlab.org/teaching/material/Structural_Bioinformatcs_Lab.pdf; https://sbcb. bioch.ox.ac.uk/users/greg/teaching/docking-2012.html. Staining of hydrophobic residues in the center. Introduction PyMOL here: https://pymolwiki.org/index.php/ Practical_Pymol_for_Beginners.

### Questions 11.14 to 11.21

Cellular communication is an essential process in eukaryotic and prokaryotic cells in order to regulate important processes or to react to an external stimulus. In prokaryotes, this is usually done by direct control, e.g. via two-component systems. A sensor activates a *responder,* which then immediately triggers transcription. In this way, a rapid response is made to an external stimulus. In eukaryotes, on the other hand, regulation is more complex

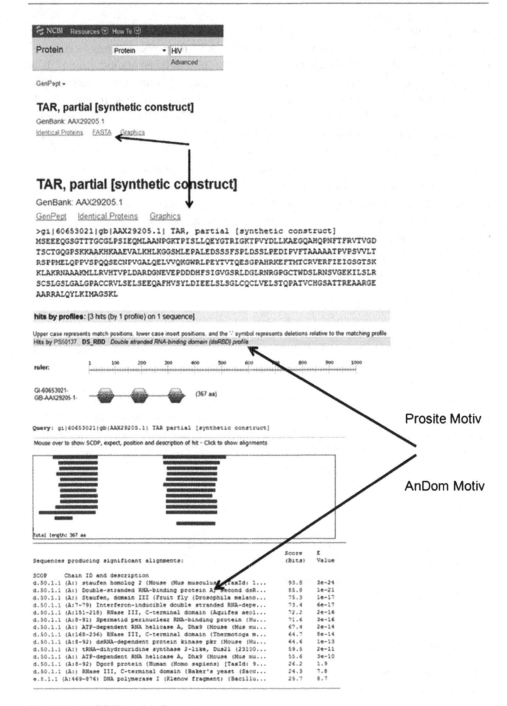

**Fig. 20.9**  PROSITE and AnDom

and usually occurs indirectly, e.g. via glucocorticoids, and is often also associated with intracellular communication. An example of cellular communication is *second messengers* that allow rapid communication, such as ATP in the energy supply in the cell (ATP is critically important for movement). It is generated in the respiratory chain after energy-rich compounds are broken down via glycolysis (anaerobic) and citric acid cycle (aerobic). The reduction equivalents (NADH, FADH) are oxidized in the respiratory chain and assembled into ATP molecules. Bioinformatically, I can look at metabolism and develop a kinetic (dynamic) model for this. Another example of cellular communication is differentiation, which is cell-to-cell communication. Here, for example, haematopoiesis (blood formation) would be interesting. For this, one can bioinformatically look at the kinase network. Important for cell differentiation is the central organizer (Speman organizer), which determines the developmental axes in the embryo, which occurs via the Wnt signaling pathway. This can also be considered bioinformatically, e.g. modeling with cellular automata or agent-based simulations. In most cases, it is therefore of interest to know the role of my protein and where it is localised, for example in the membrane or in the cell nucleus, in order to also draw conclusions about its function. For this purpose, there are already numerous databases in which I can find relevant interactions and information, e.g. PlateletWeb, KEGG, STRING and SPdb (Signal Peptide database; https://proline.bic.nus.edu.sg/spdb/). Bioinformatically, I can also predict localization, for example with SignalP (localization of signal peptides; https://www.cbs.dtu.dk/services/SignalP) or TargetP (https://www.cbs.dtu.dk/services/TargetP). Given a training dataset of proteins with known, experimentally verified localization, these programs learn to predict a particular localization from the amino acid composition. The localization in the cell can thus be determined from the protein sequence with the help of programs with hidden Markov models or neuronal networks, and new sequences to be investigated can then be assigned accordingly. Specifically, a transcription factor should be localised in the nucleus, an acid protease in the lysosome, a storage protein in the Golgi, a secreted protein in the endoplasmic reticulum and a membrane protein (prediction with TMHMM) in the membrane, and so on. A program should also predict this accordingly. If you want to write your own program, it should have an input and output part. In the middle is the processing part (prediction part). This consists of either a neural network or a hidden Markov model.

The information content of a message can be described with the Shannon entropy: One bit of information is the smallest unit of information, a "yes" or "no" decision. Words and sentences can thus be assigned their information content according to their length. In a further step, one can include the different signal sources and consider the quality, i.e. how high or low the information value is, e.g. low if the same characters are always sent. This knowledge can also be transferred to biological systems, for example if one wants to take a bioinformatic look at cell differentiation or intracellular communication, such as a signal cascade between body cells via *second messengers* (e.g. cAMP). In this way, signal transmission for cell growth and cell differentiation can be described in more detail, for example by amplification or attenuation of cellular signals by kinases and phosphatases (the quality of the signal depends on the ratio of signal to background noise). In this way, it is possible to observe and model various complex cellular processes bioinformatically. One is thus in a position to understand them better.

### Question 11.22

The TMHMM server link is: https://www.cbs.dtu.dk/services/TMHMM/. Here, any sequence can be seen by simply pasting it into the question form in terms of transmembrane helices including graphics for the extra- or intracellular loops.

### Question 11.23

The NucPred link is: https://www.sbc.su.se/~maccallr/nucpred/. Here I can determine all nuclear localization signals. There is also a database LocSigDB, from which one can derive many nuclear localization signals.

### Question 11.24

The SignalP server link is: https://www.cbs.dtu.dk/services/SignalP/. Here, different neuronal networks are combined to achieve the best possible prediction (for gram-negative and -positive bacteria and eukaryotes).

### Question 11.25

The PROSITE server link is: https://prosite.expasy.org/scanprosite/. The PROSITE motifs also specify catalytic residues, protein modifications as well as typical amino acid residue combinations for enzyme families and a range of localization motifs and interaction motifs.

### Question 11.26

The ELM server plays all this back in bundled form. It retrieves several programs that are installed there, i.e. it is a meta server (https://elm.eu.org).

### Question 11.27

The link is: https://geneontology.org. A distinction is made between molecular function (MF), biological process (BP) and cellular compartment (CC).

### Questions 11.28 and 11.29

Cytoscape can be found at: https://www.cytoscape.org. Downloadable e.g. from: https://www.cytoscape.org/download.php. Protein networks are read in and can then be further analyzed by suitable subprograms (*plugins*) (see Sect. 20.5). BiNGO (https://apps.cytoscape.org/apps/bingo) calculates overrepresentations of biological processes and signaling pathways (GO terms). Please also have a look at the tutorial section.

The PlateletWeb can be found at: https://plateletweb.bioapps.biozentrum.uni-wuerzburg.
de/plateletweb.php. For a query, you can, for example, first enter the VASP protein and
have all interactions of VASP calculated. Please also have a look at the tutorial section.

## 20.12   Life Continuously Acquires New Information in Dialogue with the Environment

The link to BLAST is: https://blast.ncbi.nlm.nih.gov/Blast.cgi.
   Now test:
   Enter random sequence: no hit.
   True biological sequence: very small *E-value* (*expected value* for a random hit). For
something very common, such as the letter "E" in the database, this value can reach
100,000 or more (if that many E's were found in an average search of the database). It is
then not a random match, but the probability that this is just a random match is very small
(e.g. less than $10^{-6}$, so less than 1 in 1 million). The larger the database, the easier it is to
get random matches, so then the *E-value* becomes higher.

The link leads to the protein blast, the database ("non-redundant protein sequences, nr, i.e.
each known protein is contained only once in the database") is found automatically:
https://blast.ncbi.nlm.nih.gov/Blast.cgi?PAGE=Proteins.

   (a) Searching in a word: DNA is in the database for protein sequences, James Watson
       fails at the J and the O.
   (b) Use only meaningful characters: Never use JUZBOX, good counter test if the one-
       letter sequence is correct.
   (c) *Wobble codons* denote several nucleotides that are possible at this position, for
       example R for purine (A or G) as well as Y for pyrimidine (C or T or in RNA U).
       *Wobble codons* for consensus, here would be optimal to recognize a good and a bad
       sequence (at wrong codons, but also at the many NNNNs, perhaps also a polyade-
       nylation site).

Check out the https://www.ncbi.nlm.nih.gov/Taxonomy/Utils/wprintgc.cgi page. Go to
the NCBI page on codons. Now you can understand how to translate all the triplets from
nucleic acid sequences to amino acids. There are also variants of the universal code, such

as in Mycoplasma. A protein translated to mycoplasma and universal code is different (STOP codon in mycoplasma means W). Or even a mitochondrial protein vividly demonstrates how cellular languages are understood and translated somewhat differently.

### Question 12.4

Get the appropriate codons from the codon table. What would be different about the *codon usage* in yeast *(yeast)* compared to the universal code?

Differences from the Standard Code:

```
Code    3Standard
AUA     Met MIle I
CUU     Thr TLeu L
CUC     Thr TLeu L
CUA     Thr Tleu L
CUG     Thr TLeu L
UGA     Trp WTer *
CGA     absentArg R
CGC     absentArg R.
```

## 20.13  Life Always Invents New Levels of Language

### Question 13.1

Domains are independent folding units in protein, ranging in size from 100 to 150 amino acids. The databases are particularly important:

| InterPro | https://www.ebi.ac.uk/interpro/ |
|----------|--------------------------------|
| **SMART** | https://smart.embl-heidelberg.de |
| **Pfam** | https://pfam.xfam.org |

Query in InterPro, SMART or even Pfam: Thousands of protein families are always stored. InterPro also has automatic annotation and collection of protein domains and proteins (fusion of previous, single databases such as ProDom). SMART assumes hand-annotated *alignments* for extracellular domains, whereas Pfam considers entire protein families (multiple domains). The recombination of protein domains during *splicing* allows the production of many different protein variants from a single muscle gene. This is an advantage in the evolution of eukaryotes. This allows a much more complex generation of new proteins than would be possible if this were not the case. This is why we have become much more easily complex multicellular organisms, while bacteria remain in a simple state without splicing. The exon boundaries/reading frames are easily seen (indicated) in the SMART database. This also indicates recombination, even for "mile-long" genes, like the one for *tittin* in the human genebank. In a word: a huge evolutionary potential.

**Question 13.2**

https://www.rcsb.org/pdb/home/home.do. This is the protein database, the large repository for all protein structures. Also relatively limited, reference to the PDB database. This has 120,642 *Biological Molecular Structures* (July 2016; 100,848 X-ray crystal structures, 10,078 NMR structures, 787 by electron microscopy). Funded by the Research Collaboratory for Structural Bioinformatics, rcsb.org (Rutgers University, UC San Diego, SDSC).

There is also the *"Molecule of the Month"* series, where one structure at a time is presented in detail in a very didactic way. The RCSB PDB Molecule of the Month: Inspiring a Molecular View of Biology; Goodsell DS, Dutta S, Zardecki C, Voigt M, Berman HM, et al. (2015) The RCSB PDB "Molecule of the Month": Inspiring a Molecular View of Biology. PLOS Biology 13(5): e1002140. https://doi.org/10.1371/journal.pbio.1002140

Important main categories:

- Health and diseases
- Essential molecules of life
- Biotechnology and nanotechnology
- Structure and structural elucidation.

SCOP: Structural Classification of Proteins (scop.mrc-lmb.cam.ac.uk/scop/). SCOP classifies all protein structures and specifies how they are built in detail, e.g. beta-sheet with a helix packed against it.

Link to *"Atlas of protein structures"*: https://www.bioinformatics.org/molvis/atlas/atlas.htm.

But with protein design, these tight limits no longer apply (see follow-up tasks). CATH: Class, Architecture, Topology/fold, Homology read here https://www.cathdb.info/.

**Question 13.3**

For example, seek out the "Reactibody" by Carletti E et al. (Released: 2011 Sep 21): https://www.rcsb.org/pdb/explore.do?structureId=2XZA (catalytic antibody).

**Question 13.4**

*Tissue plasminogen activator* and *engineering of loop structure*: An optimal response is to visualize the PDB structure 5BRR, such as with RasMol (5BRR Michaelis complex of tPA-S195A:PAI-1, Gong L, Liu M, Zeng T, Shi X, Yuan C, Andreasen PA, Huang M (2015) Crystal Structure of the Michaelis Complex between Tissue-type Plasminogen Activator and Plasminogen Activators Inhibitor-1. J. Biol. Chem. 290 p. 25795–25,804). There you can look at the loop regions of tPA (their removal would prolong the effect) and an inhibitor in complex with tPA. View the structure for this with RasMol or PyMOL.

This original paper describes exactly what you see: Hydrophilic peptides derived from the transframe region of Gag-Pol inhibit the HIV-1 protease (Louis JM, Dyda F, Nashed NT, Kimmel AR, Davies DR (1998). Hydrophilic peptides derived from the transframe region of Gag-Pol inhibit the HIV-1 protease. Biochemistry. 37(8):2105–10. https://doi.org/https://doi.org/10.1021/bi972059x).

*The HIV-1 transframe region (TFR) is between the structural and functional domains of the Gag-Pol polyprotein, flanked by the nucleocapsid and the protease domains at its N and C termini, respectively. Transframe octapeptide (TFP) Phe-Leu-Arg-Glu-Asp-Leu-Ala-Phe, the N terminus of TFR, and its analogues are competitive inhibitors of the action of the mature HIV-1 protease. The smallest, most potent analogues are tripeptides: Glu-Asp-Leu and Glu-Asp-Phe with Ki values of approximately 50 and approximately 20 microM, respectively. Substitution of the acidic amino acids in the TFP by neutral amino acids and d or retro-d configurations of Glu-Asp-Leu results in a > 40-fold increase in Ki. Protease inhibition by Glu-Asp-Leu is dependent on a protonated form of a group with a pKa of 3.8; unlike other inhibitors of HIV-1 protease which are highly hydrophobic, Glu-Asp-Leu is extremely soluble in water, and its binding affinity decreases with increasing NaCl concentration. However, Glu-Asp-Leu is a poor inhibitor (Ki approximately 7.5 mM) of the mammalian aspartic acid protease pepsin. X-ray crystallographic studies at pH 4.2 show that the interactions of Glu at P2 and Leu at P1 of Glu-Asp-Leu with residues of the active site of HIV-1 protease are similar to those of other product-enzyme complexes. It was not feasible to understand the interaction of intact TFP with HIV-1 protease under conditions of crystal growth due to its hydrolysis giving rise to two products. The sequence-specific, selective inhibition of the HIV-1 protease by the viral TFP suggests a role for TFP in regulating protease function during HIV-1 replication.*

Chellappan S, Kiran Kumar Reddy GS, Ali A et al. (2007) Design of mutation-resistant HIV protease inhibitors with the substrate envelope hypothesis. Chem Biol Drug Des. 2007 May; 69(5): 298–313.

*There is a clinical need for HIV protease inhibitors that can evade resistance mutations. One possible approach to designing such inhibitors relies upon the crystallographic observation that the substrates of HIV protease occupy a rather constant region within the binding site. In particular, it has been hypothesized that inhibitors which lie within this region will tend to resist clinically relevant mutations. The present study offers the first prospective evaluation of this hypothesis, via computational design of inhibitors predicted to conform to the substrate envelope, followed by synthesis and evaluation against wild-type and mutant proteases, as well as structural studies of complexes of the designed inhibitors with HIV protease. The results support the utility of the substrate envelope hypothesis as a guide to the design of robust protease inhibitors.*

*CARB-AD37 docked into HIV protease from crystal structure. Inhibitors were tested against wild-type HIVP and a panel of three proteases with clinically relevant mutation sets: M1 (L10I/G48V/ I54V/L63P/V82A), M2 (D30N/L63P/N88D), and M3 (L10I/L63P/ A71V/ G73S/I84V/L90M).*

**Question 13.6**

Technically solve this task by querying PubMed: https://www.ncbi.nlm.nih.gov/pubmed/?term=Baker-D+AND+Nature, i.e. query: "Baker-D AND Nature". This results in the following articles, among others:

1. Bale JB, Gonen S, Liu Y et al. (2016) Accurate design of megadalton-scale two-component icosahedral protein complexes. Science. 2016 Jul 22; 353(6297): 389–394. PubMed PMID: 27463675.

2. Hsia Y, Bale JB, Gonen S et al. (2016) Design of a hyperstable 60-subunit protein icosahedron. Nature. 2016 Jul 7; 535(7610): 136–139. PubMed PMID: 27309817; PubMed Central PMCID: PMC4945409.

3. Boyken SE, Chen Z, Groves B et al. (2016) De novo design of protein homo-oligomers with modular hydrogen-bond network-mediated specificity. Science. 2016 May 6; 352(6286): 680–687. https://doi.org/10.1126/science.aad8865. Erratum in: Science. 2016 May 20; 352(6288). pii: aag1318. https://doi.org/10.1126/science.aag1318. PubMed PMID: 27151862

4. Huang PS, Feldmeier K, Parmeggiani F et al. (2016) De novo design of a four-fold symmetric TIM-barrel protein with atomic-level accuracy. Nat Chem Biol. 2016 Jan; 12(1): 29–34. https://doi.org/10.1038/nchembio.1966. Epub 2015 Nov 23. PubMed PMID: 26595462; PubMed Central PMCID: PMC4684731.

5. Doyle L, Hallinan J, Bolduc J et al. (2015) Rational design of α-helical tandem repeat proteins with closed architectures. Nature. 2015 Dec 24; 528(7583): 585–588. https://doi.org/10.1038/nature16191. Epub 2015 Dec 16. PubMed PMID: 26675735; PubMed Central PMCID: PMC4727831.

6. Brunette TJ, Parmeggiani F, Huang PS et al. (2015) Exploring the repeat protein universe through computational protein design. Nature. 2015 Dec 24; 528(7583): 580–584. https://doi.org/10.1038/nature16162. Epub 2015 Dec 16. PubMed PMID: 26675729; PubMed Central PMCID: PMC4845728.

Now the next thing you can do is read these excellent articles as well. David Baker (and others) have come a good step closer to protein design in recent times.

**Question 13.7**

(a) This is the following link: https://gosyn.bioapps.biozentrum.uni-wuerzburg.de.

The descriptive publication can be found here: https://database.oxfordjournals.org/content/2013/bat043.full. Read the publication and/or work through the tutorial and database on the net. With the database you can actually do synthetic biology design yourself and compare technical and biological control.

(b)  *Oncolytic virus*

You can find information here:

- https://www.ncbi.nlm.nih.gov/pmc/articles/PMC4303349/
- https://www.ncbi.nlm.nih.gov/pmc/articles/PMC4105246/
- https://www.genelux.com/leadership-in-oncolytic-virotherapy/oncolytic-virotherapy/.

Best to browse the internet yourself.

### Question 13.8

Here is the link to DrumPID (https://drumpid.bioapps.biozentrum.uni-wuerzburg.de/compounds/index.php), which is a database that combines protein interactions with *drugs* (i.e. chemical compounds). Here you can easily compare main and side effects, protein interactions and pharmaceuticals.

Here is the related publication along with the tutorial: https://database.oxfordjournals.org/content/2016/baw041.full. Also work through this database and the tutorial.

Another interesting and powerful database in this direction is the STITCH database at EMBL:

https://nar.oxfordjournals.org/content/early/2015/11/19/nar.gkv1277.full.pdf
https://stitch.embl.de

### Question 13.9

Here are four beautiful breakthroughs gathered together:
Adleman LM (1994) Molecular Computation of Solutions to Combinatorial Problems. Science 266: 1021–1024. Leonard Adelman deserves credit for having recreated the traveling salesman problem for simple cases (up to six cities) in DNA molecules by ligation.

Zimmer R (1998) Patent on parallel, universal and free-programmable information system for general computing operations. WO9847077 (A1) 1998 Oct 22. Prof. Ralph Zimmer (LMU Munich) considered how to convert the lambda calculus into a universal calculating machine and implement it in living organisms.

Win MN, Smolke CD (2008) Higher-order cellular information processing with synthetic RNA devices. Science 322, 456–460. Here, RNA aptamers that measure caffeine and diazepam have been interconnected with an RNA lever in such a way that they act like mini-sensors, switching to start only when the concentrations of either have a suitable range.

Tero A, Takagi S, Saigusa T et al. (2010) Rules for biologically inspired adaptive network design. Science. 2010 Jan 22; 327(5964): 439–442. https://doi.org/10.1126/science.1177894. The latest work calculates the optimal plan for the Tokyo metro system with the help of a slime mold that has been suitably distributed food.

**Question 13.10**

Generally speaking, you first need a light driveable domain. For this, look for BLUF (blue light sensitive) or LOV *(light operate voltage channel)* domains. If you add such a domain, the protein suddenly becomes driveable by light with matching wavelength, in particular it becomes active only when this light wavelength hits, but stops otherwise.

## 20.14   We Can Think About Ourselves – The Computer Cannot

**Question 14.1**

Gödel's theorem states that in any complete mathematical system it is possible to formulate statements that the system cannot decide. Systems in which such statements do not occur, on the other hand, are incomplete. A very nice derivation is given by Douglas R. Hofstadter in his classic book "Gödel-Escher-Bach."

**Question 14.2**

All general computable problems can be reproduced with the help of the Turing machine. All non-Turing-computable problems cannot be solved by computers and remain tasks for humans.

**Question 14.3**

Test for artificial intelligence: Human and computer are hidden behind a cloth, and outside people are now supposed to guess who is who. If the computer can fool the humans, then it has passed the Babbage test and possesses artificial intelligence.

**Question 14.4**

Get to know *neuronal network:*

(a) TMHMM (https://www.cbs.dtu.dk/services/TMHMM/) is already explained in Chap. 11 and Sect. 20.11, respectively.
(b) The ELM server (https://elm.eu.org) has also already been explained in Chap. 11 and Sect. 20.11, respectively.
(c) PredictProtein secondary structure prediction uses neural networks.

Prof. Burkhard Rost has spent years working on neuronal networks and secondary structure predictions of proteins. Protein sequences are simply read into the server. A neural network then predicts whether the amino acids are able to form a helix, a second predicts the ability to form *beta strands* and a third predicts whether a loop region is present. A fourth neural network is trained to decide how best to make an overall prediction from these three predictions, for example, if *beta strand* and helix but no loop region are predicted simultaneously by the three lower-level networks.

Further tricks additionally improve the predictions of this software. In particular, many sequences with similar structure are automatically added to the question sequence (multiple *alignment*). Thus, this secondary structure prediction allows an accuracy of up to 80%. This is already very close to the theoretical optimum. The only way to become even more accurate is to predict the three-dimensional structure at the same time.

### Question 14.5

One software is MemBrain (https://www.membrain-nn.de/index.htm; https://www.membrain-nn.de/).

### Question 14.6

Please search the internet for *deep learning* and inform yourself. Helpful is also the page: https://deeplearning.net/. For AlphaGo also on the Internet (https://deepmind.com/research/alphago; https://www.youtube.com/watch?v=mzpW10DPHeQ).

### Question 14.7

Classification models are used in bioinformatics for the classification between two categories (binary), for example for the diagnosis of a disease (sick/healthy). It is important here to become familiar with a classification table (confusion matrix; TP, FP, FN, TN), but also to look at the performance metrics (sensitivity, false positive rate, specificity, PPV, NPV, accuracy, misclassification rate, prevalence, ROC, AUC) for evaluating a classification model. Here it is also important to know what are, for example, differences between sensitivity and PPV, but also between specificity and NPV. For example, let's imagine: A person gets a positive (negative) test result from a predictive test that has a sensitivity of 90%, specificity of 99%, a PPV of 80%, and a NPV of 99%. Here, the positive test result could only be trusted 80% to actually be positive (sick) (20% false positive, so fortunately healthy), whereas a negative test result could be trusted more to actually be healthy (1% false negative, so actually sick). Most diagnostic testing procedures take this into account and, in the case of a positive test result, carry out a second test to confirm the diagnosis (e.g. mammography screening). On the other hand, a test should in any case be accurate enough to identify a healthy person with a high probability (here it would be worse to send home a supposedly healthy person [negative test result] who is in fact sick [false negative] and thus does not get any helping therapy or infects other people with a virus [e.g. COVID-19]). In addition, one should think about problems (little data, etc.) in creating a classification model, but also what requirements a classification model should meet. To build a predictive model, it is advisable to use a training and test dataset (splitting 80/20%) and validate the model on at least one independent dataset to better evaluate the predictive power.

### Question 14.8

Data in biology and medicine are usually high-dimensional, i.e. they contain different variables (features), correlations, confounders, batch effects and multicollinearity. For this, machine learning methods in bioinformatics are helpful to structure the data and extract relevant features, but also to develop classification models (predictive models). PCA tries to decompose high dimensional data into principal components and reduce their complexity (dimension reduction), but also to detect group differences. Cluster analyses try to classify data into groups (clusters) with similar feature structures (characteristics), e.g. healthy group (normal blood pressure) and diseased group (high blood pressure). Regression analyses attempt to find correlations and relationships between a dependent ("response variable") and independent ("predictor variable") variable, e.g. probability of developing high blood pressure (and subsequently dying of heart failure) if one is overweight. It is important to also look again at the underlying algorithms and statistical parameters to assess model goodness of fit. Further details and information can be found in the papers Worster et al. (2007), Schneider et al. (2010), Singh and Mukhopadhyay (2011) and Zwiener et al. (2011).

## 20.15   How Is Our Own Extremely Powerful Brain Constructed?

### Question 15.1

For this purpose, please refer to the website https://www.neuron.yale.edu/neuron/ (tutorial: https://www.neuron.yale.edu/neuron/docs available).

### Question 15.2

For more information, please visit the website https://www.openworm.org/index.html.

### Question 15.3

For more information, please visit the website https://www.humanconnectomeproject.org/.

### Question 15.4

To do this, simply search the Internet, for example, with size constancy in the brain, and inform.

### Question 15.5

To do this, simply search the Internet and inform yourself (there are also nice Youtube videos about this).

**Question 15.6**

OMIM stands for Online Mendelian Inheritance in Man. Go to the website (https://www.ncbi.nlm.nih.gov/omim) and search for *"alcoholism"* and *"schizophrenia"*.

## 20.16  Bioinformatics Connects Life with the Universe and all the Rest

**Question 16.1**

You can find the digital manifesto online here: https://www.spektrum.de/news/wie-algorithmen-und-big-data-unsere-zukunft-bestimmen/1375933; https://www.spektrum.de/thema/das-digital-manifest-algorithmen-und-big-data-bestimmen-unsere-zukunft/1375924.

**Question 16.2**

Here is a brief explanation of global warming: https://www.climatehotmap.org/about/global-warming-causes.html.

**Question 16.3**

Here you can find the *Doomsday Clock* (the clock of "doom", i.e. how close people are to catastrophe), but is of course a bit exaggerated here to get people to act. Easily found at: https://thebulletin.org/timeline. *Doomsday* here refers to the general demise of humanity, quickly by nuclear weapons, slowly by global warming. Two years (2018, 2019) the situation was again so volatile that the clock was at 3–12 min. But in 2020 the clock switched to only 100 s before midnight – it's really time to act.

**Question 16.4**

Here's Plan B, a particularly carefully crafted plan (Version B 4.0) for sustainability and rebuilding our environment, promoted by the Earth Watch Institute: https://www.earth-policy.org/books/pb4.

**Question 16.5**

Here is some information on Plan C: Sustainable, highly resilient and adaptable technologies that can help us stay strong in an emergency. A recent result is the coupling of enhanced $CO_2$ fixation in plants together with an alternative pathway that minimizes $CO_2$ losses through light respiration: This would allow us to be 5 times better at removing $CO_2$ from the air while making plants more productive [6] (Naseem et al. 2020).

Other examples:

1. The internet (doesn't break due to war, lost node computers are replaced by others on the fly). Many people are working to make the internet even more resilient. The nanocellulose chip without garbage already mentioned above, in which electronics are replaced by light, could be one of several possibilities.
2. Greenhouses that would still bring food even in winter, drought, famine, but would also help against nuclear winter or destroyed UV layer.
3. Using Flettner rotor ships to keep global warming down through low clouds (very effective, could stop all global warming; "Marine Cloud Brightening"; https://en.wikipedia.org/wiki/Marine_cloud_brightening).

## Literature

Aydinli,Muharrem, Chunguang Liang, Thomas Dandekar (2022) Motif and conserved module analysis in DNA (promoters, enhancers) and RNA (lncRNA, mRNA) using AlModules. Scientific Reports. 12:17588, 1–12. https://doi.org/10.1038/s41598-022-21732-0 (2022).

Fell DA (2005) Metabolic control analysis. In: Alberghina L, Westerhoff H (eds) Systems biology. Topics in current genetics, vol 13. Springer, Berlin

Heinrich R, Neel BG, Rapoport TA (2002) Mathematical models of protein kinase signal transduction. Mol Cell 9(5):957–970

Robubi A, Mueller T, Fueller J, Hekman M, Rapp UR, Dandekar T (2005) B-Raf and C-Raf signalling investigated in a simplified model of the mitogenic kinase cascade. Biol Chem 386(11):1165–1171. https://doi.org/10.1515/BC.2005.133

Schneider A, Hommel G, Blettner M (2010) Lineare Regressionsanalyse. Dtsch Arztebl Int 107(44):776–782. https://doi.org/10.3238/arztebl.2010.0776

Singh R, Mukhopadhyay K (2011) Survival analysis in clinical trials: basics and must know areas. Perspect Clin Res 2(4):145–148. https://doi.org/10.4103/2229-3485.86872

Worster A, Fan J, Ismaila A (2007) Understanding linear and logistic regression analyses. CJEM 9(2):111–113. https://doi.org/10.1017/s1481803500014883

Zwiener I, Blettner M, Hommel G (2011) Überlebenszeitanalyse. Dtsch Arztebl Int 108(10):163–169. https://doi.org/10.3238/arztebl.2011.0163

## Further Reading

Grant B, Scarabelli G (2013) BIOINF527: Structural bioinformatics lab session. Introduction to protein structure visualization and small molecule docking. https://thegrantlab.org/teaching/material/Structural_Bioinformatcs_Lab.pdf

Plan B 4.0: Mobilizing to save civilization Lester R. Brown. Released 2009. https://www.earthpolicy.org/images/uploads/book_files/pb4book.pdf

Naseem M, Osmanoglu Ö, Dandekar T (2020) Synthetic rewiring of plant $CO_2$ sequestration galvanizes plant biomass production. Trends Biotechnol 38(4):354–359

# Overview of Important Databases and Programs and Their General Use

## Alignment/Tribes

| | |
|---|---|
| CLUSTALW/Clustal Omega | https://www.ebi.ac.uk/Tools/msa/clustalo/ |
| MUSCLE | https://www.ebi.ac.uk/Tools/msa/muscle/ |
| PHYLIP | https://evolution.genetics.washington.edu/phylip.html |

## Datasets on Biological Quantities/Biotechnology/Synthetic Biology

| | |
|---|---|
| BioNumbers | https://bionumbers.hms.harvard.edu |
| BioBricks | https://biobricks.org/ |
| GoSynthetic | https://gosyn.bioapps.biozentrum.uni-wuerzburg.de/index.php |

## Dotplot

| | |
|---|---|
| Yolk | https://sonnhammer.sbc.su.se/Dotter.html |
| GEPARD | https://mips.gsf.de/services/analysis/gepard |
| JDotter | https://athena.bioc.uvic.ca/virology-ca-tools/jdotter/ |

## Functional Databases

| | |
|---|---|
| Functional Glycomics | https://www.functionalglycomics.org/; https://ncfg.hms.harvard.edu/ |
| Gene Ontology | https://www.geneontology.org |

© Springer-Verlag GmbH Germany, part of Springer Nature 2023
T. Dandekar, M. Kunz, *Bioinformatics*,
https://doi.org/10.1007/978-3-662-65036-3

## Brain Blueprints

| | |
|---|---|
| Blue Brain Project (EU) | https://bluebrain.epfl.ch/ |
| Brain Activity Atlas | https://www.brainactivityatlas.org/ |
| Brain Activity Project (USA) | https://www.braininitiative.nih.gov/ |
| Connectome project | https://www.openconnectomeproject.org |
| Mouse Brain Connectivity Atlas | https://mouse.brain-map.org/static/atlas |
| Neuroactivity Detection | https://www.ncbi.nlm.nih.gov/pubmed/23537512 |
| Temporal lobe | https://www.temporal-lobe.com/background/connectome |
| Virtual Insect Brain Lab | https://www.neurofly.de/ |
| WormWiring | https://wormwiring.org/ |
| Sausage Atlas | https://www.wormatlas.org/ |

## Genome Annotation/Sequence Analysis/Online Libraries/ Experimental Datasets

| | |
|---|---|
| BLAST | https://blast.ncbi.nlm.nih.gov/Blast.cgi |
| GenScan | https://genes.mit.edu/GENSCAN.html |
| RepeatMasker | https://www.repeatmasker.org/ |
| ENCODE | https://www.encodeproject.org |
| Ensembl | https://www.ensembl.org/Homo_sapiens/Info/Index |
| GATK Workshop | https://software.broadinstitute.org/gatk/guide/article?id=7869#1.3 |
| Genomic Science Program | https://genomics.energy.gov |
| Human Genome Project | https://web.ornl.gov/sci/techresources/Human_Genome/index.shtml |
| UCSC | https://genome.ucsc.edu/ |
| DDBJ (DNA Data Bank of Japan) | https://www.ddbj.nig.ac.jp/ |
| EBI | https://www.ebi.ac.uk/services |
| iGEM Parts | https://igem.org/Main_Page |
| MEDLINE/NCBI/PubMed | https://www.ncbi.nlm.nih.gov/pubmed/ |
| NIH | https://www.genome.gov |
| OMIM | https://www.omim.org/ |
| Swiss Bioinformatics Institute | https://www.sib.swiss/ |
| WebDirectory | https://www.biologydir.com/over-population/p1.html |
| Computational Population Biology | https://compbio.cs.uic.edu/ |
| GENEVESTIGATOR | https://genevestigator.com/gv/ |
| GEO | https://www.ncbi.nlm.nih.gov/geo/ |

## Graphics Programs, Modeling and Network Analysis

| | |
|---|---|
| CellDesigner | https://www.celldesigner.org/ |
| CellNetAnalyzer | https://www2.mpi-magdeburg.mpg.de/projects/cna/cna.html |
| Cytoscape | https://www.cytoscape.org/ |
| COBRA | https://opencobra.github.io/ |
| COPASI | https://copasi.org/ |
| Flux balance analysis | https://systemsbiology.ucsd.edu/Downloads/FluxBalanceAnalysis |
| Jimena | https://www.bioinfo.biozentrum.uni-wuerzburg.de/computing/jimena_c/ |
| MATLAB | https://de.mathworks.com/products/matlab.html |
| Metatool | https://pinguin.biologie.uni-jena.de/bioinformatik/networks/metatool/ |
| Odefy | https://www.helmholtz-muenchen.de/icb/software/odefy/index.html |
| PLAS | https://enzymology.fc.ul.pt/software/plas/ |
| PottersWheel | https://www.potterswheel.de/Pages/ |
| SQUAD | https://www.vital-it.ch/software/SQUAD |
| YANA/YANAsquare | https://www.bioinfo.biozentrum.uni-wuerzburg.de/computing/yanasquare/ |

## Interaction Database, Drug Interaction Database

| | |
|---|---|
| catRAPID | https://s.tartaglialab.com/page/catrapid_group |
| HPRD | https://hprd.org/ |
| iHOP | https://www.ihop-net.org/UniPub/iHOP/ |
| KEGG | https://www.genome.jp/kegg/ |
| NPInter | https://www.bioinfo.org/NPInter/ |
| PlateletWeb | https://plateletweb.bioapps.biozentrum.uni-wuerzburg.de/plateletweb.php |
| Roche Pathways | https://biochemical-pathways.com/#/map/1 |
| STRING | https://string-db.org |
| DrumPID | https://drumpid.bioapps.biozentrum.uni-wuerzburg.de/compounds/index.php |
| STITCH | https://stitch.embl.de/ |
| EcoCyc | https://ecocyc.org/ |

## Localization/Motive Prediction

| | |
|---|---|
| LocP | https://ekhidna2.biocenter.helsinki.fi/LOCP/ |
| LocSigDB | https://genome.unmc.edu/LocSigDB/ |
| nucloc | https://www.nucloc.org/ |
| NucPred | https://www.sbc.su.se/~maccallr/nucpred/ |
| SignalP | https://www.cbs.dtu.dk/services/SignalP/ |
| TMHMM | https://www.cbs.dtu.dk/services/TMHMM/ |
| Functional Glycomics | https://www.functionalglycomics.org/ |
| ELM | https://elm.eu.org/ |

## Programming Languages

| | |
|---|---|
| Biojava | https://biojava.org/ |
| BioPerl | https://bioperl.org/ |
| C++ | https://www.cplusplus.com/ |
| Java | https://www.oracle.com/technetwork/java/index.html |
| Perl | https://www.perl.org/ |
| Python | https://www.python.org/ |
| R | https://cran.r-project.org/ |
| Bioconductor | https://www.bioconductor.org/ |

## Promoter Analysis

| | |
|---|---|
| ALGGEN PROMO | https://alggen.lsi.upc.es/cgi-bin/promo_v3/promo/promoinit.cgi?dirDB=TF_8.3 |
| Genomatix | https://www.genomatix.de/ |
| JASPAR | https://jaspar.genereg.net/cgi-bin/jaspar_db.pl |
| MotifMap | https://motifmap.igb.uci.edu/ |
| TESS | https://www.cbil.upenn.edu/tess/ |
| TRANSFAC | https://www.gene-regulation.com/pub/databases.html |

## Protein Analysis

| | |
|---|---|
| AnDom | https://andom.bioapps.biozentrum.uni-wuerzburg.de/index_new.html |
| CATH | https://www.cathdb.info/ |
| Conserved Domains | https://www.ncbi.nlm.nih.gov/Structure/cdd/wrpsb.cgi |
| ExPASy | https://www.expasy.org |
| InterPro | https://www.ebi.ac.uk/interpro/ |
| MODELLER | https://salilab.org/modeller/tutorial/ |
| PDB | https://www.rcsb.org/pdb/home/home.do |
| Pfam | https://pfam.xfam.org/ |
| ProDom | https://prodom.prabi.fr/prodom/current/html/home.php |
| PRODORIC | https://prodoric.tu-bs.de/ |
| PROSITE | https://prosite.expasy.org |
| PyMOL | https://www.pymol.org/ |
| QUARK | https://zhanglab.ccmb.med.umich.edu/QUARK/ |
| Ramachandran plot | https://mordred.bioc.cam.ac.uk/~rapper/rampage.php |
| RasMol | https://www.openrasmol.org/ |
| SCOP (old) | https://scop.mrc-lmb.cam.ac.uk/scop/ |
| SCOP updated | https://scop.berkeley.edu/ |
| SMART | https://smart.embl-heidelberg.de/ |

| SWISS-MODEL | https://swissmodel.expasy.org |
|---|---|
| UniProt/Swiss-Prot | https://www.uniprot.org/ |

**RNA Analysis**

| ITS2 | https://its2.bioapps.biozentrum.uni-wuerzburg.de/ |
|---|---|
| LNCipedia | https://www.lncipedia.org/ |
| microRNA.org/miRanda | https://www.microrna.org/microrna/home.do |
| miRBase | https://www.mirbase.org/ |
| regRNA | https://regrna2.mbc.nctu.edu.tw/ |
| Rfam | https://rfam.xfam.org/ |
| Riboswitch Finder | https://riboswitch.bioapps.biozentrum.uni-wuerzburg.de/ |
| RNAAnalyzer | https://rnaanalyzer.bioapps.biozentrum.uni-wuerzburg.de/ |
| RNAfold web server | https://rna.tbi.univie.ac.at/cgi-bin/RNAWebSuite/RNAfold.cgi |
| TargetScan | https://www.targetscan.org |
| tRNAscan | https://lowelab.ucsc.edu/tRNAscan-SE/ |
| Vienna Package | https://www.tbi.univie.ac.at/RNA/ |

Printed in the United States
by Baker & Taylor Publisher Services